Structure and
Dynamics of RNA

NATO ASI Series

Advanced Science Institutes Series

A series presenting the results of activities sponsored by the NATO Science Committee, which aims at the dissemination of advanced scientific and technological knowledge, with a view to strengthening links between scientific communities.

The series is published by an international board of publishers in conjunction with the NATO Scientific Affairs Division

A	**Life Sciences**	Plenum Publishing Corporation
B	**Physics**	New York and London
C	**Mathematical and Physical Sciences**	D. Reidel Publishing Company Dordrecht, Boston, and Lancaster
D	**Behavioral and Social Sciences**	Martinus Nijhoff Publishers
E	**Engineering and Materials Sciences**	The Hague, Boston, and Lancaster
F	**Computer and Systems Sciences**	Springer-Verlag
G	**Ecological Sciences**	Berlin, Heidelberg, New York, and Tokyo

Recent Volumes in this Series

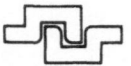

Series A: Life Sciences

Structure and Dynamics of RNA

Edited by

P. H. van Knippenberg

University of Leiden
Leiden, The Netherlands

and

C. W. Hilbers

University of Nijmegen
Nijmegen, The Netherlands

Plenum Press
New York and London
Published in cooperation with NATO Scientific Affairs Division

Proceedings of a NATO Advanced Research Workshop on
3D Structure and Dynamics of RNA,
held August 21–24, 1985,
in Renesse, The Netherlands

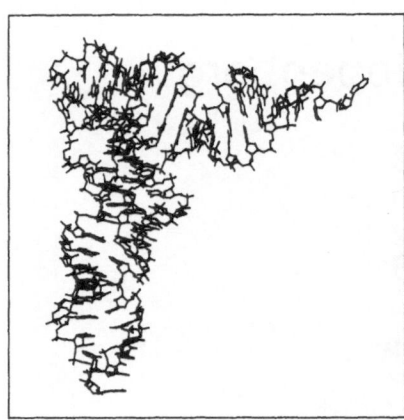

Library of Congress Cataloging in Publication Data

NATO Advanced Research Workshop on 3D Structure and Dynamics of RNA
  (1985: Renesse, Netherlands)

  Structure and dynamics of RNA.

  (NATO ASI series. Series A, Life sciences; v. 110)
  "Proceedings of a NATO Advanced Research Workshop on 3D Structure and
Dynamics of RNA, held August 21–24, 1985, in Renesse, The Netherlands"—
T.p. verso.
  "Published in cooperation with NATO Scientific Affairs Division."
  Includes bibliographies and index.
  1. Ribonucleic acid—Structure—Congresses. 2. Molecular dynamics—
Congresses. I. Knippenberg, P. H. van. II. Hilbers, C. W. III. North Atlantic Treaty
Organization. Scientific Affairs Division. IV. Title. V. Series. [DNLM: 1. RNA—
analysis—congresses. 2. RNA—physiology—congresses. QU 58 N279s 1985]
QP623.N38   1985   574.87'3283   86-15111

ISBN 978-1-4684-5175-7   ISBN 978-1-4684-5173-3 (eBook)
DOI 10.1007/978-1-4684-5173-3

© 1986 Plenum Press, New York
Softcover reprint of the hardcover 1st edition 1986
A Division of Plenum Publishing Corporation
233 Spring Street, New York, N.Y. 10013

PREFACE

This volume contains contributions from the speakers at the NATO Advanced Research Workshop on "3D Structure and Dynamics of RNA", which was held in Renesse, The Netherlands, 21 - 24 August, 1985.

Two major developments have determined the progress of nucleic acid research during the last decade. First, manipulation of genetic material by recombinant DNA methodology has enabled detailed studies of the function of nucleic acids in vivo. Second, the use of powerful physical methods, such as X-ray diffraction and nuclear magnetic resonance spectroscopy, in the study of biomacromolecules has provided information regarding the structure and the dynamics of nucleic acids. Both developments were enabled by the advance of synthetic methods that allow preparation of nucleic acid molecules of required sequence and length.

The basic understanding of nucleic acid function will ultimately depend on a close collaboration between molecular biologists and biophysicists. In the case of RNA, the ground rules for the formation of secondary structure have been derived from physical studies of oligoribonucleotides. Powerfull spectroscopic techniques have revealed more details of RNA structure including novel conformations (e.g. left-handed Z-RNA).

A wealth of information has been obtained by studying the relatively small transfer RNA molecules. A few of these RNAs have been crystallized, enabling determination of their three-dimensional structure. It has become apparent that "non-classical" basepairing between distal nucleotides gives rise to tertiary interactions, determining the overall shape of the molecule. Independent evidence for the 3D folding stems from high resolution proton NMR studies of dissolved molecules. Newly started molecular dynamics calculations promise to provide us with a detailed knowledge of the atomic motions in these molecules. Details of the structures and of the interaction with ligands are also derived from data obtained by a variety of spectroscopic techniques. When these are combined with results of (bio)-chemical analysis it is possible to arrive at a clear picture of structure and function of this class of RNA molecules.

In most cells the bulk of RNA is present in ribosomes. The three classes of ribosomal RNA, 5S RNA, 16S RNA and 23S RNA have been studied extensively, although not in such detail as transfer RNA. For all three classes "consensus" secondary structures have been derived These are primarily based on phylogenetic data, but are supported by experiments using (bio)-chemical approaches. Unfortunately, it has as yet not been possible to crystallize a ribosomal RNA (or part of it) and the molecules are too large to be studied by NMR at its current state of the art. However, some progress has been made with fragments of ribosomal RNA. Similarly, through the use of a variety of techniques, including recombinant DNA methods, functional areas in ribosomal RNAs have been mapped.

Important information has also been obtained regarding the folding of viral RNA. Here, some novel folding principles have been introduced which might also play a role in other RNA molecules.

A recent finding that RNA molecules, like enzymes, have catalytic properties, has attracted much attention. The auto-catalytic selfsplicing of RNA appears to depend on a precise folding pattern of the RNA near the splice junction.

NATO Scientific Affairs Division is gratefully acknowledged for granting an award that made the organization of the workshop possible. Generous financial support was obtained from the Royal Netherlands Academy of Arts and Sciences. Contributions were also made by Amersham Nederland BV, Beckmann Instruments Nederland BV, Boehringer Mannheim BV, Bruker Spectrospin NV, Gibco-BRL, Gist-Brocades BV, Salm & Kipp and Westburg BV (Anglian Biotechnology).

Finally it should be mentioned that the success of the meeting and the high scientific standard of this volume are the result of the enthousiastic co-operation of the participants. There is clearly a need for meetings devoted to RNA at regular intervals in the future.

P.H. van Knippenberg
C.W. Hilbers

CONTENTS

IMPROVED PARAMETERS FOR PREDICTION OF RNA SECONDARY STRUCTURE

AND INSIGHTS INTO WHY RNA FORMS DOUBLE HELIXES

D. H. Turner,[1] S. M. Freier,[1] N. Sugimoto,[1]
D. R. Hickey[1], J. A. Jaeger,[1] A. Sinclair,[2] D. Alkema,[2]
T. Neilson,[2] M. H. Caruthers,[3] and R. Kierzek[4]

[1]Department of Chemistry
University of Rochester
Rochester, New York 14627
U.S.A.

[2]Department of Biochemistry
McMaster University
Hamilton, Ontario
Canada L8N 3Z5

[3]Department of Chemistry and Biochemistry
University of Colorado
Boulder, Colorado 80309
U.S.A.

[4]Institute of Bioorganic Chemistry
Polish Academy of Sciences
Noskowskiego 12/14, 61-704 Poznan
Poland

ABSTRACT

   Thermodynamic parameters for double helix formation have been
measured for a large number of oligoribonucleotides. These data have
been analyzed to provide free energy changes associated with base
pairs, dangling ends, and base mismatches. The results suggest base
stacking and base pairing are important determinants of RNA stability,
but that hydrophobic bonding is not. The improved thermodynamic
parameters are applied to predict secondary structures for the self
splicing intervening sequence from the ribosomal RNA precursor of
Tetrahymena thermophila.

INTRODUCTION

   Knowledge of the forces directing nucleic acid chemistry is
important for understanding the structure and dynamics of RNA. There
is considerable controversy, however, over the relative contributions
of interactions such as hydrophobic bonding, base stacking, and
hydrogen bonding. Empirical measures for the magnitudes of these
contributions can be obtained from optical studies of double helix

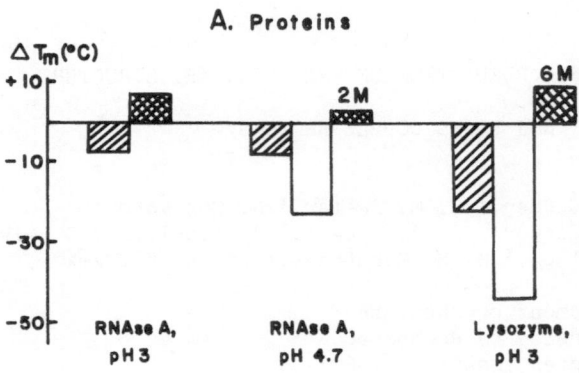

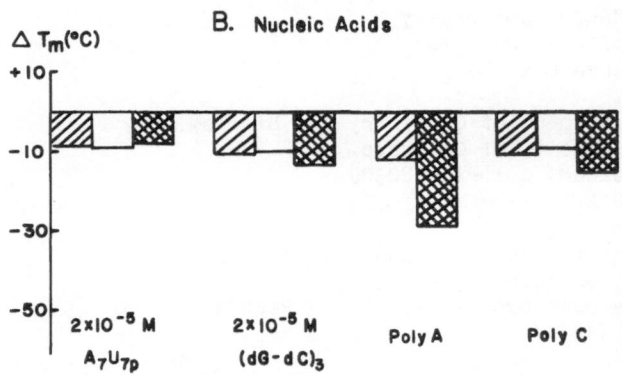

Fig. 1. Cosolvent-induced changes in T_m relative to water. A.
Proteins: ribonuclease A (RNAse A) at pH 3 (Brandts and
Hunt, 1967; Gekko and Timasheff, 1981b) and pH 4.7 (Schrier
and Scheraga, 1962; Schrier et al., 1965; Gerlsma and Stuur,
1974), lysozyme at pH 3 (Parodi et al., 1973; Back et al.,
1979). T_m's in H_2O are 45, 61, and 67°C, respectively. B.
Nucleic Acids at pH 7: 2×10^{-5} M A_7U_7p (Hickey and Turner,
1985a), 2×10^{-5} M $(dG-dC)_3$ (Albergo and Turner, 1981; Freier
et al., 1983b), poly A (Dewey and Turner, 1980), poly C
(Freier et al., 1981). T_m's in H_2O are 37, 48, 35, and 49°C,
respectively. Cosolvents are 10 mol % except where noted and
are represented by: ▨ , ethanol;   ☐ , 1-propanol;
        ▦ , glycerol.

formation by oligonucleotides. Absorbance versus temperature melting curves are analyzed to provide thermodynamic parameters for the single strand to double helix transition. Such studies also provide parameters for improving predictions of RNA structure from sequence. In this paper, we review the results of several such studies.

HYDROPHOBIC BONDING

One possible source of free energy driving helix formation by RNA is classical hydrophobic bonding. Many studies indicate hydrophobic bonding is important for protein folding (Brandts and Hunt, 1967; Kauzmann, 1959; Cantor and Schimmel, 1980), and it has been suggested that it is also important for folding of nucleic acids. One indication that hydrophobic bonding stabilizes the folded form of proteins is the decrease in protein melting temperature induced by addition of aliphatic alcohols. Typical results are shown in Figure 1. Presumably, this effect is due to the favorable interactions between the hydrophobic groups of the alcohols and proteins. The observation that propanol is a stronger denaturant than ethanol (see Figure 1) is consistent with this interpretation because the longer aliphatic chain of propanol makes it more hydrophobic. Nucleic acids, however, do not follow this trend. As illustrated in Figure 1, ethanol and propanol have similar effects on the coil to helix transition for double stranded A_7U_7p and $(dG-dC)_3$, and for single stranded poly (cytidylic acid). Glycerol is a cosolvent that enhances hydrophobic bonding (Gekko and Timasheff, 1981a,b). There is an unfavorable interaction between CHOH and CH_2 groups (Okamoto et al., 1978) that is presumably responsible for this effect. Thus glycerol raises the melting temperatures of proteins, as illustrated in Figure 1. The opposite effect is observed for the coil to helix transition in both single and double strand nucleic acids (see Figure 1). Thus solvent effects on stability differ for proteins and nucleic acids. This should not be surprising since the chemical structures of proteins and nucleic acids are quite different. The interiors of proteins contain many non-polar aliphatic groups, while the buried parts of nucleic acids are largely aromatic and polar. Recent computer simulations of solute-solvent interactions indicate water-water structure increases around non-polar solutes (Geiger et al., 1979; Pangali et al., 1979; Swaminathan et al., 1978) but not around benzene (Linse et al., 1984), and that hydrophobic bonding is not responsible for stacking of purines (Langlet et al., 1980).

STACKING

An empirical measure of the free energy change associated with stacking can be obtained by comparing the stabilities for completely complementary double helixes and double helixes containing terminal unpaired nucleotides (dangling ends) (Petersheim and Turner, 1983a; Freier et al., 1983a, 1985a, 1986a). For example, half the difference of free energy changes for helix formation by CCGG and CCGGA provides the stacking free energy change for a 3' A on a GC base pair. Many such free energy increments have been measured using CCGG, GGCC, and GCGC as core double helixes (Petersheim and Turner, 1983a; Freier et al., 1983, 1985a, 1986a). These are listed in Table 1 for 37°C, and several are shown in Figure 2. Some trends are apparent. Free energy increments for 3' dangling ends are much larger than for 5' dangling ends. In fact, in 1 M NaCl, a 5' dangling end adds essentially the same stability increment as a 5' phosphate. This suggests the base of a 5' dangling nucleotide interacts little with the adjacent base pair.

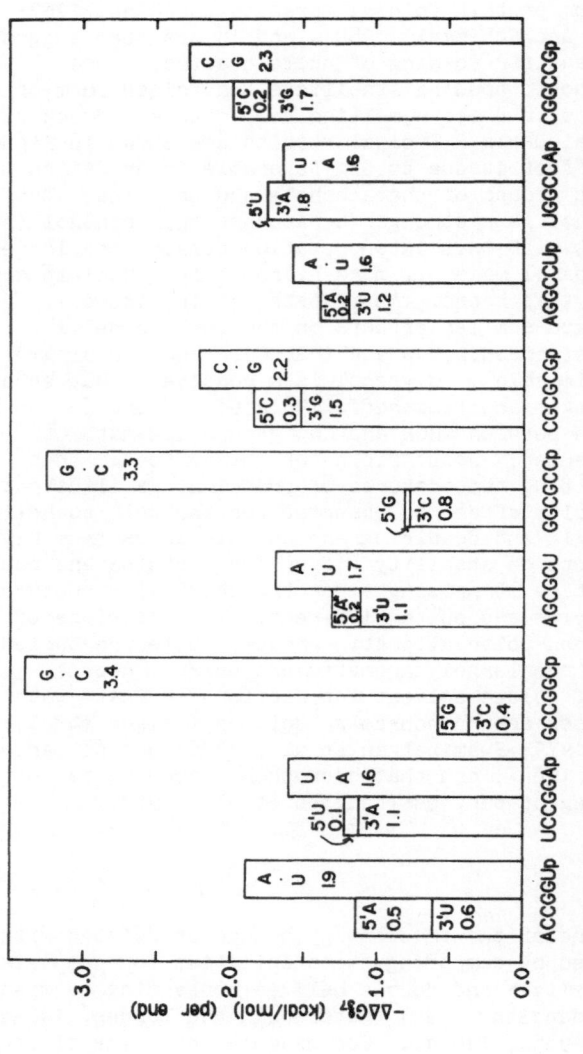

Fig. 2. Free energy increments at 37°C for adding a terminal base pair or dangling end to a CCGG, GCGC, or GGCC core. The left hand bars in each set represent free energy increments for 5' and 3' dangling ends. The right hand bars represent the free energy increments for base pair formation. (*) This free energy increment for a 5' dangling A was measured on AGGCCp.

Table 1. Stability Increments for Adding Terminal Phosphates, Dangling Ends, Terminal Base Pairs and Terminal Mismatches to GGCC, CCGG, and GCGC in 1 M NaCl.[a]

	$-\Delta\Delta G°(37°C)$, kcal/mol		
core helix:	GGCC	CCGG	GCGC
added terminus			
terminal phosphates:			
5' phosphate (5'p)	0.2	0.3	
3' phosphate (3'p)	-0.1	-0.2	
5' dangling ends:			
5'Ap + 3'p	0.2	0.5	
5'Cp + 3'p	0.2		0.3
5'Gp + 3'p		0.2	0.0
5'Up + 3'p	0.0	0.1	
3' dangling ends:			
3'pAp	1.8	1.1	1.7
3'pCp	0.8	0.4	0.8
3'pC	0.9		
3'pGp	1.7	1.3	1.5
3'pUp	1.2	0.6	1.1
Watson-Crick pairs:			
5'Ap + 3'pUp	1.6	1.9	
5'Cp + 3'pGp	2.3		2.2
5'Gp + 3'pCp		3.4	3.3
5'Up + 3'pAp	1.6	1.6	
5'Ap + 3'pU	1.5		1.7
GU pairs:			
5'Up + 3'pGp	1.5	1.4	
5'Gp + 3'pUp		2.3	
5'Gp + 3'pU			1.9
Mismatches			
5'Ap + 3'pAp		1.1	
5'Ap + 3'pCp		1.1	
5'Ap + 3'pGp		1.6	
5'Gp + 3'pAp		1.3	
5'Gp + 3'pGp		1.5	
5'Up + 3'pU			1.2

[a] $\Delta\Delta G°(37°C)$ is half the difference between the $\Delta G°(37°C)$ of helix formation for the molecule containing the core helix plus the added termini and the $\Delta G°(37°C)$ of helix formation for the tetramer core. For example, for the dangling end, 3'pAp on a CCGG core: $\Delta\Delta G°(37°C)$ = 0.5 [$\Delta G°$(CCGGA) − $\Delta G°$(CCGG)]. Results are from Petersheim and Turner, 1983a; Freier et al., 1983a; Freier et al., 1985a; Freier et al., 1986a,d.

NMR chemical shifts as a function of temperature for ACCGGp are consistent with this interpretation (Petersheim and Turner, 1983b). This lack of stacking is also consistent with standard A-form RNA geometry (Freier et al., 1985a). For 3' dangling ends, the order of additional stability is A,G>U>C, and they add more stability when adjacent to a C than to a G. The range of free energy increments is -0.4 to -1.8 kcal/mole for 3' dangling ends. For full base pairs, this range is -0.9 to -3.3 kcal/mole at 37°C (see Tables 1 and 3), indicating stacking is an important determinant of nucleic acid stability.

PAIRING

We define pairing as the interactions between nucleotides within a base pair. Quantum mechanical calculations indicate hydrogen bonding dominates these interactions (Pullman and Pullman, 1968; 1969). An empirical measure of the free energy increment for pairing is the difference between the free energy change for adding a base pair to a helix and the sum of the free energy changes for adding the corresponding dangling ends (Petersheim and Turner, 1983; Freier et al., 1985a, 1986a). Free energy changes for terminal base pairs and dangling ends are listed in Table 1, and presented as bar graphs in Figure 2. Presumably, stacking of dangling ends provides an upper limit for the stacking contribution to base pair formation since a dangling end has more freedom to adopt an optimum stacking geometry. It is even possible that stacking interactions interfere with base pair formation if they must be disrupted to allow a geometry appropriate for hydrogen bonding. In seven of the nine cases shown in Figure 2, the sums of the free energy increments from 5' and 3' dangling ends are more than half the increments for adding the corresponding base pairs. In two cases, GCCGGCp and GGCGCCp, the sums of the dangling ends are less than one quarter of the increment for the full base pair. This indicates pairing is also an important determinant of nucleic acid stability. In making this comparison, it should be realized that the free energy changes associated with 3' dangling end stacking and base pair formation contain contributions from unfavorable configurational entropy. In principle, this term must be factored out to derive measures of the pure attractive forces driving base pair formation. Unfortunately, there is currently uncertainty about the magnitude of the configurational entropy.

PREDICTING RNA STRUCTURE

Current computer algorithms (Zuker and Stiegler, 1981; Jacobson et al., 1984; Nussinov et al., 1982; Pipas and McMahon, 1975; Salser, 1977, Papanicolaou et al., 1984) for predicting RNA secondary structure from sequence are based on a nearest neighbor model (Tinoco et al., 1971; Gralla and Crothers, 1973). Such a model is reasonable if stacking and hydrogen bonding drive helix formation, since both involve short range forces. The results in Table 1 provide some direct evidence supporting the nearest neighbor model. Specifically, the free energy increments for adding 3' dangling ends, a CG or AU base pair to GGCC are the same, within experimental error, as for adding each to GCGC.

Another test of the nearest neighbor model is shown in Table 2 which lists melting temperatures and free energy changes measured for helix formation by pairs of oligomers that have the same nearest neighbors, but different sequences (Freier et al., 1986b). The

Table 2.   Thermodynamic Parameters of Helix Formation for Oligonucleotides with Identical Nearest Neighbors, but Different Sequences[a]

Oligomer	$-\Delta G°(37°C)$ (kcal/mol)	T_m (°C) (at 1×10^{-4}M)
AGAUAUCU	6.58	41.4
AUCUAGAU	7.20	45.1
AACUAGUU	7.17	45.7
AGUUAACU	6.36	41.1
ACUUAAGU	6.16	40.2
GAACGUUC	9.30	52.3
GUUCGAAC	8.76	50.4
UCUAUAGA	6.96	43.6
UAGAUCUA	7.25	45.3
GUCGAC	7.09	45.4
GACGUC	7.35	46.2
GCCGGCp[b]	11.24	67.2
GGCGCCp[c]	11.33	65.2
ACUAUAGU	6.98	44.0
AGUAUACU	6.80	43.7

[a] Parameters derived from plots of reciprocal melting temperature (T_m^{-1}) vs. log(concentration). Data from Freier et al., 1986b.
[b] Data from Freier et al., 1985a.
[c] Data from Freier et al., 1986a.

Table 3.   Free Energy Parameters for Nearest Neighbor Interactions in 1M NaCl at 37°C.[a,b]

	3' Nucleotide			
5' Nucleotide	A	C	G	U
A	-0.9	-2.2	-1.7	-1.0
C	-1.9	-3.0	-2.2	-1.7
G	-2.2	-3.3	-3.0	-2.2
U	-1.2	-2.2	-1.9	-0.9

[a] Free energy parameters obtained by a multiple linear regression to thermodynamic parameters for single strand to double helix equilibria for 36 oligonucleotides. For each oligonucleotide, thermodynamic parameters were derived from plots of inverse melting temperature vs. log(concentration). Values of $\Delta H°$ and $\Delta S°$ from these plots were within 15% of the $\Delta H°$ and $\Delta S°$ derived by averaging fits of the melting curves to a two-state model.

[b] Sources of oligonucleotide data were: Petersheim and Turner, 1983a; Nelson et al., 1981; Freier et al., 1983a, 1985a,b, 1986a,b,c; Groebe, D. R., Uhlenbeck, O. C., Freier, S. M., and Turner, D. H., unpublished experiments.

melting temperatures of the pairs differ from 0.3 to 5.5°C with an average difference of 2.4°C, corresponding to about 1% on the Kelvin scale. The free energy differences range from 0.8 to 15.2% with an average difference of 6%. These results indicate the nearest neighbor model should provide reasonable predictions for helix stabilities.

In order to make the best possible predictions of helix stability with the nearest neighbor model, it is necessary to measure free energy parameters for all ten nearest neighbor interactions (Borer et al., 1974). This has not been previously possible due to limitations in methods for making RNA. Recent advances in synthetic methods have eliminated this constraint (Kierzek et al., 1986; England and Neilson, 1976; Uhlenbeck and Gumport, 1982; Romaniuk and Uhlenbeck, 1983; Beckett and Uhlenbeck, 1984). Table 3 lists free energy changes at 37°C for all ten nearest neighbors as derived from melting studies on 36 oligonucleotides (Freier et al., 1986c).

The largest differences between these parameters and those used most often for prediction of RNA secondary structure (Borer et al., 1974) are for nearest neighbors containing two GC base pairs (Freier et al., 1985b). The values of $\Delta G°(37°C)$ for CG, GG, and GC stacks are -2.2, -3.0, and -3.3 kcal/mole, respectively. If helix stability was determined only by hydrogen or hydrophobic bonding, then $\Delta G°(37°C)$ should be the same for all these stacks. Thus these values also provide additional evidence for the importance of stacking.

In addition to standard AU and GC base pairs, GU pairs are known to occur in RNA. We have measured the thermodynamic parameters for several oligomers containing GU base pairs in order to improve GU parameters. Results from four oligomers containing terminal GU pairs are listed in Table 1. Evidently, a terminal GU pair is essentially equivalent to a terminal AU pair. This confirms previous suggestions based on the binding of the codons AUG and GUG to formylmethionine tRNA (Uhlenbeck et al., 1970; Gralla and Crothers, 1973). Preliminary measurements have also been made on oligonucleotides containing internal GU base pairs. Together with previously published data (Uhlenbeck et al., 1971; Gralla and Crothers, 1973), the results suggest an internal nearest neighbor stack containing GU is roughly 0.1 to 0.4 kcal/mole less stable than the corresponding AU stack (N. Sugimoto, S. M. Freier and D. H. Turner, unpublished experiments).

The thermodynamic parameters for dangling end stacking suggest that terminal mismatches other than GU should also significantly stabilize helixes. Stability increments for six such mismatches are listed in Table 1 (Hickey and Turner, 1985b; Freier et al., 1986d). They range from -1.1 to -1.6 kcal/mole. In the nearest neighbor model, there are a total of 48 different terminal mismatches. Thus it would be useful to have rules for approximating the free energy increments of terminal mismatches without directly measuring each one. One possible approximation is to sum the increments from the corresponding 3' and 5' dangling ends. Figure 3 shows that this is adequate for pyrimidine-pyrimidine and pyrimidine-purine mismatches, but not for A·A mismatches. This might be due to geometrical considerations. Purine-purine mismatches are better approximated by the stability increment of the appropriate 3' dangling end made more favorable by 0.2 kcal/mole for the effect of the 5' phosphate.

To determine the effect of revised nearest neighbor parameters on prediction of RNA structure, the parameters in Table 3 were used with the computer program of Zuker and Stiegler (1981) to predict the structure of the self splicing intervening sequence from the RNA

8

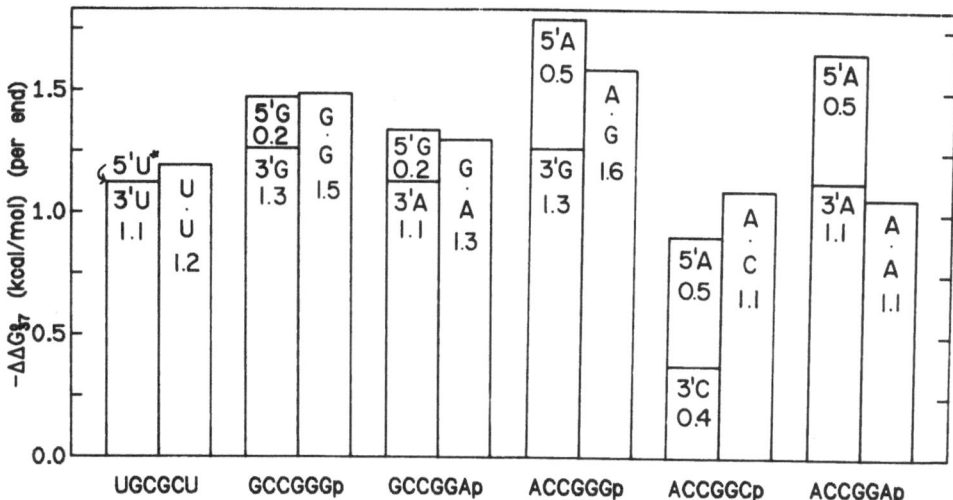

Fig. 3. Free energy increments at 37°C for adding a terminal mismatch
to tetramer cores compared with increments for the
corresponding 3' and 5' dangling ends. (*) This free energy
increment for a 5' dangling U was measured on UGGCCp.

precursor of Tetrahymena thermophila (Cech et al., 1983). In these
calculations, internal and terminal GU stacks were given free energy
increments 0.4 kcal/mol less favorable than the corresponding AU
stacks. Certain nucleotides were forced to be unpaired based on
nuclease sensitivity (Cech et al., 1983) or chemical reactivity data
(Inoue and Cech, 1985). The results for a linear and circular form of
the intervening sequence are shown in Figures 4 and 5, respectively.
The structures are very similar to the one suggested by Cech et al.
(1983), except the conserved sequences 9R and 9R' are paired in a
secondary rather than tertiary interaction. In addition, the pairing
between conserved sequences 9L and 2 is broken. This may be an
artefact, however, due to the inability of the algorithm to consider
tertiary interactions. Structures with tertiary pairing of 9L and 2
would have free energies similar to the structures in Figures 4 and 5.
Other tertiary interactions are also possible. To test the
sensitivity of the structures to the parameters for GU pairs,
calculations were also done with GU increments 0.2 and 0.3 kcal/mol
less favorable than the corresponding AU stacks. In these
calculations, the structures were considerably different from the 3'
side of box 9R to the 5' side of box 2 (nucleotides 279 to 306). This
suggests this region could exhibit conformational diversity.

SUMMARY

     Experimental evidence indicates stacking and hydrogen bonding are
important determinants of RNA stability, but that hydrophobic bonding
is not. The nearest neighbor model appears to be adequate for summing
up the interactions determining RNA structure. Recent advances in the
synthesis of RNA oligonucleotides make it possible to improve on and
expand the parameters used with the nearest neighbor model. These new
parameters should help improve predictions of RNA structure from
sequence.

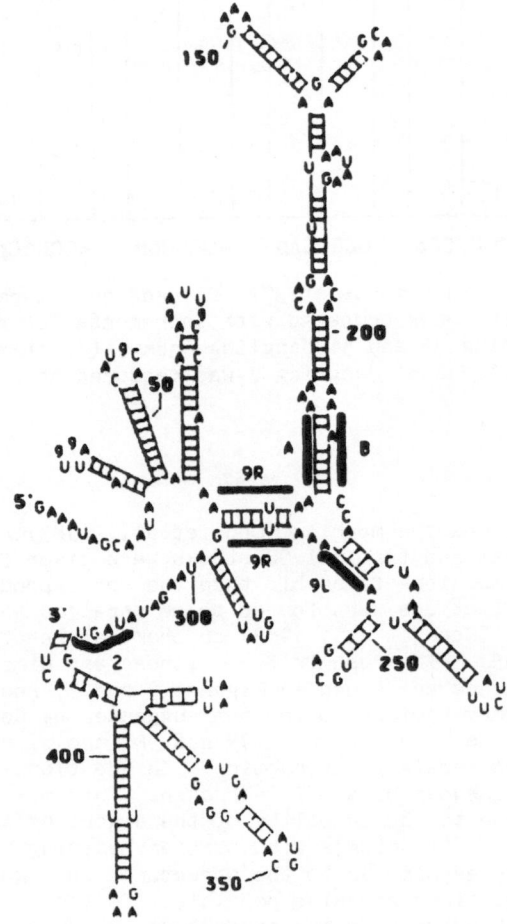

Fig. 4.   Secondary structure predicted for excised intervening
sequence from rRNA precursor of Tetrahymena thermophila.   The
computer program of Zuker and Stiegler (1981) was used with
the free energy parameters from Table 3, with GU stacks 0.4
kcal/mol less favorable than the corresponding AU stacks, and
with the T1 ribonuclease sensitivity data of Cech et al.
(1983).   (g) sites of T1 sensitivity.   Dark lines denote
conserved regions.

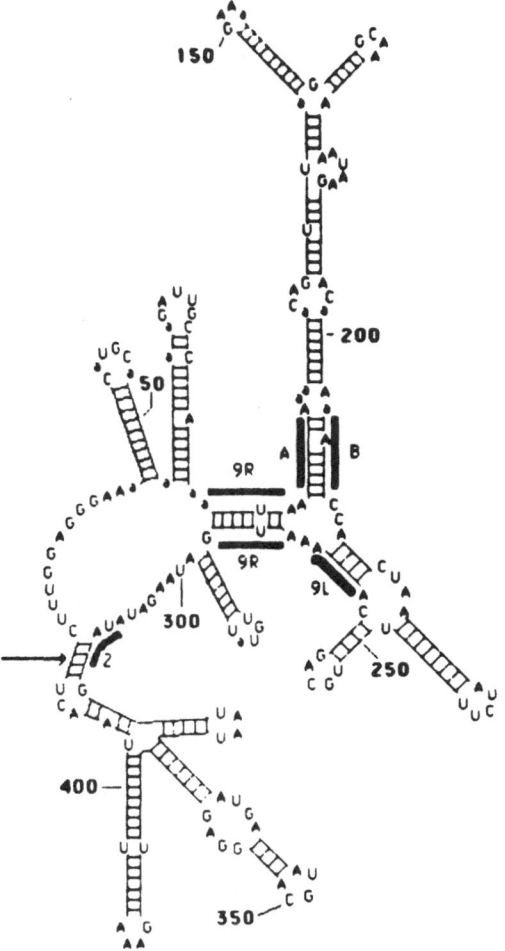

Fig. 5. Secondary structure predicted for circular form of the
Tetrahymena rRNA intervening sequence. The computer program
of Zuker and Stiegler (1981) was used with the free energy
parameters from Table 3, with GU stacks 0.4 kcal/mol less
favorable than the corresponding AU stacks, and with the
methylation data of Inoue and Cech (1985). The circle form
was approximated by a linear sequence with A(152) as the 5'
end. (a) sites of strong methylation by reaction with
dimethyl sulfate. Dark lines denote conserved regions.
Arrow denotes site of cyclization.

REFERENCES

Albergo, D. D., and Turner, D. H., 1981, Biochemistry, 20:1413.
Back, J. F., Oakenful, D., and Smith, M. B., 1979, Biochemistry,
    18:5191.
Beckett, D. and Uhlenbeck, O. C., 1984, Enzymatic Synthesis of
    Oligoribonucleotides, in: "Oligonucleotide Synthesis: A
    Practical Approach," M. J. Gait, ed., IRL Press, Oxford,
    pp. 185-197.
Borer, P. N., Dengler, B., Tinoco, I., Jr., and Uhlenbeck, O. C.,
    1974, J. Mol. Biol., 86:843.
Brandts, J. F. and Hunt, L., 1967, J. Am. Chem. Soc., 89:4826.
Cantor, C. R. and Schimmel, P. R., 1980, "Biophysical Chemistry
    Part I: The Conformation of Biological Macromolecules,"
    W. H. Freeman, San Francisco, California.
Cech, T. R., Tanner, N. K., Tinoco, I., Weir, B. R., Zuker, M., and
    Perlman, P. S., 1983, Proc. Natl. Acad. Sci. U.S.A., 80:3903.
Dewey, T. G. and Turner, D. H., 1980, Biochemistry, 19:1681.
England, T. E. and Neilson, T., 1976, Can. J. Chem., 54:1714.
Freier, S. M., Hill, K. O., Dewey, T. G., Marky, L. A., Breslauer,
    K. J., and Turner, D. H., 1981, Biochemistry, 20:1419.
Freier, S. M., Burger, B. J., Alkema, D., Neilson, T., and Turner,
    D. H., 1983a, Biochemistry, 22:6198.
Freier, S. M., Albergo, D. D., and Turner, D. H., 1983, Biochemistry,
    22:1107.
Freier, S. M., Alkema, D., Sinclair, A., Neilson, T., and Turner,
    D. H., 1985a, Biochemistry, 24:4533.
Freier, S. M., Sinclair, A., Neilson, T., and Turner, D. H., 1985b,
    J. Mol. Biol., 184:in press.
Freier, S. M., Sinclair, A., Alkema, D., Neilson, T., Kierzek, R.,
    Caruthers, M. H., and Turner, D. H., 1986a, in preparation.
Freier, S. M., Turner, D. H., Caruthers, M. H., and Kierzek, R.,
    1986b, in preparation.
Freier, S. M., Sugimoto, N., Jaeger, J. A., Neilson, T., Kierzek, R.,
    Caruthers, M. H., and Turner, D. H., 1986c, in preparation.
Freier, S. M., Neilson, T., Kierzek, R., Caruthers, M. H., and
    Turner, D. H., 1986d, in preparation.
Geiger, A., Rahman, A., and Stillinger, F. H., 1979, J. Chem. Phys.,
    70:263.
Gekko, K. and Timasheff, S. N., 1981a, Biochemistry, 20:4667.
Gekko, K. and Timasheff, S. N., 1981b, Biochemistry, 20:4677.
Gerlsma, S. Y. and Stuur, E. R., 1974, Int. J. Pept. Protein Res.,
    6:65.
Gralla, J. and Crothers, D. M., 1973, J. Mol. Biol., 73:497.
Hickey, D. R. and Turner, D. H., 1985a, Biochemistry, 24:2086.
Hickey, D. R. and Turner, D. H., 1985b, Biochemistry, 24:3987.
Inoue, T. and Cech, T. R., 1985, Proc. Natl. Acad. Sci., U.S.A.,
    82:648.
Jacobson, A. B., Good, L., Simonetti, J., and Zuker, M., 1984,
    Nucleic Acids Res., 12:45.
Kauzmann, W., 1959, Adv. Protein Chem., 14:1.
Kierzek, R., Freier, S. M., Swinton, D., Turner, D. H., and Caruthers,
    M. H., 1986, in preparation.
Langlet, J., Giessner-Prettre, C., Pullman, B., Claverie, P., and
    Piazolla, D., 1980, Int. J. Quantum Chem., 18:421.
Linse, P., Karlström, G., and Jönsson, B., 1984, J. Am. Chem. Soc.,
    106:4096.
Nelson, J. W., Martin, F. H., and Tinoco, I., Jr., 1981, Biopolymers,
    20:2509.
Nussinov, R., Tinoco, I., Jr., and Jacobson, A. B., 1982, Nucleic
    Acids Res., 10:341.

Okamoto, B. Y., Wood, R. H., and Thompson, P. T., 1978, J. Chem. Soc., Faraday Trans. 1, 74:1990.

Pangali, C., Rao, M., and Berne, B. J., 1979, J. Chem. Phys., 71:2982.

Papanicolaou, C., Gouy, M., and Ninio, J., 1984, Nucleic Acids Res., 12:31.

Parodi, R. M., Bianchi, E., and Ciferri, A., 1973, J. Biol. Chem., 248:4047.

Petersheim, M. and Turner, D. H., 1983a, Biochemistry, 22:256.

Petersheim, M. and Turner, D. H., 1983b, Biochemistry, 22:269.

Pipas, J. M. and McMahon, J. E., 1975, Proc. Natl. Acad. Sci., U.S.A., 72:2017.

Pullman, A. and Pullman, B., 1968, Adv. Quantum Chem., 4:267.

Pullman, B. and Pullman, A., 1969, Prog. Nucleic Acid Res. Mol. Biol. 9:327-402.

Romaniuk, P. J. and Uhlenbeck, O. C., 1983, Methods Enzymol., 100:52.

Salser, W., 1977, Cold Spring Harbor Symp. Quant. Biol., 42:985.

Schrier, E. E., Ingwall, R. T., and Scheraga, H. A., 1965, J. Phys. Chem., 69:298.

Schrier, E. E. and Scheraga, H. A., 1962, Biochem. Biophys. Acta, 64:406.

Swaminathan, S., Harrison, S. W., and Beveridge, D. L., 1978, J. Am. Chem. Soc., 100:5705.

Tinoco, I., Jr., Uhlenbeck, O. C., and Levine, M. D., 1971, Nature (London), 230:362.

Uhlenbeck, O. C., Baller, J., and Doty, P., 1970, Nature, 225:508.

Uhlenbeck, O. C., Martin, F. H., and Doty, P., 1971, J. Mol. Biol., 57:217.

Uhlenbeck, O. C. and Gumport, R. I., 1982, The Enzymes, 15:31.

Zuker, M. and Stiegler, P., 1981, Nucleic Acids Res., 9:133.

ANOMALOUS CONFORMATIONS OF RNA CONSTITUENTS :

2D NMR AND CALCULATIONAL STUDIES

Cornelis Altona

Gorlaeus Laboratory of the State University
P.O. Box 9502
2300 RA Leiden, The Netherlands

ABSTRACT

Short single-stranded RNA fragments normally display a strong tendency to favour a right-handed helical conformation. Relatively small changes in enthalpy and/or entropy of stacking, induced by minor structural variations may cause a large difference in stacking behaviour in aqueous solution, but do not appear to affect the detailed geometry of the stacked state. Recently developed NMR methods that allow for a determination of the sugar-phosphate backbone geometry along β (O5'-C5'), γ (C5'-C4'), δ (C4'-C3'), and ε (C3'-O3') are surveyed. At the level of trimers and higher oligomers conformational transmission factors, such as next-nearest-neighbour inter-actions, may come into play. For example, the trimer U-$\overline{\text{A}}$-U ($\overline{\text{A}}$ = m$_2^6$A) be-haves in a fashion that can be predicted from the known stacking proper-ties of its dimer constituents U-$\overline{\text{A}}$ and $\overline{\text{A}}$-U, whereas the trimer $\overline{\text{A}}$-U-$\overline{\text{A}}$ be-haves in an entirely different way. In the latter compound the two purines engage in a 1-3 stacking interaction. At the same time the central pyri-midine residue is pushed outside the purine-purine interaction zone (bulge-out).

Several interesting properties of bulges have come to light : (i) a longer alternating pu-py sequence displays multiple bulges, witness $\overline{\text{A}}$-U-$\overline{\text{A}}$-U-$\overline{\text{A}}$; (ii) a strong stacking interaction at its 3'-end does not affect the bulge, for example in $\overline{\text{A}}$-U-$\overline{\text{A}}$-U; (iii) in contrast, the bulge is abandoned in favour of a normal right-handed stacking pattern by a strong stacking interaction at its 5'-end : $\overline{\text{A}}$-$\overline{\text{A}}$-U-$\overline{\text{A}}$; (iv) a self-comple-mentary alternating pu-py sequence, e.g. (A-U)$_3$, is able to convert from a bulged single strand at elevated temperatures into a regular A-type duplex at low temperature. Thus, bulge-out structures may occur either in loops or under conditions where a duplex is forced to open up.

Another interesting conformation is shown by the 3'-terminal $\overline{\text{C}}$-$\overline{\text{A}}$ in $\overline{\text{C}}$-$\overline{\text{C}}$-$\overline{\text{A}}$ ($\overline{\text{C}}$ = m$_2^4$C), a chemically modified 3'-acceptor of tRNAs. In contrast to the usual right-handed parallel stacking pattern favoured by the $\overline{\text{C}}$-$\overline{\text{C}}$ part, the 3'-terminal $\overline{\text{A}}$ residue prefers to adopt a left-handed antiparal-lel stacking.

With the aid of molecular-mechanics calculations (AMBER program) various plausible A-U-A and C-C-A models could be generated.

INTRODUCTION

Geometrical details and dynamic properties of nucleic acid constituents in aqueous solution can be investigated in considerable depth by a combination of various techniques. The judicious use of high-resolution proton NMR spectroscopy provides insight into the magnitude of proton-proton and proton-phosphorus torsion (rotation) angles.[1] In addition, ^{13}C NMR yields carbon-phosphorus rotation angles.[2-4] This information, by inference, allows one to construct an accurate spatial configuration of the preferred form or forms of the sugar ring and of part of the nucleotide backbone.[1] At the same time this knowledge serves to exclude a large number of theoretically allowed conformational combinations, thus enabling one to carry out a well-oriented computer search for the relative merits of the remaining possibilities. Moreover, temperature-dependent ^{31}P NMR spectroscopy can yield quantitative information on the rotamer distribution about the two elusive P-O ester torsion angles, a distribution that thus far escapes direct monitoring by NMR methods.[5] Circular dichroism (CD) of oligonucleotides constitutes a probe for base-base interaction. The spectral intensity and band shape reflect general features of this interaction; the changes with temperature yield quantitative information on the important overall thermodynamic parameters ΔH and ΔS that govern the unstack-stack equilibrium.[6-10] NMR chemical-shift versus temperature profiles[10-13] (^{1}H, ^{13}C) in principle contain the same information — but on a structurally more localized scale — and the comparison of CD with NMR results may serve as a useful tool to distinguish between overall and local conformational changes.

The above remarks pertain in particular to oligonucleotides that display "normal" stacking behaviour, i.e. molecular systems showing well-defined nearest-neighbour base-base interactions.[14] This "vertical" base-base interaction usually is accompanied by a large change of the enthalpy ($-\Delta H_{stack} \sim 5$ to 7 kcal/mole per interaction in a single-helical array, the exact value of ΔH depending on the chemical nature of the bases in question). The extremely favourable enthalpy factor does not necessarily mean that nucleic acids are permanently locked into a base-stacked array, for at ambient temperature it is balanced by a counter factor: a large and unfavourable entropy term: $- \Delta S_{stack} \sim 15$ to 25 cal/mole/K. At 300 K the value of $- T\Delta S$ thus ranges from ~ 4.5 to ~ 7.5 kcal/mole, in favour of the random coil state. This delicate balance means that relatively small changes in ΔH and/or ΔS, caused by minor structural variations, may result in a large difference in stacking behaviour at ambient temperatures in aqueous solution.

At this point it is well to remember that the midpoint T_m of a two-state stack/unstack equilibrium follows from the relation $T_m = \Delta H/\Delta S$, i.e. a moderate 10% increase in ΔH (or 10% decrease in ΔS) would cause an appreciable (30°C) increase in T_m. Well-documented cases in point are the changes in T_m caused by N^6-dimethylation of the adenines in A-A and A-U to give $\overline{A}$-$\overline{A}$ and $\overline{A}$-U ($\overline{A} = m_2^6$A): T_m increases from 295 K to 317 K ($\overline{A}$-$\overline{A}$) and from 295 K to 310 K ($\overline{A}$-U)[8] In both cases cited meticulous CD and NMR studies[8,10,12] revealed that the substantial increase in T_m in aqueous solution (more stack at all temperatures between 0 - 100°C) is due to the fact that both ΔH and ΔS become less negative. However, the change of the entropy term outweighs the change in enthalpy. A more general phenomenon may be involved, because dimethylation of the N^4-amino groups in C-C also increases intramolecular stacking (H.P.M. de Leeuw and C. Altona, unpublished observations). Moreover, it is long known that the self-association constant (stacking) of free adenine bases in H_2O increases on N^6-dimethylation.[14] It should be emphasized that the increase in stacking propensity on N^6-(or N^4)-dimethylation is not accompanied by detectable structural or conformational changes of the sugar-phosphate backbone in any of the compounds examined thus far. The effect is perhaps related to changes in hydrophobicity of the methylated base (D. Riesner, private communication). Little is known of the influence of other N^6 or N^4 substituents. The T_m of m_2^6A-U is 300 K,[8] i.e. between A-U and $\bar{A}$-U, whereas our older work[15] suggests that the isopentenyl group actually slightly <u>decreases</u> the stacking proclivity of i^6A-U compared to A-U itself.

At the level of trimers and higher oligomers conformational transmission factors, such as next-nearest-neighbour interactions, may come into play. Optical studies[16,17] on a large number of trinucleoside diphosphates purport to show that the assumption of additivity of dimer properties works well in a number of cases, but fails completely in others. This should not come as a surprise, because longer-distance interactions — absent in dinucleoside monophosphates — may easily interfere with the delicate balance between ΔH and ΔS mentioned above.

The present paper reviews the conformational properties of some small oligoribonucleotides that are anomalous in the sense that these properties could not have been predicted from knowledge of the behaviour of the constituent dimers. The compounds in question have been studied at Leiden in recent years by means of ^1H, ^{13}C and ^{31}P NMR and circular dichroism. Some preliminary results of molecular mechanics calculations will be discussed. Since it was our explicit desire to analyse the behaviour of pure single strands, any possibility for Watson-Crick base-pairing was removed by chemical modification: N^6-dimethylation of the A bases ($\bar{A}$) and N^4-dimethylation of the C bases ($\bar{C}$). Moreover, this particular modification enhances stacking proclivity (*vide supra*) and it was conjectured that anomalous stacking interactions — if present — would stand out more sharply, as indeed proved to be the case.

CONFORMATIONAL ANALYSIS

Notation

Recent recommendations on nucleotide conformational nomenclature[18] are adopted in this paper, Fig. 1. The nucleotide units are indexed in parentheses, from the 5' to the 3' terminal: A(1)-B(2)-C(3)-D(4) ; the atoms,

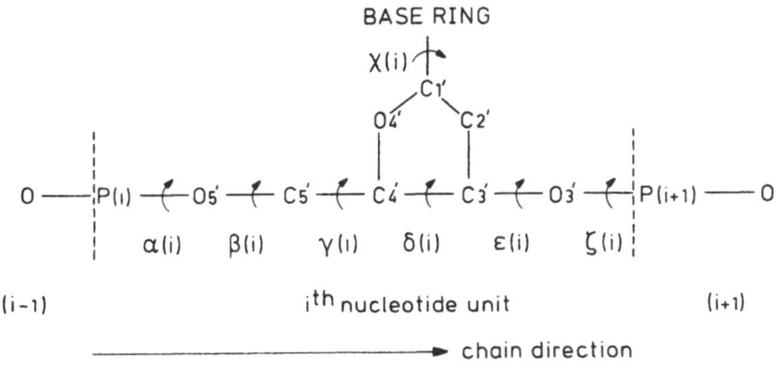

Fig. 1. α - ζ Notation for important torsion angles of nucleotide residue (i).

torsion angles and the corresponding NMR coupling constants etc. are designated with the same index. The torsion angles along the backbone are denoted
in sequence: α, β, γ, δ, ε, ζ, starting with P-O5' as the central bond
for α. The signs of the torsion angles are defined according to the usual
Klyne-Prelog rule.[19] Staggered conformational ranges correspond to g+
($\sim 60°$), t ($\sim 180°$) and g⁻ ($\sim 300°$ or $-60°$). The operational definitions
for g and t includes non-classical geometries, e.g. ε (220) implies ε^t.
The torsion angle about the glycosyl bond C1'-N, which gives the orienta-
tion of the base with respect to the sugar, is denoted by χ . For $\chi = 0°$,
the O4'-C1' bond eclipses the N9-C4 bond in purines and the N1-C2 bond in
pyrimidines.

The Sugar Ring

In biochemical literature the two distinct D-ribose conformations are
commonly denoted as C3'-endo and C2'-endo, respectively. One should keep
in mind that these terms carry a precise meaning: C3'-endo denotes a ^3E
(envelope) form with the phase angle of pseudorotation[20] P = 18° , while
C2'-endo stands for a ^2E form with P = 162°. However, X-ray and NMR
studies have shown that pure ^3E and ^2E conformers are extremely rare oc-
currences.[20,21] Moreover, these terms tend to give the false impression
that sugar conformations are rigid, whereas in reality the pseudorotational
movement of the furanose ring constitutes one of the major degrees of freedom of the backbone of nucleic acids (in DNA even more outspoken than in
RNA). For this reason we prefer to distinguish between the N-type genus of
forms (P = 0° ± 90°) and the S-type genus (P = 180° ± 90°), Fig. 2. The
torsion angles ϕ_j in an — approximately — equilateral 5-membered ring
are mutually related by a 2-parameter equation[20], Eq. 1:

$$\phi_j = \phi_m \cos [P + 4\pi(j-2)/5] \qquad j = 0, 1, 2, 3, 4 \qquad (1)$$

in which ϕ_m represents the maximum amplitude of pucker, P the phase angle
and the endocyclic torsions are numbered clockwise, starting with
C4'-O4'-C1'-C2' (j=0). A recent survey of 178 crystal structures[22] of
nucleosides and nucleotides amply confirmed our earlier finding[20] that the
normally observed P values can be grouped into two fairly narrow ranges:
0°-18° (N-type) and 144°-180° (S-type), with occasional structures found
outside these limits. Knowledge of two or three endocyclic torsion angles,
e.g. from X-ray crystallography or from proton NMR coupling constants,
enables one to uniquely deduce the pseudorotation parameters P and ϕ_m
(Eq. 1) from which not only the remaining endocyclic torsions but also the
important backbone angle δ can be calculated, Eq. 2.

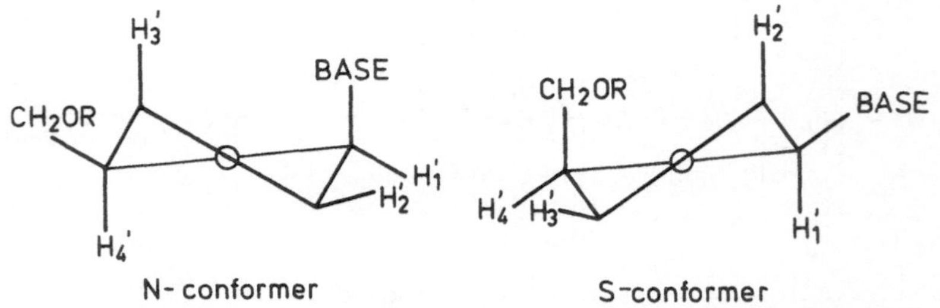

Fig. 2. Idealized projections of the two basic conformations of the
ribose ring: N-type (C3'-endo) and S-type (C2'-endo), see
ref. 20 for the first pseudorotation analysis of ribosides.

$$\delta = 120.6 + 1.1 \, \phi_m \cos(P + 145.2) \tag{2}$$

Eq. 2 is based upon a corrected pseudorotation equation[23]. The constants of Eq. 2 were determined from least-squares fit to the crystal structures data set[23].

Given the fact that in (aqueous) solution of nucleic acids the N- and S-type conformers most often occur side-by-side in fast equilibrium, the NMR spectroscopist is faced with the simultaneous determination of four geometrical parameters (P_N, ϕ_N, P_S, ϕ_S) and one equilibrium constant K. The Leiden group has spent considerable effort in order to provide a dependable and routine solution to this problem. Two keys to success were developed: (i) de Leeuw et al.[22] derived a set of empirical equations which relate proton-proton torsion angles to the endocyclic furanose torsions; (ii) Haasnoot et al.[24,25] extended the classical Karplus equation by the introduction of terms which account for the electronegativity and orientation of substituents. The successful application of these findings to the conformational analysis of nucleic acids and prolines has been documented extensively[9,10,26,33] and will not be treated here. Suffice it to say that an iterative least-squares computer program PSEUROT[30,34] yields the pseudorotation parameters, i.e. the geometry, of the stable furanose conformers and the S/N molar ratio, given a complete set of $^3J_{HH}$ couplings along the bonds C1'-C2', C2'-C3' and C3'-C4'. Some representative couplings are shown in Table I. The accuracy of the procedure is greatly enhanced in case the N/S equilibrium constant displays a strong shift with temperature, as is usually found for oligomers. In fact, the geometry of the sugars, and thus also backbone angle δ (Eq. 2), of nucleic acid fragments in solution now often can be determined with an accuracy that is at least equal to that of a high-resolution X-ray crystal structure analysis.

Backbone Angles β, γ, ϵ in Solution

β(P - O5' - C5' - C4'). The conformational preference about β is best monitored by means of the vicinal proton-phosphorus couplings $J_{H5'P}$ and $J_{H5''P}$, Fig. 3. In 5'-mononucleotides, the β^t (trans $\sim$ 180°) form is invariably preferred (> 70%). Nearest-neighbour stacking in oligomers requires the adoption of a pure β^t state and for this reason the behaviour of $J_{H5'P}$ and $J_{H5''P}$ deserves special attention. A Karplus-type relation, which correlates J_{HP} with the torsion angle ϕ_{HP}, was recently reparametrized[2], Eq. 3.

$$^3J_{(HCOP)} = 15.3 \cos^2\phi_{HP} - 6.1 \cos\phi_{HP} + 1.6 \tag{3}$$

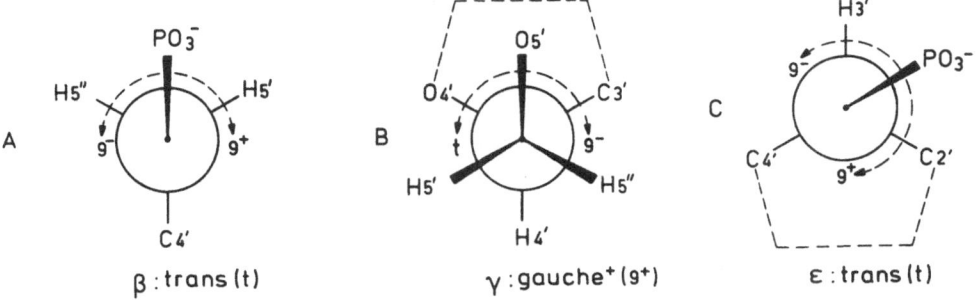

$$\beta : trans\,(t) \qquad \gamma : gauche^+\,(g^+) \qquad \epsilon : trans\,(t)$$

Fig. 3. Newman projections of the preferred backbone rotamers: β, viewed along the O5'-C5' bond; γ, along the C5'-C4' bond: ϵ, along the O3'-C3' bond. Possible rotations into less favoured regions are indicated by arrows. The ϵ^+ region appears to be avoided completely.

Table I. Predicted coupling constants[a] (Hz) of a few ribose N-type and S-type conformations (corrected for the through-space transmission effect[b])

	P_N	ϕ_N	δ_N[c]	$J_{1'2'}$	$J_{2'3'}$	$J_{3'4'}$	$J_{1'2'}+J_{3'4'}$
${}^3_2T$	0°	37°	87.2°	1.01	4.78	8.34	9.35
	9	37	84.0	1.05	4.84	8.67	9.72
${}^3E$	18	37	81.6	1.20	5.04	8.89	10.09
	P_S	ϕ_S	δ_S[c]				
${}^2_1T$	144°	37°	134.0°	7.93	5.80	2.18	10.11
	153	37	139.8	7.89	5.38	1.60	9.49
${}^2E$	162	37	145.2	7.76	5.07	1.19	8.95

[a]In practice the use of program PSEUROT[34] is recommended to derive P_N, ϕ_N and δ from measured sets of couplings. [b]de Leeuw et al.[35] [c]Eq. 2.

Under the additional assumption that in unstacked states an equilibrium exists between three classical rotamers (β^t, 180°; β^+, 60°; β^-, 300°) a sum rule is easily derived, Eq. 4.

$$p(\beta^t) = (25.4 - \Sigma') / 20.6 \qquad (4)$$

where p stands for fractional population and $\Sigma' = J_{H5'P} + J_{H5''P}$. It can be shown that Eq. 4 holds well (± 3%) even when β^t deviates up to 10° from the classical 180° value. Such deviations are revealed by the difference $J_{H5''P} - J_{H5'P}$ [10].

A recalculation of a series of β^t values of stacked (single-helical) RNA species (Table VII in ref. 1), with the aid of the new $J_{(HCOP)}$ Karplus parameters[2], gives a range of β^t 172°-180°, mean 174°. This number perfectly agrees with that obtained from X-ray crystallography: (i) mean[1] of stacked and base-paired dimers AU, GC and AA[+]: β^t, 174° ± 8°; (ii) RNA-11 duplex fibre[36]: β^t, 175°. At this point it should be noted that the H5', H5'' region in the NMR spectra of larger oligonucleotides is densely packed with overlapping resonances and an accurate spectral analysis in terms of $J_{H5'P}$, $J_{H5''P}$ of double helices is currently quite impractical. A similar remark applies to $J_{H4'H5'}$ and $J_{H4'H5''}$, which couplings monitor backbone angle γ.

γ(O5' - C5' - C4' - C3'). Three staggered rotamers exist and all three are known from X-ray crystallography, Fig. 3: γ^+ ($\sim$ 60°), γ^t ($\sim$ 180°), γ^- ($\sim$ 300°). A survey of crystal structures[22] showed that the γ^+ region is most richly populated (76%), followed by γ^t (18%) and by the rare γ^- (6%). This distribution is fairly accurately mirrored by the distribution found for monomeric nucleosides and nucleotides in aqueous solution at ambient temperatures. The crystallographic data moreover revealed outspoken trends towards non-classical angles, especially for γ^+ ($\sim$ 53°) and γ^- ($\sim$ 290°).

The γ^+ population $p(\gamma^+)$ of nucleic acids in solution can be deduced[1] from the sum $\Sigma = J_{H4'H5'} + J_{H4'H5''}$, Eq. 5.

$$p(\gamma^+) = (13.3 - \Sigma) / 9.7 \qquad (5)$$

A right-handed stack requires a pure γ^+ conformation and thus knowledge of the percentage of stack at a series of temperatures often allows reasonably accurate (± 0.3 Hz) extrapolation of observed couplings $J_{H4'H5'}$ and $J_{H4'H5''}$ to values corresponding to the fully stacked state. With the aid of the extended Karplus equation[24] it is then possible (provided the H5', H5'' assignment is secure, see ref. 1) to derive torsion angle γ^+ from NMR data. A survey (Table VII in ref. 1) revealed that γ^+ in right-handed single helical arrays remains surprisingly constant: 49° ± 4° in the RNA fragments studied thus far. This figure again agrees with crystalline RNA miniduplexes[1]: 55° ± 5° and with the RNA-11 duplex[36]: 49°.

ϵ(C4' - C3' - O3' - P). The conformational situation about ϵ is usually given in terms of an equilibrium between two, distinctly non-classical, rotamers ϵ^t and ϵ^-, the ϵ^+ region appears "forbidden". Fig. 3C reveals that ϵ can in principle be monitored by three couplings: $J_{H3'P}$, $J_{C4'P}$ and $J_{C2'P}$. Unfortunately, from $J_{H3'P}$ only limited information can be obtained since the ϵ^t and ϵ^- regions both position the ^{31}P gauche to H3'. The situation is much improved when ^{13}C NMR is employed: the ϵ^t region (by definition) has C4' *trans* to ^{31}P (ϵ 216 8.0 Hz) and C2' *gauche* to ^{31}P (ϵ 216 1.1 Hz) ; in the ϵ^- region these magnitudes are reversed.

The successful application of ^{13}C NMR to the study of the geometry of nucleic acid constituents was in the past hampered by several factors: (i) the lack of a fully reliable assignment procedure; (ii) the lack of a good set of calibration points from which Karplus-type parameters can be extracted. Limitation (i) has been removed[13] by the advent of a two-dimensional (2 D) NMR technique. Heteronuclear chemical shift correlation spectroscopy provides a one-to-one correspondence between the ^1H spectrum and the ^{13}C signals of C-H coupled carbons. Of course, the method presupposes the correct assignment of the ^1H spectrum, but in an era where so many excellent 2D homonuclear techniques are available this has become almost routine.

Limitation (ii) has been removed by Lankhorst et al.[2] A careful study of several oligoribonucleotides, measured at a number of temperatures, led to the determination of six Karplus parameters and four torsion angles for $J_{(CCOP)}$ and $J_{(HCOP)}$ from 17 experimental couplings. This was achieved by means of a least-squares computer analysis. The resulting Karplus parameters are given in Eq. 3 (above) and in Eq. 6.

$$^3J_{(CCOP)} = 6.9 \cos^2\phi_{CP} - 3.4 \cos \phi_{CP} + 0.7 \qquad (6)$$

It was noted[2] that applications of Eq. 6 in cases where $0° < \phi_{CP} < 60°$ should be treated with caution, because a directly measured coupling in this range is as yet unavailable. The fit for the 17 data points $60° < \phi_{CP} < 360°$, rms deviation 0.2 Hz, was quite satisfactory. Similarly gratifying was the comparison[2] of ϵ^t values in four fully stacked RNA fragments in solution (ϵ^t_{mean} 219°) with the mean of 10 X-ray determinations (ϵ^t 218°). For the more elusive (minor) ϵ^- rotamer a value of 277° (± 5°) was estimated.

In summary, it is concluded that a careful ^1H and ^{13}C NMR analysis of RNA fragments in aqueous solution in principle may produce reliable geometrical parameters (torsion angles) along four out of six sugar-phosphate backbone bonds, β, γ, δ, and ε. In practice, limitations are encountered. In the first place, single-stranded fragments do not attain a state of conformational purity (100% stack) in aqueous solution within the accessible temperature range. Extrapolation procedures therefore are required and these necessitate either the quantitative determination of ΔH and ΔS or — at least — knowledge of rotamer populations as function of temperature. In the second place, 2D NMR spectra of duplexes — which compounds satisfy the requirement of conformational purity — as yet defy the requirement that accurate couplings (± 0.3 Hz) need to be extracted in order to carry out a satisfactory geometrical analysis. Undoubtedly, further hardware and software developments in the 2D NMR field will soon lift the limitations and allow for a detailed insight into the conformational behaviour of RNA duplexes. The groundwork has already been laid.

The anomalous RNA sequences that are treated in the next section partly defy a precise geometrical analysis of the NMR data, because these compounds appear to exist as a complex mixture of stacked conformations. Nevertheless, the $^3J_{HH}$, $^3J_{HP}$ and $^3J_{CP}$ data obtained serve to exclude a large number of *a priori* plausible theoretical models.

ANOMALOUS RNA CONFORMATIONS

Alternating Purine-Pyrimidine-Purine Sequences

A recent investigation[37] of the behaviour of the chemical shifts of the base and H1' protons of thirteen RNA trimers revealed that most of the shifts could be predicted from the constituent dimers. The purine bases, through the well-known ring-current effect, strongly shield the protons (upfield shift) of nearest-neighbour residues when these are stacked "above" or "below". To a good approximation these shieldings are additive. In contrast, the three R-Y-R trimers that were present in the collection[37], A-C-G, A-U-G, and G-U-G, behaved anomalously. The protons of the central pyrimidine residues were less shielded, whereas those of the flanking pur-ines were more shielded than expected. The authors[37] therefore proposed the existence of a conformation in which the interior residue is "bulged-out" and the two terminal bases stack upon each other.

The work by the Leiden group has modified and extended our knowledge of the bulging phenomenon. Tables II and III show the differential shield-ings featured by regular nearest-neighbour stacks and by bulged-out frag-ments, respectively. Take for example the UH1' proton shifts. In U-Ā and A-Ū the UH1' signals move upfield by ∿ 0.4 ppm. One therefore predicts that an upfield shift of ∿ 0.8 ppm would occur in an Ā-U-Ā single helix; the ex-perimental value amounts to only 0.14 ppm. The UH5 and UH6 resonances follow a similar pattern. Therefore, the internal U residue of Ā-U-Ā spends at most 15-20% of its time between the two flanking Ā residues. In contrast, the protons of the terminal Ā residues of Ā-U-Ā show an surprisingly large upfield shift of 0.13-0.24 ppm compared to expectations: in a regular stack (witness Ā-U and U-Ā) as well as in random-coil forms the differential shieldings should be close to zero. Instead, the shifts found for e.g. AH1'(1) and AH1'(3) amount to roughly 50% of those in the well-stacking dimer Ā-Ā, those of AH8 even amount to ∿ 100%. It must be concluded, follow-ing Lee and Tinoco[37], that the flanking Ā bases stack upon each other to a large extent at the expense of the internal U residue, which is squeezed out.

Bulged-out triplets share another characteristic property: compared to classical RNA helices, which invariably prefer to have the sugars adopt the N form (Table IV), the bulges definitely prefer S (Table V).

Taking the data of Tables III-V together, as well as other results (not shown), the following picture unfolds: (i) neither the Ā-U nor the U-Ā sequence by itself impediments stacking, witness U-Ā-U; (ii) Ā-U-Ā-U displays normal stacking at its 3'-terminus as well as a bulge at its 5'-terminus[27]. Evidently, the bulge structure allows 3'-stacking and this has important consequences for computer model building; (iii) in contrast, the bulge structure does not seem to allow 5'-stacking, as Ā-Ā-U-Ā largely reverts to a normal helix at least as far as the first three residues are concerned; (iv) the double bulge[39] displayed by the pentamer Ā-U-Ā-U-Ā probably can be interpreted to mean that the bulges retain sufficient

Table II. Differential shieldings[a,b] (10^{-2} ppm, 20°C, 5 mM, pH 7) of U and Ā residues of some RNA fragments that partake in a nearest-neighbour stack

proton	5'-terminal		3'-terminal			internal
	UĀ	UĀU	ĀU	UAU	AUAU	ĀAUĀ
UH5	16	24	55	48	56	41
UH6	17	5	35	32	41	39
UH1'	39	34	40	29	37	41

	5'-terminal			3'-terminal			internal	
	ĀU	ĀA	ĀAUĀ	UĀ	ĀA	ĀAUĀ	UAU	AAUĀ
AH2	-1	31	17	0	9	1	-1	24
AH8	-7	13	13	0	25	17	3	30
AH1'	6	35	28	-4	24	8	7	34

[a] Refs 27, 38, 39 and unpublished observations by P.Gijsman, C.M. Groeneveld, Y.Th. van den Hoogen, S.J.Treurniet, G.M.Visser and C.Altona. [b] Shieldings are relative to the corresponding monomeric methyl phosphate esters: mpU, Upm, mpUpm, mpA, Apm, mpApm. Values were obtained by graphical interpolation in cases where a measurement at 20°C was unavailable.

Table III. Differential shieldings[a] (10^{-2} ppm, 20°C, 5 mM, pH 7) of U and Ā residues of some RNA fragments that partake in the bulging phenomenon

proton	internal			
	ĀUĀ	ĀUAU	ĀUĀUĀ	ĀUĀUĀ
UH5	15	11	13	25
UH6	10	7	10	17
UH1'	14	12	14	24

	5'-terminal			3'-terminal		internal
	ĀUĀ	ĀUĀU	ĀUĀUĀ	ĀUĀ	ĀUĀUĀ	ĀUĀUĀ
AH2	18	14	25	13	10	19
AH8	18	15	22	24	17	32
AH1'	14	15	15	13	11	32

[a] Footnotes a and b, Table II.

flexibility to adapt to conflicting demands; (v) the non-methylated hexamer (A-U)$_3$, although bulged at high temperature, reverts to a normal A-RNA-like duplex at low temperature[40]; (vi) nearest-neighbour stacked structures, by virtue of the large TΔS term, characteristically unstack rapidly on increasing temperature, as revealed by large changes in N-type population and in chemical shifts. In contrast, all residues partaking in bulges — except perhaps Ā(5) of the pentamer — characteristically show virtually no

Table IV. Populations[a] of N-type conformers (%) of some RNA residues that partake in a nearest-neighbour stack as function of temperature[b]

T (°C)	5'-terminal						internal			
	$\overline{A}U$	$\overline{AA}$	$\overline{AAUA}$	$U\overline{A}U$	AA	$U\overline{A}$	$U\overline{A}U$	$AU\overline{A}U$	$\overline{AA}U\overline{A}$	$AAUA$
0	91	88	80	87	73	63	100	89	79	74
20	74	83	75	71	62	54	84	80	71	54
45	57	64	64	56	--	--	68	67	66	51
60	49	--	56	--	--	45	--	51	58	40

3'-terminal

T (°C)	$\overline{A}U$	$AU\overline{A}U$	$\overline{AA}$	AA	$U\overline{A}$	$\overline{AAUA}$[c]
0	92	85	83	66	59	51
20	79	75	75	57	56	49
45	64	65	--	--	53	46
60	--	56	--	--	53	47

[a]Footnote a, Table II. [b]Temperatures are rounded off to the nearest 5°C. [c]This residue appears to be largely unstacked.

Table V. Populations[a] of N conformer (%) for U and $\overline{A}$ residues that partake in bulging structures as function of temperature[b]. Monomer data are shown for purposes of comparison

T (°C)	5'-terminal				
	$\overline{A}pm$	Upm	$\overline{A}UA$	$A\overline{U}AU$	$AUAUA$
0	11	40	26	34	30
20	15	41	27	30	30
45	18	--	31	28	30
60	21	41	32	29	30

internal

T (°C)	$mp\overline{A}pm$	$mp\overline{U}pm$	$\overline{A}UA$	$A\overline{U}AU$	$\overline{A}UAUA$	$AU\overline{A}UA$	$AUA\overline{U}A$
0	14	24	30	34	25	43	40
20	16	25	30	32	25	41	--
45	--	25	32	32	25	39	41
60	22	25	33	32	25	38	--

3'-terminal

T (°C)	mpA	$iprp\overline{U}$	$\overline{A}UA$	$AU\overline{A}UA$
0	34	45	49	56
20	34	42	49	51
45	--	46	49	46
60	--	--	49	46

[a]Footnote a, Table II. [b]Footnote b, Table IV.

geometrical changes with temperature: the N/S equilibrium (Table V) as well as the differential shieldings of UH5, UH6 and UH1' (Fig. 9 in ref. 39) remain approximately constant over the temperature range 0°C - 75°C. This finding does not exclude the possibility of a strong 1,3-stacking inter-action between purine bases; it must mean, however, that ΔS of the 1,3-interaction is much smaller than ΔS of 1,2 interactions. A theoretical four-state thermodynamic computer analysis[41] of the trimer $\overline{A}$-U-$\overline{A}$ showed that — under assumption of normal ΔH and ΔS for the $\overline{A}$-U and U-$\overline{A}$ constitu-ent parts — the bulge population will remain high and approximately con-stant (> 70%) from 0°C to 100°C for $\Delta H_{1,3}$ 6 kcal/mole and $\Delta S_{1,3} \leq 14$ cal/mole/K. A smaller entropy factor suggests either greater conformational freedom of the bulge (i.e. the side-by-side existence of various bulged states) or an increase in low-lying vibrational states.

The backbone geometry of the series of bulged molecules has been the subject of in-depth NMR investigations at our laboratory. The β^t and γ^+, γ^t, γ^- populations have been tabulated elsewhere[27,38,39]. The main results can be summarized as follows. The β^t populations of the bulges (77 ± 5%) equals that of the constituent monomers mpA and iprpU (78-81%). No prefer-ence for either β^- or β^+ can be detected. The bulges also prefer the well-known γ^+ rotamer; the internal U residues display γ^+ populations ranging from 77-86% (compare the monomers 78-81%); the internal and 3'- terminal A residues, however, display a relatively high γ^t content (24-33%) at the

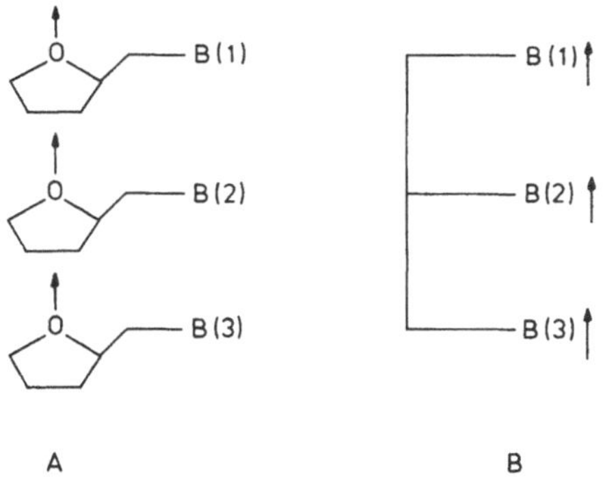

Fig. 4. Schematic representation of a right-handed stacked array. The furanose O4' all point "up", i.e. in the 5' direction.

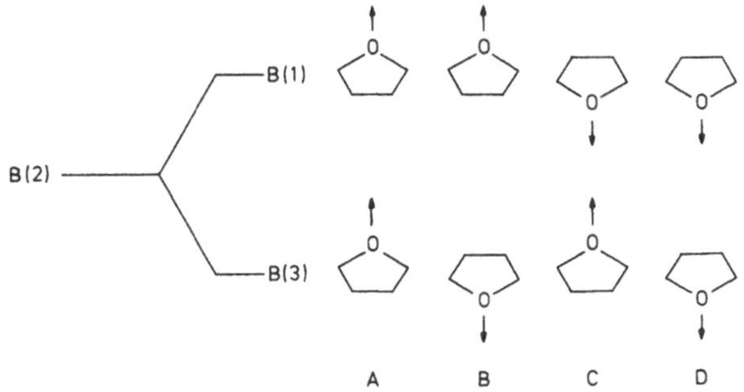

Fig. 5. Schematic representation of the four possible furanose O4' orientations in bulged structures that have a 1,3 base-base interaction. Sample models for cases B, C, and D are shown in Fig. 6, see also Table VI.

Table VI. Backbone angles and energy of the calculated A-U-A 'best' models[a]

no[b]	E	5' st	res.	α	β	γ	δ	ε	ζ	χ	3' st
2	-109	yes	A(1)	-	-	59	85	194	277	-119	
(B)			U(2)	183	179	56	145	213	39	-158	
			A(3)	75	191	67	102	-	-	- 73	no
3	-108	no	A(1)	-	-	58	152	292	69	-115	
(C)			U(2)	65	182	61	153	189	23	-134	
			A(3)	169	175	171	165	-	-	-176	yes
4	-112	no	A(1)	-	-	69	153	279	61	-105	
(D)			U(2)	73	179	60	152	213	69	-127	
			A(3)	64	189	56	92	-	-	-127	no
5	-106	yes	A(1)	-	-	60	84	199	288	-139	
(E)			U(2)	280	168	64	86	190	283	-142	
			A(3)	288	181	66	153	-	-	-125	yes

[a] Energies are given in kcal/mole. Molecular mechanics calculations were performed by means of the AMBER program[43], 1984 force field[44]. The dielectric constant $\varepsilon(diel)= 2,5$ was chosen instead of the usual $\varepsilon(diel)= 1R$. [b] The lettering corresponds to that indicated in Fig. 6. Models B, C, and D correspond to the sugar orientations depicted in Fig. 5; Model E represents a classical right-handed single helix.

Table VII. Backbone angles and energy of some calculated C-C-A models[a,b]

no	E	5' st	res.	α	β	γ	δ	ε	ζ	χ	3' st
1	-106	yes	C(1)	-	-	57	80	189	274	-134	
			C(2)	285	184	56	86	178	49	-177	
			A(3)	88	194	69	153	-	-	-155	no
2	-104	yes	C(1)	-	-	58	82	189	273	-113	
			C(2)	280	172	57	89	176	45	-118	
			A(3)	90	190	67	88	-	-	-162	no
3	-101	yes	C(1)	-	-	58	80	189	274	-131	
			C(2)	280	176	58	85	197	305	-154	
			A(3)	282	172	58	90	-	-	-163	yes

[a] Footnote a, Table VI. [b] Models number 1 and 2 represent a classical right-handed stack of the C(1)-C(2) sequence and an antiparallel stack at the C(2)-A(3) step, N-N-S and N-N-N, respectively. Model number 3 represents an N-N-N single helix.

expense of γ^+ (63-76%). The conformational situation about ε in $\overline{A}$-U-$\overline{A}$ appears to differ significantly from that in normally stacked compounds[2]. In the latter, typical couplings $J_{C4'P}$ (7.2-7.5 Hz) and $J_{C2'P}$ (1.3-1.9 Hz) signify preference for ε^t; in $\overline{A}$-U-$\overline{A}$ $J_{C4'P}$ (3.9 Hz) < $J_{C2'P}$ (5.1 Hz in $\overline{A}$(1) and 4.4 Hz in U(2)). It is concluded that the ε^- rotamer — which correlate: with S-type sugar conformation — plays an important role in the overall structure of the bulges. This conclusion accords with the observation[38,39,41] of large (W-path) long-range $J_{H2'P}$ couplings in all bulges (0.5-1.3 Hz).

The ^{31}P resonances of $\overline{A}$-U-$\overline{A}$ have been assigned[41] by means of selective decoupling experiments; at 1°C P(2) 0.85 ppm and P(3) 0.70 ppm, compare $\overline{A}$-U P(2) 0.44 ppm, relative to A>p. Since it has been estimated[5] that the ^{31}P nucleus in ζ/α combinations of the type g/t or t/g resonates about 2.5 ppm downfield of A>p, it follows that the bulge strongly prefers one or more of the following sequences: ζ^-/α^+, ζ^+/α^-, or ζ^+/α^+ (ζ^-/α^- does not apply, since it leads to a regular 1,2 stack).

The glycosyl angle χ (O4'-C1'-N1-C4 in purines, O4'-C1'-N1-C2 in pyrimidines) escapes direct determination by NMR. A qualitative estimate of *anti* and *syn* conformations appears possible, however, by means of T_1 and Nuclear Overhauser Effect (NOE) measurements. Preliminary NOE results[4l] indicate an *anti* base for residues U(2) and A(3) in $\overline{A}$-U-$\overline{A}$, whereas both *anti* and *syn* seem present in residue A(1). Further work is needed to clear up this point.

A Chemically Modified tRNA Acceptor End; $\overline{C}$-$\overline{C}$-$\overline{A}$

The trimer $\overline{C}$-$\overline{C}$-$\overline{A}$ and its constituent dimers were extensively studied[42] by means of ^{1}H, ^{13}C, and ^{31}P NMR spectroscopy and by CD. Only the main findings will be reiterated here. At low temperature the trimer displays a normal right-handed stacking pattern for the $\overline{C}$(1)-$\overline{C}$(2) step: strong preferences for N(1), ε^t(1), β^t(2), γ^+(2), N(2) (all ≥ 84% at 24°C), both bases *anti* (NOE and T_1) and a ζ^-/α^- combination (δ ^{31}P(2) < 0.7 ppm at low temperature). The dimer C-Cpm exhibits much the same conformational parameters as does the C(1)-C(2) step in the trimer. Moreover, this compound displays an excellent two-state behaviour (stack-unstack) in CD and NMR as the temperature is raised. In contrast, the C(2)-A(3) step of the trimer displays novel features, although part of the backbone behaves in a normal fashion: N(2), ε^t(2), β^t(3), γ^+(3), S(3). The outspoken preference for S-type ribose (64% at 0°C) varies little with temperature. It would be wrong, however, to conclude from the above data that the $\overline{A}$(3) residue behaves as a non-stacked (random-coil) entity. First, it was observed that several protons of residue $\overline{C}$(2) "feel" the shielding power of $\overline{A}$(3). Second, if the $\overline{C}$-$\overline{A}$ part of the trimer were unstacked, the CD spectrum of $\overline{C}$-$\overline{C}$-$\overline{A}$ would show a close resemblance to that of dimer $\overline{C}$-Cpm, which is not the case (Fig. 6 in ref. 42). Instead, the shape of the CD difference spectrum — C-Cpm subtracted from $\overline{C}$-$\overline{C}$-$\overline{A}$ — is the inverse of the usual dinucleotide spectrum. Third, the resonance signal of the ^{31}P(2) nucleus is found at low field (1.1-1.2 ppm from A>p) and its temperature conduct is anomalous. These and other data, taken together, suggest the approximately simultaneous melting out of (at least) two different antiparallel stacked states, one characterized by a ζ/α gg combination and another by a t,g or g,t combination.

Molecular Mechanics Calculations

A right-handed helical array naturally has all sugar-ring oxygens (vectors from the midpoint of the C2'-C3' bond to O4') pointing towards the 5' direction of the helix (arrows in Fig. 4). However, 1,3 stacking, as envisaged for bulges, in principle can occur in four different fashions (Fig. 5). It is easily appreciated that the "up" situation of the 5' residue (Fig.5A,B) would allow normal 1,2 stacking "on top", provided that other geometrical conditions (sugar N, base anti) are met. Similarly, the "up" direction of the 3' residue (Fig. 5A,C) in principle accommodates a 1,2 stacking "below". The bulge plus 3' stack displayed by $\overline{A}$-U-$\overline{A}$-U thus immediately rules out the situation depicted in Fig. 5B,D, at least for this particular molecule. In order to see whether sterically acceptable models of 1,3 stacks can be built, energy minimizations were carried out with the aid of program AMBER[43] and use of the 1984 force field[44]. Some preliminary results for (non-methylated) A-U-A and C-C-A are presented in Tables VI and VII, respectively. The calculated A-U-A structures of Table VI are visualized in Fig. 6. The approximately parallel purine base planes in the various A-U-A bulge models are at

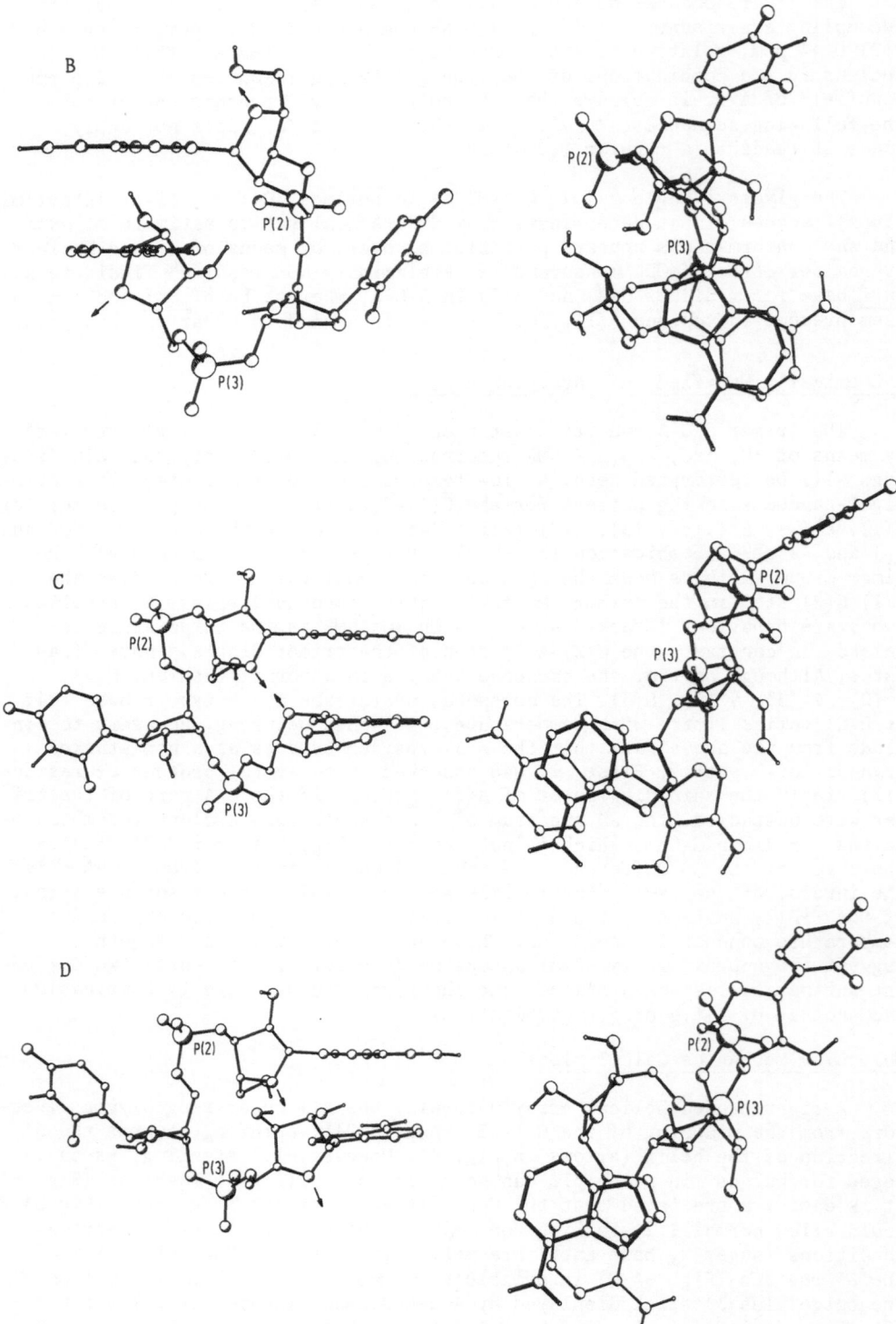

Fig. 6. Calculated models (Table VI) of three possible bulged A-U-A conformations, cases B, C, and D of Fig. 5. According to our calculations the A(1) base of B, C, and D can adopt a *syn* conformation at little cost, whereas A(1) *syn* in a calculated helix significantly raises the energy.

a distance of 3.2-3.4 Å and the overlap is good. Further model-building studies are currently being planned.

Acknowledgements

This work was supported by the Netherlands Foundation of Chemical Research (SON) with financial aid from the Netherlands Organization for the Advancement of Pure Research (ZWO). The synthetic efforts of Professor Dr. J.H. van Boom, Dr. G.A. van der Marel and Mrs. G. Wille are gratefully acknowledged. The following persons participated in this research; Bulge structures: A.J. Hartel, C.M. Groeneveld, P. Gijsman, Y.Th. van den Hoogen, Dr. P.P. Lankhorst, S.J. Treurniet and G.M. Visser; C.C.A.: H.P.M. de Leeuw and Dr. J. Doornbos; molecular mechanics; A.A. van Beuzekom. Dr. E. Westhof kindly provided a copy of the AMBER program with permission from Professor Dr. P. Kollman.

REFERENCES

1. C. Altona, Recl. Trav. Chim. Pays-Bas 101:413 (1982).
2. P. P. Lankhorst, C. A. G. Haasnoot, C. Erkelens, and C. Altona, J. Biomol. Struct. Dyns. 1:1387 (1984).
3. P. P. Lankhorst, C. A. G. Haasnoot, C. Erkelens, and C. Altona, Nucleic Acids Res. 12:5419 (1984).
4. P. P. Lankhorst, C. A. G. Haasnoot, C. Erkelens, H. P. Westerink, G. A. van der Marel, J. H. van Boom, and C. Altona, Nucleic Acids Res. 13:927 (1985).
5. C. A. G. Haasnoot, and C. Altona, Nucleic Acids Res. 6:1135 (1979).
6. J. T. Powell, E. G. Richards, and W. B. Gratzer, Biopolymers 11:235 (1972).
7. M. J. Lowe, and J. A. Schellman, J. Mol. Biol. 65:91 (1972).
8. C. S. M. Olsthoorn, C. A. G. Haasnoot, and C. Altona, Eur. J. Biochem. 106:85 (1980).
9. C. S. M. Olsthoorn, L. J. Bostelaar, J. H. van Boom, and C. Altona, Eur. J. Biochem. 112:95 (1980).
10. C. S. M. Olsthoorn, J. Doornbos, H. P. M. de Leeuw, and C. Altona, Eur. J. Biochem. 125:367 (1982).
11. C. Altona, A. J. Hartel, C. S. M. Olsthoorn, H. P. M. de Leeuw, and C. A. G. Haasnoot, in:"Nuclear Magnetic Resonance Spectroscopy in Molecular Biology", B. Pullman, ed. p. 87, D. Reidel Publishing Co. Dordrecht, Holland (1978).
12. A. J. Hartel, P. P. Lankhorst, and C. Altona, Eur. J. Biochem. 129:343 (1982).
13. P. P. Lankhorst, C. Erkelens, C. A. G. Haasnoot, and C. Altona, Nucleic Acids Res. 11:7215 (1983).
14. P. O. P. Ts'o, in:"Basic Principles of Nucleic Acid Chemistry", P. O. P. Ts'o ed., Academic Press, New York (1974).
15. C. Altona, in: "Structure and Conformation of Nucleic Acids and Protein-Nucleic Acid Interactions", M. Sundaralingam and S. T. Rao, eds, p. 613, University Park Press, Baltimore (1975).
16. D. M. Gray, and I. Tinoco jr, Biopolymers 11:1235 (1972).
17. D. Frechet and J. Gabarro-Arpa, Biochim. Biophys. Acta 609:1 (1980).
18. IUPAC-IUB Joint Commission on Biochemical Nomenclature, Eur. J. Biochem. 131:9 (1983)
19. W. Klyne, and V. Prelog, Experientia 16:521 (1960).
20. C. Altona, and M. Sundaralingam, J. Am. Chem. Soc. 94:8205 (1973).
21. C. Altona, and M. Sundaralingam, J. Am. Chem. Soc. 95:2333 (1973).
22. H. P. M. de Leeuw, C. A. G. Haasnoot, and C. Altona, Isr. J. Chem. 20:108 (1980).
23. F. A. A. M. de Leeuw, P. N. van Kampen, C. Altona, E. Diez, and A. L. Esteban, J. Mol. Struct. 125:67 (1984).

24. C. A. G. Haasnoot, F. A. A. M. de Leeuw, and C. Altona, Tetrahedron 36:2783 (1980).
25. C. A. G. Haasnoot, F. A. A. M. de Leeuw, and C. Altona, Bull. Soc. Chim Belg. 89:125 (1980)
26. C. A. G. Haasnoot, F. A. A. M. de Leeuw, H. P. M. de Leeuw, and C. Altona, Org. Magn. Reson. 15:43 (1981).
27. A. J. Hartel, G. Wille, J. H. van Boom, and C. Altona, Nucleic Acids Re: 9:1405 (1981).
28. C. A. G. Haasnoot, F. A. A. M. de Leeuw, H. P. M. de Leeuw, and C. Altona, Biopolymers 20:1211 (1981).
29. F. A. A. M. de Leeuw, and C. Altona, J. Chem. Soc. Perkin II, 375 (1982).
30. F. A. A. M. de Leeuw, and C. Altona, J. Comp. Chem. 4:428 (1983).
31. F. A. A. M. de Leeuw, and C. Altona, Int. J. Pept. Protein Res. 20:120 (1982).
32. J.-R. Mellema, A. K. Jellema, C. A. G. Haasnoot, J. H. van Boom, and C. Altona, Eur. J. Biochem. 141:165 (1984).
33. J.-R. Mellema, J. M. L. Pieters, G. A. van der Marel, J. H. van Boom, C. A. G. Haasnoot, and C. Altona, Eur. J. Biochem. 143:285 (1984).
34. F. A. A. M. de Leeuw and C. Altona, Quant. Chem. Progr. Exch. No 463 (1983).
35. F. A. A. M. de Leeuw, A. A. van Beuzekom, and C. Altona, J. Comp. Chem. 4:438 (1983).
36. S. Arnott, P. J. Campbell Smith, and R. Chandrasekan, in: "CRC Handbook of Biochemistry and Molecular Biology", p. 411 (1975).
37. C.-H. Lee and I. Tinoco Jr, Biophys. Chem. 11:283 (1980).
38. P. P. Lankhorst, C. M. Groeneveld, G. Wille, J. H. van Boom, C. A. G. Haasnoot, and C. Altona, Recl. Trav. Chim. Pays-Bas 101:253 (1982).
39. P. P. Lankhorst, G. Wille, J. H. van Boom, C. A. G. Haasnoot, and C. Altona, Nucleic Acids Res. 11:2839 (1983).
40. P. P. Lankhorst, G. A. van der Marel, G. Wille, J. H. van Boom, and C. Altona, Nucleic Acids Res. 13:3317 (1985).
41. C. M. Groeneveld, P. P. Lankhorst, and C. Altona, unpublished observations.
42. J. Doornbos, H. P. M. de Leeuw, C. S. M. Olsthoorn, G. Wille, H. P. Westerink, J. H. van Boom, and C. Altona, Nucleic Acids Res. 11:7517 (1983).
43. P. Kollman, P. Weiner, and A. Dearing, Biopolymers 20:2583 (1981)
44. S. J. Weiner, P. A. Kollman, D. A. Case, U. Chandra Singh, C. Ghio, G. Alagona, S. Profeta Jr, and P. Weiner, J. Am. Chem. Soc. 106:765 (1984).

NMR STUDIES OF BASE-PAIR KINETICS OF NUCLEIC ACIDS

Jean Louis Leroy, Daniel Broseta, Nicolas Bolo, and
Maurice Guéron

Groupe de Biophysique du Laboratoire* de Physique de la
Matière Condensée. Ecole Polytechnique, 91128 Palaiseau,
France

*G.R. 050038 du Centre National de la Recherche Scientifique

1. INTRODUCTION

The rate of exchange with water of the labile protons of a macromole-
cule may be connected with the rate of fluctuations which expose such protons to the solvent. The exchange process can be monitored by proton
magnetic resonance which has the potential capability of identifying each
exchangeable proton and of measuring its exchange rate.

Poly(rA).poly(rU) and similar homopolymers have been the subject of
extensive studies and may be considered as a reference for the description
of proton exchange in nucleic acids (1). Exchange times longer than 10
seconds have been studied by proton-tritium exchange monitored by tritium
release (2,3).

Proton-deuterium exchange has been measured in stopped-flow, monitored
by optical absorption with a time resolution of 3 milliseconds (4). Typically, two rates were found in such studies. The interpretation was based
on measurements of base catalysis and on the analysis of a detailed
mechanism of exchange. In A-U compounds the slow exchange rate was assigned
to the adenine amino protons, even though they are exposed to the solvent
in the double helix structure. The fast exchange rate was assigned to the
uracil imino proton, even though it is inaccessible in the double helix.
It was argued that exchange of all three protons occurs only from an open
state, in which the A-U base-pair is broken. This analysis suggested that
the lifetimes of the closed base-pair and of the open state are respectively
500 and 25 ms at 25°C (4).

These pioneering studies were followed by numerous others, using
similar methods and proton magnetic resonance as well. The systems studied
range from synthetic polymers and oligodeoxynucleotides to natural products,
including different species of tRNA.

In many studies, base catalysis of proton exchange was investigated
in order to determine if chemical exchange or base-pair opening is rate-
limiting. For base-pairs far from the double-strand extremities, the imino
proton exchange rates were mostly insensitive to the buffer (often phosphate)
used as a catalyst. This has lead to the commonly held assumption that,
failing contrary evidence, the exchange rate of hydrogen-bonded imino
protons may be identified to the opening rate of the base-pair.

Recently, we have investigated the exchange process of imino protons
in tRNAs (5). In these molecules, some protons exchange very slowly, in

minutes or hours. We found that their exchange rates are often strongly sensitive to buffer. This led us to discover similar effects in the exchange of rapidly exchanging imino protons of tRNAs. These effects have been missed in earlier work, perhaps because phosphate is a poor catalyst for imino exchange in these nucleic acids. Our results, obtained with base catalysts such as tris and imidazole lead to a drastic revision of the lifetimes of tRNA base-pairs.

At this point it had become clear that in many cases, imino proton exchange rates are limited by exchange catalysis. A new investigation of the model compounds poly(rA).poly(rU) and poly(rI).poly(rC) seemed in order.

This investigation has been reported recently (6), and forms the basis of the present contribution. We shall see that it supports some conclusions of the earlier work, in particular the assignment of the slower exchange rate to the amino protons. On the other hand, the base-pair lifetimes are about 100 times shorter than the earlier values.

2. MATERIALS AND METHODS

Sample Preparation. Poly(rA).poly(rU) and poly(rI).poly(rC) were broken by sonication (7) and chromatographed on Sephadex G100. Fragments in the range of 50 $\pm$ 15 base-pairs were pooled and used in the present work.

NMR Methods. Proton spectra were obtained at 276 MHz on a home-built Fourier transform spectrometer (8). We used the SJR sequence for water suppression (9). For NOE experiments, selective inversion of a line was obtained using a $(\theta, \tau, -\theta)_n$ so-called DANTE sequence (10) of strong pulses, with $\theta = 1.8°$ and n = 50. For magnetization transfer from water, the water line was selectively inverted by a $(\theta, \tau)_n$ sequence with $\theta = 1.8°$, $\tau = 100$ microseconds, n = 100; transverse magnetization was destroyed by a z-gradient pulse lasting 5 to 10 ms. The free precession was routinely broadened by exponential multiplication, and resolution-enhanced by sine-cosine multiplication (11).

Determination of the proton exchange time. Most exchange times were determined by the line broadening induced by addition of catalyst, according to $1/\tau_{ex} = \pi\Delta$, where Δ is the increase of the full linewidth at half-height (12). The increase in the longitudinal relaxation rate could also have been used, but this measurement is more time-consuming than the broadening method and its interpretation is complicated by the phenomenon of spin diffusion (13). The method of saturation transfer gives the exchange time directly. However, it has the same draw-backs as the longitudinal relaxation method. It was used in some cases as a control, and the results agreed with those obtained from line-broadening.

3. SPECTRAL ASSIGNMENTS OF EXCHANGEABLE PROTONS

Apart from the imino proton which has been previously assigned (14,15), two amino protons (from A and from C respectively) and two ribose 2'OH protons are expected to exchange with water. Indeed, the spectra in H2O and D2O differ by the loss of peaks corresponding to a total intensity of five protons. The 2'-OH protons should exchange much faster with water than the amino protons. The latter, alone among exchangeable protons, should be NOE-coupled to the imino proton.

In poly(rI).poly(rC) the 2'OH protons are assigned at -6.7 and -6.33 ppm by magnetization transfer from water (Fig. 2). In this experiment the exchange rate at 20°C, pH 6.9, is found to be 66 s^{-1}. The alternate assignment, to amino protons, is excluded by the observation that this rate is much larger than the exchange rate reported for the amino protons of poly(rC) (16).

In poly(rA).poly(rU), the -6.61 ppm peak is exchangeable and integrates to three protons in low buffer (Fig. 3). At moderate buffer concentrations (10 mM tris, pH 7), intensity corresponding to two protons is lost. These must be the two ribose 2'OH protons.

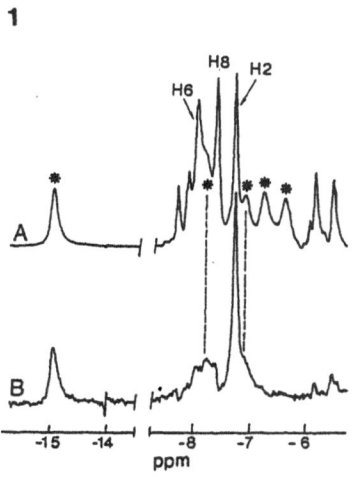

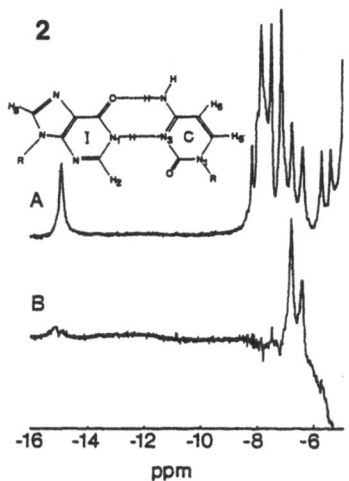

Figure 1. (A). Proton nmr spectrum of poly(rI).poly(rC) in H_2O at 30°C. (B). Difference between the spectrum in (A) and that obtained 60 ms after inversion of the resonance at -14.95 ppm. By this NOE experiment, inosine H2 is assigned at -7.19 ppm. The broad NOEs at -7.74 and -7.05 ppm are attributed to the cytosine amino protons. Concentration in base-pairs: 30 mM; 0.1 M NaCl, pH 6.9.

Figure 2. Saturation transfer from H_2O to exchangeable protons of poly(rI). poly(rC) at 20°C (0.1 M NaCl, pH 7). (A) Reference spectrum. (B) Difference between the spectrum in (A) and that obtained 10 ms after inversion of the H_2O magnetization. The rapidly exchanging protons ($t_{ex} \approx 15$ ms) are identified as the ribose 2'OH. Exchange of the imino proton is slower, hence its low intensity. The amino protons are not observed, hence their exchange is even slower. The broad line around -12.5 ppm corresponds to non base-paired imino protons.

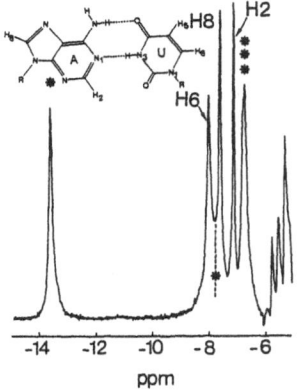

Figure 3. Proton nmr spectrum of poly(rA).poly(rU) (38-65 base-pairs) in H_2O at 30°C. Resonances which disappear in D_2O are marked with one star per proton. The suppression of water excitation results in spectral intensities which are chemical-shift dependent. Resonances at -5.7 and -5.5 ppm correspond to one and two protons respectively. Same conditions as in Figure 1.

TABLE I : Spectral assignments

	poly(rA).poly(rU)	poly(rI).poly(rC)
Imino (U,I)	-13.5	-14.95
Amino (internal) (A,C)	-7.81	-7.74
Amino (exposed) (A,C)	-6.61	-7.05
$C_2'OH$	$-6.61, -6.61$	$-6.70, -6.33$
CH5 (U,C)	-5.57	-5.45
CH6 (U,C)	-7.79	-7.89
CH2 (A,I)	-7.15	-7.19
CH8 (A,I)	-7.61	-7.52
CH1'	$-5.77, -5.57$	$-5.77, -5.45$

Lastly we turn to the amino protons. In poly(rI).poly(rC), the spectrum displays two poorly resolved resonances at -7.74 and -7.05 ppm, which disappear in D_2O. These resonances are NOE-coupled to the imino proton (Fig. 1). They are attributed to the cytosine amino protons.

In the poly(rA).poly(rU) spectrum, there are no resolved lines which we can attribute to amino protons. However, as with poly(rI).poly(rC), two broad resonances are NOE-coupled to the imino proton, at -7.81 and -6.61 ppm (not shown). They are attributed to the amino protons. (The -6.61 ppm line lies at the same position as the 2'OH lines).

4. THEORY OF PROTON EXCHANGE

Exchange of imino protons with water proceeds in most cases via formation of a hydrogen-bonded complex with a base catalyst B (19):

$$AH + B \rightarrow (AH...B)$$

The complex dissociates to give either (AH + B) or $(A^- + BH^+)$, in proportions corresponding to equilibrium between these two couples. The yield for transferring the proton from AH is therefore $1/(1 + 10^{-\Delta pK})$, where:

$$\Delta pK = pK_B - pK_A$$

The transfer rate k_{tr} is then:

$$k_{tr} = k_D/(1 + 10^{-\Delta pK}) \qquad (1)$$

where k_D is the collision rate. If there are no energetic or structural barriers, k_D is diffusion-controlled and is given by:

$$k_D = k_R x(B) \qquad (2)$$

where (B) is the base concentration; k_R, the diffusion rate, is usually in the range of 10^{10} $(M.s)^{-1}$. In particular, this value is appropriate for catalysis of NH exchange in uridine by phosphate, trifluoroethylamine and OH^- (20).

Exchange via the open state of a duplex. In double-stranded poly(rA).poly(rU), the uracil imino proton is hydrogen-bonded to N_1 of adenosine. It is therefore considered to be protected from attack by a base, and cannot exchange as long as the A-U base-pair remains closed.
Transient opening is usually discussed in terms of a two-state model, with the base-pair either closed or open:

$$Ad - Ur \underset{k_{cl}}{\overset{k_{op}}{\rightleftharpoons}} Ad + Ur$$

Two cases occur depending on whether k_{cl} is smaller or larger than the transfer rate k_{tr} from the open state. In the first case transfer occurs every time the base-pair opens. Transfer then measures k_{op}, or its inverse the base-pair lifetime: $k_{ex} = \tau_0^{-1}$ (3)

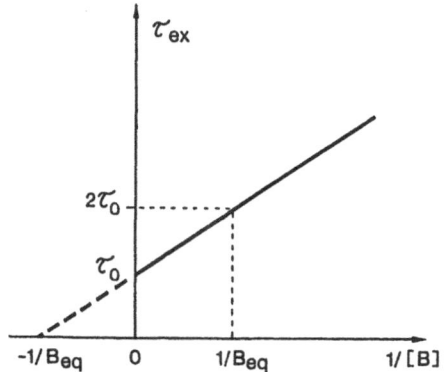

Figure 4. Graphic determination of B_{eq}, the base catalyst concentration at which the transfer rate from the open state is equal to the base-pair closing rate k_{cl}. The base-pair lifetime is τ_0. The net time for exchange is τ_{ex}, and it is a linear function of the inverse catalyst concentration (equations 1,2 and 5). It extrapolates to the base-pair lifetime τ_0 for infinite catalyst concentration. For a base catalyst concentration B_{eq}, the exchange time is equal to $2 \times \tau_0$.

In the second case, the net transfer rate is equal to that from the open state, multiplied by the fraction of the time that the base-pair is open:

$$k_{ex} = k_{tr}/(\tau_0 \, k_{cl}) \qquad (4)$$

The general formula for exchange time is:

$$t_{ex} = \tau_0 \, (1 + k_{cl}/k_{tr}) \qquad (5)$$

The base-pair lifetime τ_0 is a parameter of the closed state. A parameter related only to the open state is the catalyst concentration, denoted B_{eq}, for which $k_{cl} = k_{tr}$. This concentration may be measured: it is the catalyst concentration such that the exchange time τ_{ex} is twice the base-pair lifetime τ_0. A graphic determination of B_{eq} is pictured in Figure 4. One may then write:

$$\tau_{ex} = \tau_0 \, (1 + B_{eq}/(B)) \qquad (6)$$

Rate of Transfer from the Open State. The transfer rate k_{tr} from the open state of a base-pair in a polynucleotide would, in the simplest model, be equal to that for the isolated nucleotide. However, the rate of exchange from the open state could be changed, due to such factors as steric hindrance to access of the catalyst, or electrostatic effects for a charged catalyst. These effects can be lumped together in a factor α, the ratio of transfer rates from the open state and from an isolated nucleotide. The factor α is unkown, buffer-dependent, presumably smaller than one. It may be named "accessibility".

According to the definition of B_{eq}, and using equations 1 and 2, we have:

$$k_{cl} = \alpha \times 10^{10} \times B_{eq}/(1 + 10^{-\Delta pK}) \qquad (7)$$

If one compares catalysts having different accessibilities, their relative efficiencies should differ for the polynucleotide and for the isolated nucleotide. Table III gives evidence for this phenomenon.

5. RESULTS

Imino Proton Exchange and Base-Pair Lifetime. The basic observation is that the imino proton linewidth is sensitive to buffer concentration. This may be seen in Figure 5, where two ranges of temperature can be distinguished.

Below 30°C and in the absence of buffer (open symbols) the line-width is dominated by magnetic interactions, independently of exchange. It diminishes slightly as the temperature is raised up to 30°C, due to the decreasing viscosity of water. Above 30°C, the broadening is attributed to exchange, in agreement with the observation of buffer-induced broadening, to which we now turn.

In the presence of base catalyst (full symbols), much larger linewidths are found, reflecting larger exchange rates. The influence of the buffer shows that the exchange process is generally not opening-rate limited.

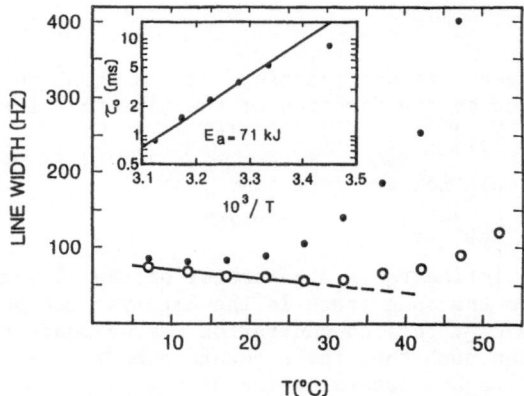

Figure 5. Imino proton linewidth in poly(rI).poly(rC) versus temperature. Open symbols: the solution is 0.1M NaCl, and 1 mM sodium cacodylate, at pH 6.7. Full symbol: 0.1M tris chloride has been added, and the pH is 7.8. Apart from exchange, the linewidth should diminish as the temperature is raised (dashed line). The increased linewidth at higher temperatures is due to exchange. (In the 1 mM cacodylate solution, the principal catalyst is probably the terminal phosphate of the polynucleotide chains). Insert: The base-pair lifetimes τ_0 are obtained from this type of data, taken at different buffer concentrations and extrapolated to infinite buffer concentration (Fig. 6). The variation of τ_0 vs. the inverse of temperature yields the activation energy for base-pair opening. Its value suggests individual base-pair opening.

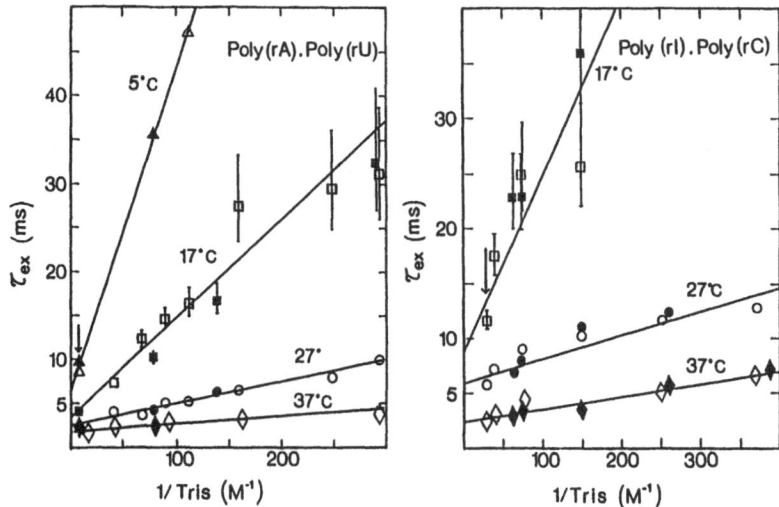

Figure 6. Exchange time, τ_{ex}, of the imino protons of poly(rA).poly(rU) and poly(rI).poly(rC), as a function of the inverse of base catalyst concentration: $1/[tris\ base] = (1 + 10^{pK-pH})/[total\ tris]$. Extrapolation to infinite catalyst concentration gives the lifetime, τ_0 of the closed base-pair. The pH was 7.15 ± 0.1, except for data indicated by arrow, for which the pH was 7.8; $NaCl = 0.1$ M. Exchange times are the same in the presence of magnesium (full symbol, 0.5 Mg/nucleotide) and in its absence (open symbol). See also legend of Figure 8.

TABLE II: Base-pair lifetimes (ms)

Temperature (Celsius)	5	17	27	37
poly(rA).poly(rU)	7	3.4	2.5	2
poly(rI).poly(rC)		8.8	5.9	2.5

At each temperature one may plot the exchange broadening as a function of the inverse of concentration of the catalysing base (Fig. 6). A straight line is obtained, in agreement with equations 1,2 and 5. Extrapolation to infinite base concentration gives the lifetime τ_0 of the closed base-pair. This is plotted in the insert of Fig. 5 to give the activation enthalpy for base-pair opening, $E_a = 71$ kJ/M. This value suggests that the opening event involves only a small number of base-pairs.

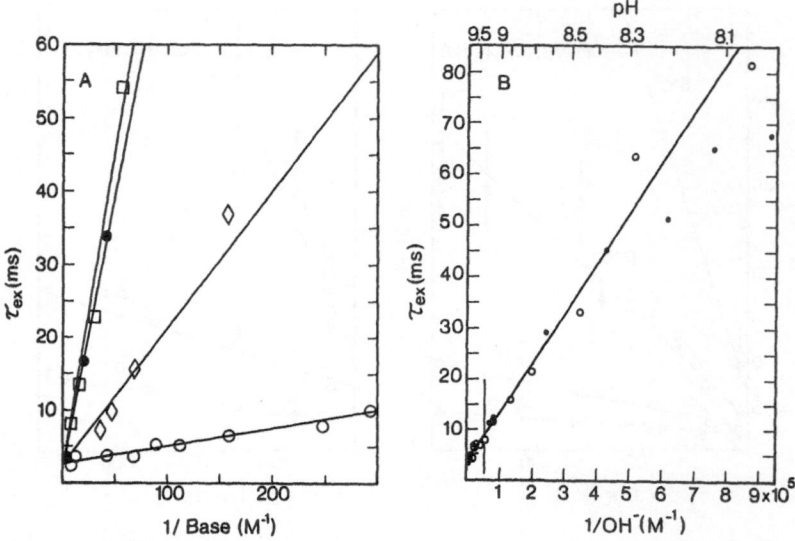

Figure 7. (A) Catalysis of exchange of the poly(rA).poly(rU) imino proton, as a function of the inverse of the base catalyst concentration (as in Figure 6). The buffers are: tris, pH 7.15, (o); triethanolamine, pH 7.1 (◇); phosphate, pH 6.8 (□); imidazole, pH 6.9 (●). The base-pair lifetime t_0 obtained by extrapolation to infinite catalyst concentration is the same for all buffers, $t_0 = 2.5$ ms.

(B) Catalysis of exchange of the poly(rA).poly(rU) imino proton by OH^-. The full symbols correspond to increasing pH, up to 10.3. The pH was then returned to 7.5 (open symbols). Aromatic signals corresponding to single stranded poly(rA) and poly(rU) were observed only for pH> 9.4 (indicated by the vertical line). T = 27°C; 0.1M NaCl; nucleic acid concentration: 30 mM, in base-pairs.

Another feature demonstrated in Figure 6 is the lack of effect of magnesium, either on the exchange process of the imino proton or on the base-pair lifetime. In Figure 7, different catalysts are considered. The base-pair lifetime obtained by extrapolation is the same for all buffers, in agreement with the model.

This observation supports the proposition that exchange of the imino proton follows the simple two-state model described above. The base-pair lifetimes are given in table II. They are in the range of milliseconds for both polynucleotides.

Imino Proton Exchange and the Open State. Whereas the proton exchange time in infinite buffer is equal to the lifetime τ_0 of the closed base-pair, the base concentration B_{eq} for which the exchange time is equal to $2\tau_0$ is characteristic of events taking place in the open state: it is the concentration for which the proton transfer rate from the open state, k_{tr}, is equal to the closing rate k_{cl}. Unfortunately, we cannot derive these rates from the measured values of B_{eq}: the quantities derived from equation 7, k_{cl}/α and $\alpha K_{op} = (\tau_0 k_{cl}/\alpha)^{-1}$, include the unknown accessibility α.

The data for exchange of the imino proton of poly(rA).poly(rU) by different catalysts and at different temperatures are given in Tables III and IV. Since k_{cl} and K_{op} are independent of the catalyst in the two-state model, the variations in the two rightmost columns reflect the variation of α as a function of the catalyst. It appears that α is much larger for OH^- than for the other catalysts. The values of α remain unknown.

TABLE III: Catalysis of imino proton exchange
in poly(rA).poly(rU); 0.1 M NaCl, T = 27°C.

Catalyst	Catalyst pK	B_{eq} (mM) (a)	k_{cl}/α (s^{-1}) (b)	αK_{op} (c)
Imidazole	6.95	300	13×10^6	3.0×10^{-5}
$PO_4H^{(2-)}$	6.8	350	11×10^6	3.8×10^{-5}
Triethanolamine	7.72	76	21×10^6	2.1×10^{-5}
Tris	8.08	10	5.9×10^6	6.9×10^{-5}
OH^-	15.7	0.033	0.33×10^6	110×10^{-5}

(a) Data of Fig. 7; The pK of the uridine imino proton is 9.3 (21); (b)
From equation (7); ($\alpha<1$) is the unknown ratio of the transfer rate from the
open state to transfer rate from isolated uridine; (c) $\alpha K_{op} = (\alpha/k_{cl}) \times (1/\tau_0)$.

The restriction ($\alpha \leqslant 1$) provides an upper boundary for k_{cl} and a lower
boundary for K_{op}. The strictest limits are those from the data obtained
with OH^-: $k_{cl} < 0.33 \times 10^6$ and $K_{op} > 1.1 \times 10^{-3}$.

TABLE IV: Open-state parameters at various temperatures (a,b)

T°C	poly(rA).poly(rU) k_{cl}/α s^{-1}	αK_{op}	poly(rI).poly(rC) k_{cl}/α	αK_{op}
5	32×10^6	0.4×10^{-5}		
17	17×10^6	1.6×10^{-5}	30×10^6	0.4×10^{-5}
27	5.9×10^6	6.9×10^{-5}	4.7×10^6	3.7×10^{-5}
37	1.9×10^6	25×10^{-5}	6.2×10^6	6.3×10^{-5}

(a) From imino proton exchange rates in tris buffer, using pK values of
8.08 for tris (21), 9.3 for uridine and 8.9 for inosine (22) at 27°C. The
same pK differentials ($pK_{tris}-pK_{nucleoside}$) were used at all temperatures.
(b) Data derived as in Table III; the value appropriate for tris base is
implied.

If α is independent of temperature, the enthalpy ΔH for base-pair
formation may be obtained from αK_{op}. For poly(rA).poly(rU), we find
$\Delta H = -86.1$ kJ/M. This is approximately twice the stacking enthalpy of one
AU base-pair in the growing process of the ply(rA).poly(rU) double helix;
$\Delta H = -45.4$ kJ/M (23). For poly(rI).poly(rC), we find $\Delta H = -100$ kJ/M. These
data should be interpreted with caution. Note for instance that ($-\Delta H$) is
larger than the activation enthalpy E_a given above.

Exchange kinetics of amino protons. In the duplex form, only poly(rI).
poly(rC) was studied since the amino protons of poly(rA).poly(rU) are not
directly resolved in the spectrum.

At 30°C, catalysis by OH^- was observed between pH 8 and pH 8.8 by line
broadening and by magnetization transfer from water (not shown). The pH
dependence suggests exchange from the closed duplex.

At pH 7, the exchange rate was too slow for us to measure, even in the
presence of tris buffer. Broadening of amino protons by tris buffer was
easily observed above the melting temperature in both polynucleotides.

Exchange kinetics of the ribose hydroxyl protons. The resonances
assigned to 2'OH are strongly sensitive to buffer catalysis. They were

examined in the absence of buffer: they are exchange-broadened above 20°C and disappear above 40°C.

Their exchange rate is much smaller ($k_{ex} < 10$ s^{-1} at 5°C, pH 6.9) than that of the 2'OH proton of poly(rU) measured previously by Young and Kallenbach (24) as $k_{ex} = 150$ s^{-1} in the same conditions. We found that the exchange of the duplex hydroxyls is strongly accelerated (at least 10 times) by magnesium. Association of magnesium to phosphate might make a phosphate to hydroxyl hydrogen-bond more labile. These observations suggest that the 2'OH protons of the duplex are protected against base catalysis, due maybe to hydrogen bonding to phosphate, as observed in tRNA (25).

6. DISCUSSION

We summarize our main results:

a) The exchange of the imino proton is buffer catalyzed. This process provides a direct determination of the base-pair lifetime and a lower limit for the open state lifetime.

b) The observations on the imino proton are well interpreted in the framework of the two-state model, with exchange occuring from the open state only.

c) The exchange of the imino proton is strikingly similar in the two duplexes poly(rA).poly(rU) and poly(rI).poly(rC), even though the imino proton resides on the pyrimidine in the first case, and on the purine in the second case. This suggests that the open state involves exposure of both bases of the base-pair.

d) The imino proton exchange kinetics and the base-pair lifetimes of the duplexes are both independent of magnesium concentration.

The base-pair dissociation constant would be expected to be salt-insensitive if it does not involve a large strand separation. This suggests a model where base-pairs open more or less independently, rather than one where many neighboring base-pairs are open together, as in the soliton model (2, 26).

Magnesium-independent exchange need not be the rule for all polynucleotides. For instance, some base-pairs of tRNA are magnesium-sensitive (5, 27)

Comparison with Previous Studies. The same duplexes have been studied by slow tritium exchange and stopped-flow deuterium exchange, as reviewed in reference (1), and more recently by NMR (14, 29).

The main quantitative result of the tritium exchange studies (2,3) was the evaluation of the base-pair dissociation constant K_{op}, computed as the ratio of the exchange rate assigned to the amino protons to the same rate for the mononucleotide. This ratio was 1/50 at 0°C. This is a surprisingly large value for K_{op}. We think that it may be questioned because it is based on the assumption that amino proton exchange occurs only from the open state.

The deuterium experiments on poly(rA).poly(rU) (4) provided two further results. First they demonstrated a fast, buffer independent, rate which was ascribed to imino proton exchange. The base-pair lifetime derived from this measurement was 1 s at 20°C, hundreds of times larger than the value obtained in the present work!

The origin of the discrepancy lies in the nature of the experimental system. The deuterium study just described was carried out in the presence of magnesium. Using the same conditions, we have found that poly(rA).poly (rU) disproportionates to form the triple helix poly(rA) (poly(rU))$_2$ (Fig. 8). The two imino protons of the triple helix exchange much more slowly than the imino proton of the duplex, and they are hardly sensitive to buffer.

The base-pair kinetics of poly(rI).poly(rC) (14) and of poly(rA).poly(rU) (29) have been studied by NMR on the basis of imino proton exchange times. The exchange times measured on both duplexes are much longer than the base-pair lifetime determined in the present work, indicating imperfect exchange catalysis.

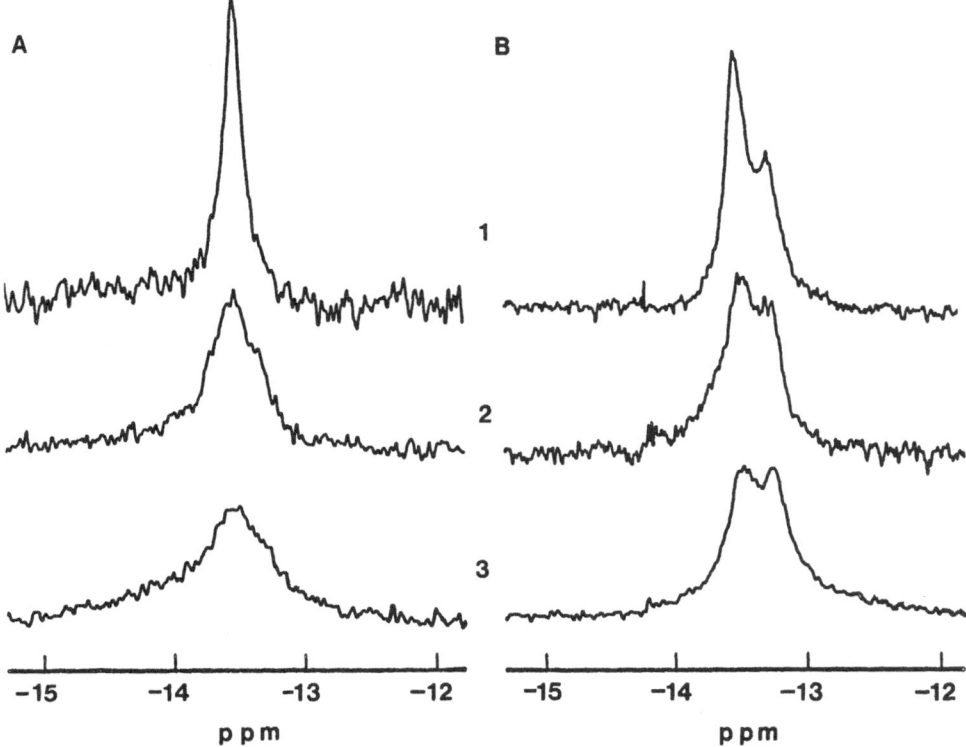

Figure 8. Formation of the triple helix poly(rA)·(poly(rU))$_2$ in a solution containing magnesium. Side A: the solutions contain NaCl, 0.1 M, and no magnesium. Side B: the solutions contain NaCl, 0.1 M, and MgCl$_2$, 10 mM. The pH is 7.2 except as noted; T = 27°C. Spectrum 1A is that of the duplex, at -13.6 ppm (cf Fig. 1). The spectrum in magnesium (1B) exhibits the same duplex peak, together with two broad peaks at the position of the Watson-Crick (-13.5 ppm) and reverse Hoogsteen (-13.2 ppm) imino protons of the triple helix (Geerdes and Hilbers, 1977). Addition of tris (final base concentration 11 mM) results in spectra 2A and 2B. The duplex imino resonance is exchange-broadened in both spectra. The triple helix peaks (in spectrum 2B) are not changed by the buffer. For spectra 3A and 3B, the tris base concentration is increased to 32 mM. The double helix imino resonance is further broadened. The triple helix peaks (in spectrum 3B) still exhibit no exchange broadening. They have equal intensities. Poly(rA) and poly(rU) were 50 base-pair long, and their concentrations were 1 mM in nucleosides. For spectrum 1B, the pH is 5.3. The spectrum was the same at pH 5.3 and 7.4, for polymers of 50 or 80 base-pairs, and for polymer concentrations up to 30 mM. The salt and pH conditions for spectrum (1B) and the polymer concentration were chosen with reference to those of the stopped-flow deuterium exchange measurements of Mandal et al. (1979), which were 10 mM free MgCl$_2$, pH 5.3, nucleoside concentration ca. 0.1 mM. The ratio of the duplex imino intensity to that of the triplex reverse Hoogsteen imino is 0.5/1 in Figure 1B. Hence only 1 uridine out of 5 is in the duplex. By contrast, the data of Figure 6 were obtained with lower Mg concentrations, and the spectra showed that 4 out of 5 uridines were in the duplex.

Other polynucleotides. The exchange of the imino proton has been investigated by NMR in various ribonucleotide (28) and deoxyribonucleotide duplexes (28, 30) as well as in tRNA (32, 33).

Buffer effects are often much weaker than those we observe. This may be due in some cases to inefficient exchange catalysis by such buffers as phosphate or cacodylate (pK = 6.3), and in other cases to genuine differences in the kinetic properties of base-pairs of different duplexes.

The example of tRNA is particularly illustrative (5). In moderately buffered solutions, for example in the presence of 10 mM phosphate, the exchange times of imino protons of yeast tRNA^{Phe} range between milliseconds and days. Investigation of the effect of exchange catalysts shows that exchange is generally limited by the catalytic step. Base-pair lifetimes derived from extrapolation to infinite tris and imidazole concentration are summarized in Figure 9.

The clearest feature is that the long-lived base-pairs are clustered in the dihydro-uracil arm which is therefore a particularly sturdy structure. This region is limited on both sides by fast opening base-pairs (U8-A14 and m$_2^2$G26-A44). Another slowly opening base-pair (τ_0 = 3 mn at 17°C), whose imino proton is not resolved, is located in another region. It may be CG2 from acceptor stem or CG49, GC53 or G19-C58 from the TΨC arm or loop. Short-lived base-pairs have been found in two regions: the acceptor stem (GU4, AU5 and UA6), and the junction between the dihydrouracil loop and the TΨC loop. The majority of base-pairs for which lifetime is not provided have lifetime between 3 mn and 0.3 s at 17°C.

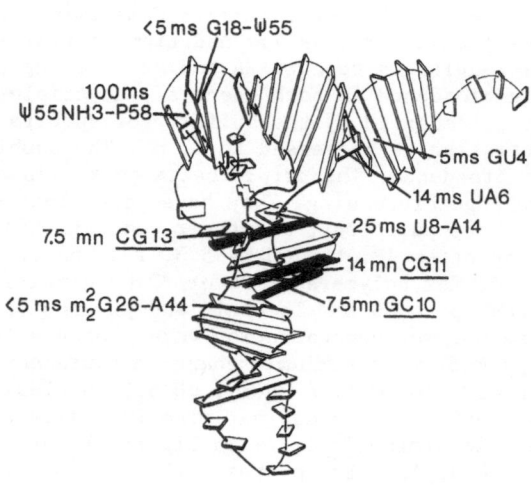

Figure 9. Base-pair lifetimes of yeast tRNA^{Phe} at 17°C in the absence of magnesium, NaCl 0.1 M. (Magnesium has moderate effects on the base-pair lifetimes: at 30 Mg/tRNA, 17°C, the lifetime of CG11 is 42 mn and that of base-pairs CG13 and GC10 is 32 mn).

CONCLUSION

This investigation of proton exchange in poly(rA).poly(rU) and poly(rI).poly(rC) supports the two-state model in which imino proton exchange occurs only from the open state. It shows that exchange is not opening-limited except in the presence of large concentrations of base-catalysts. A direct result of the experiments is the base-pair lifetime which is found to be short (10^{-2} s).

We are presently unable to determine either the lifetime of the open state or the base-pair dissociation constant. Lower limits at $27^{\circ}C$ are 3 microseconds and 10^{-3} respectively. The picture of the two duplexes which results from the present study is quite different from the earlier ones: the base-pairs open much more frequently. It may well be that they close faster and are more stable (depending however, on the unknown value of the accessibility α).

The two duplexes studied here behave very similarly. However, one should not generalize our results to all nucleic acids. In a limited number of cases we have observed imino protons whose exchange is insensitive to catalysts and therefore controlled by the opening of the base-pair. These exceptions concern two CG base-pairs localized in the TΨC arm of yeast tRNA^{Asp}, a non-assigned base-pair of yeast tRNA^{Phe} and possibly the triple helix of poly(rA)(poly(rU))$_2$.

More recently, we have studied imino proton exchange of short deoxyribonucleotides containing 6 to 10 base-pairs. We found that, in the B form as in the Z form, the exchange rate is increased when a base catalyst is added.

This gives us further arguments to believe that, in a number of studies on nucleic acids, proton exchange controlled by incomplete catalysis may have been mistaken for base-pair opening. The exchange times measured earlier at low catalyst concentrations are often in the range of hundreds of milliseconds, and they are weakly dependent on the nature of the base-pair or on sequence. It may be that the base-pair lifetimes are much shorter and also more diverse. The structural fluctuations related to base-pair opening would then become possible candidates for sequence-dependent recognition processes.

REFERENCES

1 Englander, S.W., Kallenbach, N.R. (1983) Q. Rev. Biophys. 16, 521-655.
2 Teitelbaum, H. and Englander, S.W. (1975a) J. Mol. Biol. 92, 55-78.
3 Teitelbaum, H. and Englander, S.W. (1975b) J. Mol. Biol. 92, 79-92.
4 Mandal, C., Kallenbach, N.R. and Englander, S.W. (1979) J. Mol. Biol. 135, 391-411.
5 Leroy, J.L., Bolo, N., Figueroa, N., Plateau, P. and Guéron, M. (1985) J. Biomol. Struct. and Dynam. 2, 915-939.
6 Leroy, J.L., Broseta, D. and Guéron, M. (1985) J. Mol. Biol. 184, 165-178.
7 Granot, J., Assa-Munt, N. and Kearns, D.R. (1982) Biopolymers 21, 873-883.
8 Caron, F., Guéron, M., Nguyen Ngoc Quoc, T. and Herzog, R.F. (1980) Rev. Phys. Appl. 15, 1267-1274.
9 Plateau, P. and Guéron, M. (1982) J. Am. Chem. Soc. 104, 7310-7311.
10 Morris, G.A. and Freeman, R. (1978) J. Mag. Res. 29, 433-462.
11 Guéron, M. (1978) J. Mag. Res. 30, 515-520.
12 Crothers, D.M., Cole, P.E., Hilbers, C.W. and Shulman, R.G. (1974) J. Mol. Biol. 87, 63-88.
13 Johnston, P.D. and Redfield, A.G. (1978) Nucleic Acids Res. 5, 3913-3927.
14 Mirau, P.A. and Kearns, D.R. (1984) J. Mol. Biol. 77, 207-221.
15 Geerdes, H.A.M. and Hilbers, C.W. (1977) Nucleic Acids Res. 4, 207-221.
16 Nakanishi, M. and Tsuboi, M. (1978) J. Mol. Biol. 124, 61-71.
17 Neumann, J.M. and Tran-Dinh, S. (1981) Biopolymers 20, 89-106.

18 Neumann, J.M. and Tran-Dinh, S. (1982) Biopolymers 21, 383-402.
19 Eigen, M. (1964) Angew. Chem. Int. ed. Engl. 3, 1-19.
20 Fritzsche, H., Kan, L.S., Ts'o, P.O.P. (1981) Biochemistry 20, 6118-6122.
21 Izatt, R.M., Christensen, J.J. and Rytting, J.H. (1971) Chem. Rev. 71, 439-480.
22 Handbook of Biochemistry and Molecular Biology, Physical and Chemical Data, 3rd edit, ed. Fassmann, G.D. (Cleveland, Ohio: Chemical Rubber Co.), p. 360.
23 Pörschke, D. (1971) Biopolymers 10, 1989-2013.
24 Young, P.R. and Kallenbach, N.R. (1978) J. Mol. Biol. 126, 467-479.
25 Quigley, G.J. and Rich, A. (1976) Science 194, 796-805.
26 Englander, S.W., Kallenbach, N.R., Heeger, A.J., Krumhamsl, J.A. and Litwin, S. (1980) Proc.Natl. Acad. Sci. USA 77, 7222-7226.
27 Figueroa, N., Keith, G., Leroy, J.L., Plateau, P., Roy, S. and Guéron, M. (1983) Proc. Natl. Acad. Sci. USA 80, 4330-4333.
28 Fritzsche, H., Lou-Sing Kan and Ts'o, P.O.P. (1983) Biochemistry 23, 277-280.
29 Mirau, P.A. and Kearns, D.R. (1985) Biopolymers 24, 711-724.
30 Haasnoot, C.A.G., de Bruin, S.H., Berendsen, R.G., Janssen, H.G.J.M., Binnendijk, T.J.J., Hilbers, C.W., Van der Marel, G.A. and Van Boom, J.H. (1983) J. of Biomolecular Structure and Dynamics 1, 115-129.
31 Assa-Munt, N., Granot, J., Behling, R.W. and Kearns, D.R. (1984) Biochemistry 23, 944-955.
32 Hurd, R.E. and Reid, B.R. (1980) J. Mol. Biol. 142, 181-193.
33 Tropp, J.S. and Redfield, A.G. (1983) Nucleic Acids Res. 11, 2121-2133.

PICOSECOND TIME DOMAIN SPECTROSCOPY OF STRUCTURE AND DYNAMICS IN NUCLEIC ACIDS

Rudolf Rigler and Flora Claesens

Department of Medical Biophysics, Karolinska Institutet

Box 60400, S-104 01 Stockholm, Sweden

INTRODUCTION

The generation of laser pulses in the picosecond and recently in the femtosecond time domain (Auston and Eisenthal, 1984) has made possible studies of vibrational and rotational modes as well as of short lived excited states in nucleotides and nucleic acids. With the tunability of dye lasers selective excitation of purines and pyrimidines is available and information on the distribution and dynamics of conformational states as well as on internal motions of nucleic acid chains and individual nucleotides can be obtained.

Fluorescence spectroscopy of nucleic acids at room temperature has been limited by short lifetimes of the excited states of purines and pyrimidines (Daniels, 1973). Due to the low quantum yield and the difficulties in detection of weak luminescence our knowledge about luminescence properties of nucleic acids has been fragmentary (Ballini et al., 1982; Ballini et al., 1983). With the use of time windows permitting collection of photons due to excited states of nucleic acids and rejection of others, these difficulties can be overcome. Time gated fluorescence spectroscopy with high sensitivity has become possible by the development of single photon detection techniques with ps time resolution (Rigler et al., 1984).

In this contribution we present data on luminescence properties and dynamics of excited states of synthetic and natural nucleic acids. Also the possibilities in using modified nucleosides such as deoxy-2-aminopurine in DNA or wybutine in tRNA for elucidating the dynamic behaviour and conformations in nucleic acids are demonstrated.

EXPERIMENTAL

Laser excitation: Generation of ps laser pulses was performed by mode locking a large frame Kr-ion laser (Coherent 3000 K) with an acousto-optic Bragg cell (Coherent 467) and synchronously pumping Rhodamine 6G at 530.9 nm in a dye laser (Coherent 699-4). In order to increase the pulse power the output of the dye laser was cavity dumped and frequency doubled with an angle tuned KDP crystal (Fig. 1).

Signal detection: The time dependent emission after pulsed excitation was recorded by time correlated single photon counting with a time to amplitude converter (TAC, Ortec 457) and pulse height analysis in a

multichannel analyzer (MCA, ND66). A system response time of 43 ps FWHM
to a 6 ps laser pulse was obtained by the use of multichannel plate
detectors (Hamamatsu R1564-U) and constant fraction discriminators with
minimal walk times (Tennelec TC453) (Rigler et al., 1984). With a split
beam arrangement allowing simultaneous detection of the (attenuated)
excitation pulse together with the polarized components of the fluorescence
variations in intensity and timing of the exciting pulse and the convoluted
response signal can be eliminated. Deconvolution of these data and non
linear parameter estimations yield a variance in lifetimes and rotational
relaxation times of a few ps. The rotational relaxation of cresyl-violet
in acetone could be determined to 63±2 ps (Rigler et al., 1984).

   Time gated emission spectra: Time windows of various width and timing
positions were obtained by discriminating the output of the TAC by a
window discriminator and gating the input in the MCA operating in the
multichannel scaling mode. A Zeiss MQIII monochromator was used in order
to minimize the time spread due to chromatic dispersion (Δt=8 ps at
400 nm).

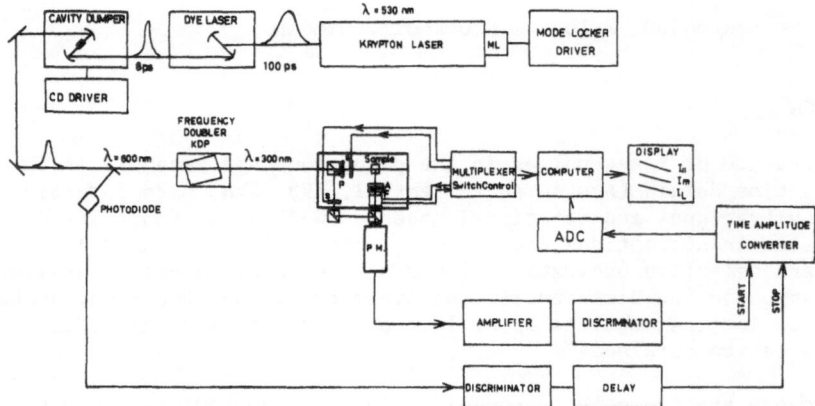

Fig. 1. Setup for time correlated single photon spectroscopy with mode
   locked synchronously pumped and cavity dumped dye laser.

RESULTS

Luminescence of purines, synthetic and natural DNAs

   When excitation with frequency double dye laser pulses was performed
at 290 nm time gated emission spectra shown in Fig. 2 could be recorded.
Contributions of Raman scattering were eliminated with a time window
discriminating against the exciting pulse and instantaneous scattering
processes. For purine nucleosides ATP and GDP emission maxima around
400 nm with a pronounced shoulder at 330 nm for GDP were found. A similar
behaviour was observed for GpC. Analogous spectra were observed for
poly[d(A-T)], poly[d(G-C)] and natural DNA (Adenovirus 2) with the short
wave emission band accentuated.

   Analysis of the excited states (Table 1) shows that with the exception
of deoxy-2-aminopurine and 7-methyl-guanosine the decay of purines contains
more than one component. While for ATP a prevalence of longlived state
(4.2 ns) is found, the opposite is true for poly[d(A-T)] where a short
lived state (90 ps) is the dominating one. Similarly in poly[d(G-C)]
the shortest lived state is most populated.

46

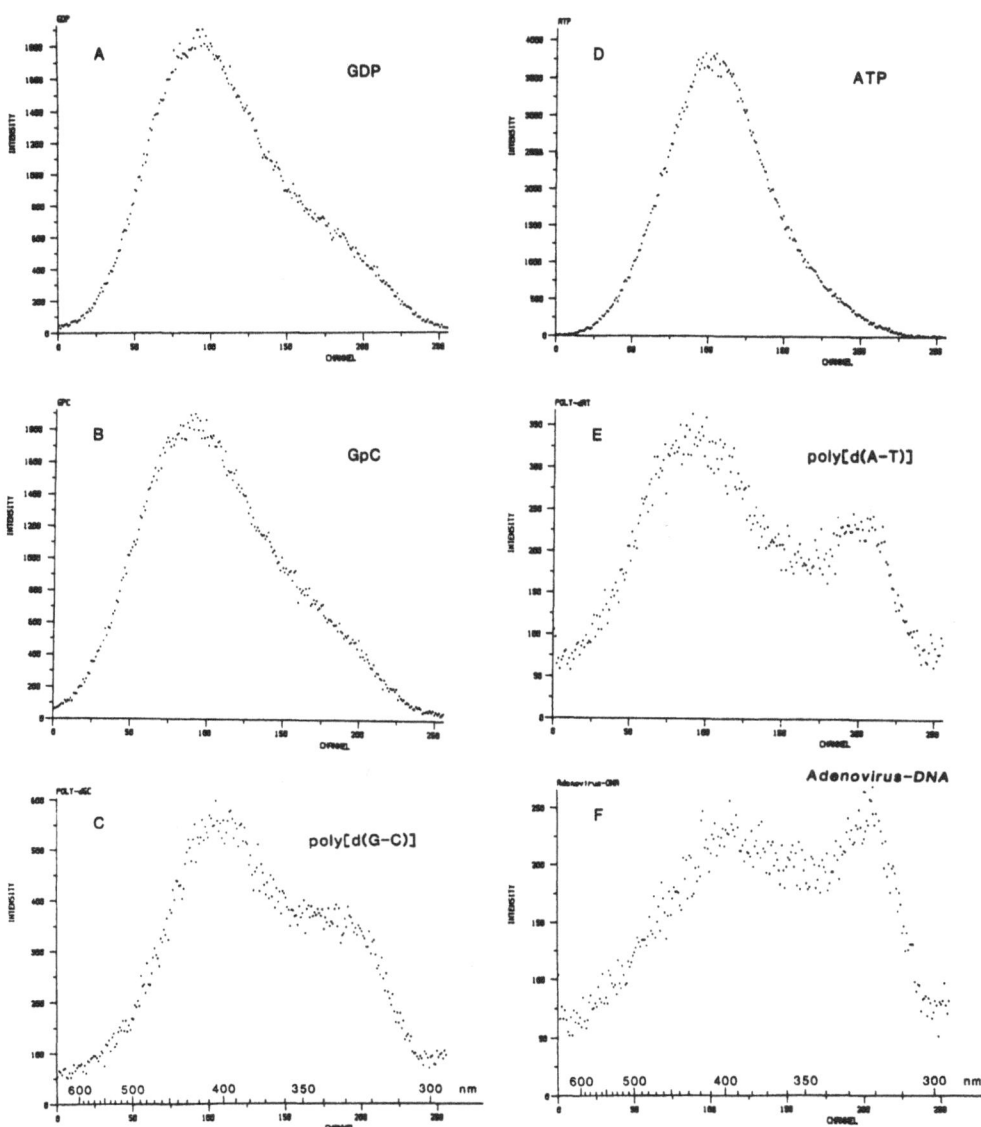

Fig. 2. Time gated fluorescence spectra of GDP (A), GpC (B), poly[d(G-C)] (C), ATP (D), poly[d(A-T)] (E) and Adenovirus 2-DNA (F) in H_2O. Excitation at 290 nm, Zeiss MQIII monochromator, bandwidth 15 nm. Time window 2 ns.

Deoxy-2-aminopurine containing DNAs

The dynamics of the DNA helix comprises a spectrum of kinetic phenomena ranging from fast motions in the ps domain of bases within the nucleotide stack and of the phosphate backbone to motions involving stacking-unstacking (ns-µs), hydrogen bonding (µs-ms) and finally pair opening of individual base pairs (s) (Pörschke, 1977). The development of algorithms in order to simulate the time dependence of molecular trajectories from the knowledge of atomic positions and elementary forces (molecular dynamics simulations and normal mode analysis (Karplus and McCammon, 1983; McCammon, 1983)) has led to interesting possibilities in analyzing the dynamics of the DNA helix with the X-ray structure as starting point (Tidor et al., 1983).

Table 1. Excited states lifetimes of natural and modified purines, tRNA, DNAs and 2-aminopurine containing DNA

	τ_1 [ps]	τ_2 [ps]	τ_3 [ps]	τ_4 [ps]	A_1	A_2	A_3	A_4	χ^2
Purines									
ATP [x]	290	4170			0.36	0.64			1.1
GdP [x]	94	870			0.97	0.03			1.4
Y-base	640	1480			0.27	0.73			1.4
d-2AP		10300				1.00			2.5
m[7]GMP	260				1.00				1.0
RNA [+]									
tRNA[Phe]	210	2800	6250		0.14	0.06	0.80		1.3
tRNA[Phe]+codon	100	1030	5940		0.76	0.1	0.14		1.6
DNA									
poly[d(A-T)]	90	870	5640		0.96	0.3	0.01		2.0
poly[d(G-C)]	250	2930	7810		0.74	0.22	0.04		1.4
d-[GCGC2APC]$_2$									
0.1 M NaCl	114	730	2780	6880	0.55	0.23	0.19	0.03	1.7
5 M KCl	91	550	1490	5650	0.55	0.22	0.16	0.07	1.6

x) data from Rigler et al., 1984
+) data from Claesens and Rigler, 1986

Amongst the presently available spectroscopic techniques laser excitation provides the best time resolution and the existence of torsional motions of the DNA helix has been concluded from the fluorescence anisotropy decay of intercalated ethidiumbromide (Millar et al., 1982).

In order to be able to follow the kinetics of the DNA helix at a localized position we have introduced a modified purine nucleoside, 2-aminopurine (2-AP) which has been prepared as deoxyribonucleoside and incorporated in synthetic DNA sequences by phosphotriester synthesis on a solid matrix (Bäumert et al., 1985).

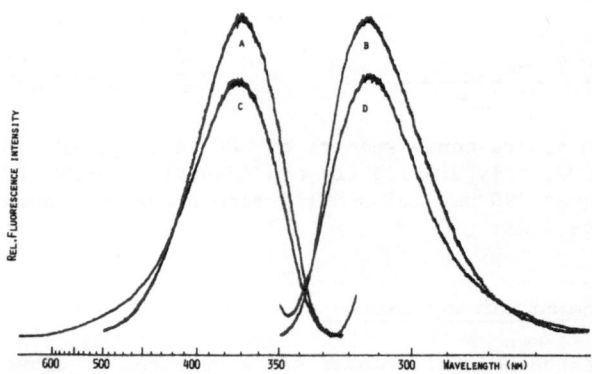

Fig. 3. Excitation (B,D) and emission (A,C) spectra of deoxy-2-amino-purine (A,B) and d-[GCGC2APC]$_2$ (C,D) in 0.01 M Tris, 0.1 M NaCl, pH 7.4, 20°C.

EMISSION AND EXCITATION SPECTRA

The selection of 2-aminopurine as local DNA probe was made because of various favorable properties: (i) an emission spectrum differing

from natural nucleotides, (ii) long lifetime (high quantum yield) and
(iii) base pairing with cytosine (as guanosine analogue) or with thymidine
(as adenine analogue).

When incorporated in a G-C containing DNA hexamer in penultima position
2-AP can be verified easily by its emission and excitation spectrum
(Fig. 3) with peaks at 375 and 315 nm.

LIFETIME OF EXCITED STATES

While no major differences in both emission and excitation spectra
are found for d-2AP in its monomeric as well as stacked form the single
lifetime of about 10 ns of the monomer is split up into several (at
least 4) terms with a strong weighting around 100 ps (Figs. 4A and 5A,D,
Table 1). While the longest lifetime is populated only to a few percent
its weighting is significantly decreased on heating (Rigler et al., 1985)
and is increased under high salt conditions (Table 1) which promote
the transition of G-C DNAs from the B-form into the Z-form (Pohl and
Jovin, 1972; Pohl, 1983).

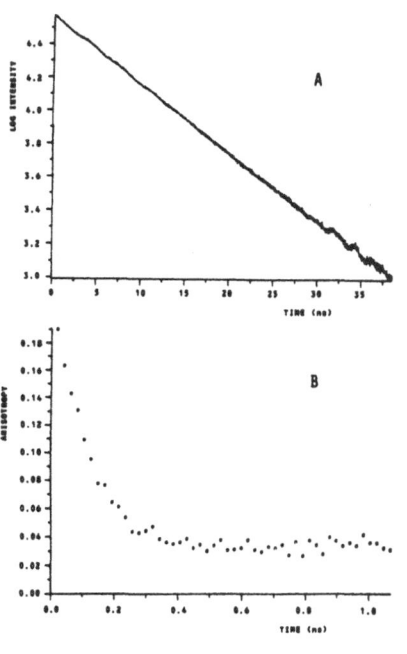

Fig. 4. Fluorescence (A) and anisotropy
(B) decay of deoxy-2-aminopurine
[1 µM] in H_2O, 20°C. Excitation
at 300 nm, emission KV380.

ANISOTROPY DECAY

Analysis of rotational tumbling from
the decay of the anisotropy of fluorescence
yields a rotational relaxation time
(τ_r) of about 150 ps for d-2AP (Fig. 4B
and Table 2) a value typical for dye
molecules of similar size (cresyl-violet)
in water (Rigler et al., 1984). When
incorporated in the G-C hexamer the
anisotropy decay can no longer be described
by a single rotational relaxation time
as is obvious from Figs. 5C and F. An
analysis of the initial part of the
anisotropy decay yields τ_r values of about
400 ps in low salt and of 700 ps in
high salt conditions, which are a factor
4 and 2 faster than would be calculated for the rotational tumbling of
the G-C hexamer (τ_r=1.6 ns).

Table 2. Rotational relaxation times of deoxy-2-aminopurine and
wybutine in d-[GCGC2APC]$_2$ and yeast tRNA[Phe] at 20°C.

	τ_1^{rot} [ps]	τ_2^{rot} [ps]
d-2AP	148	
d-[GCGC2APC]$_2$ 0.1 M NaCl	400	
5 M KCl	717	
tRNA[Phe]		19000
tRNA[Phe]+codon		16200

49

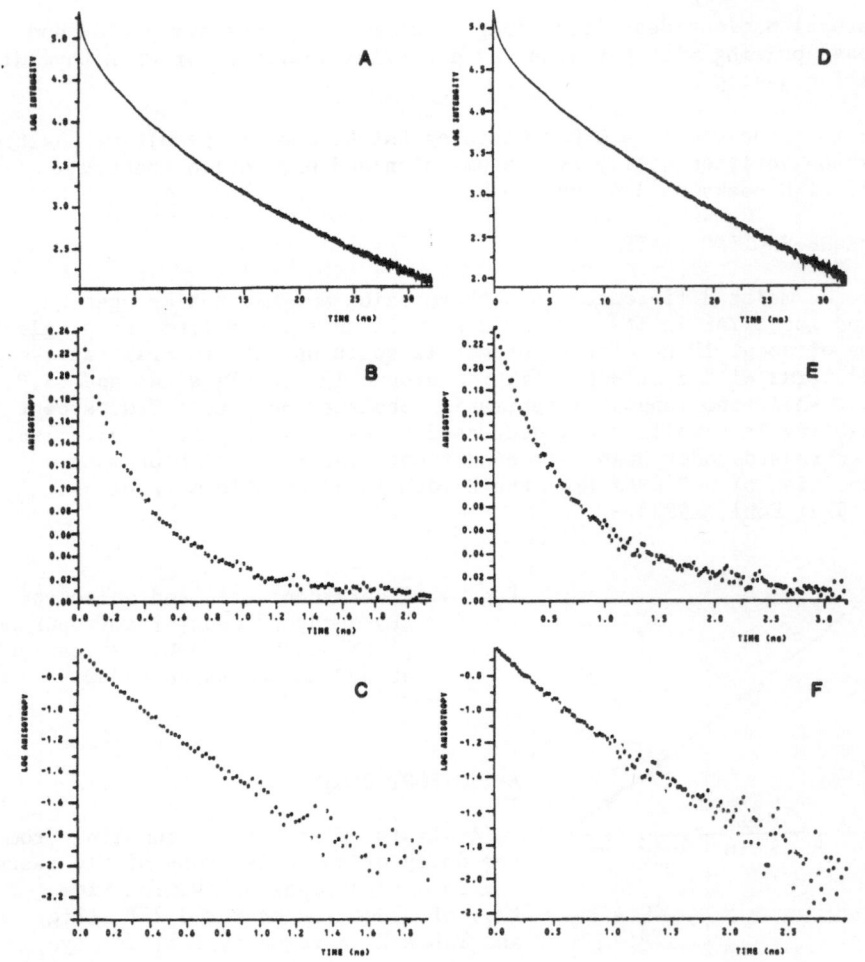

Fig. 5. Fluorescence (A,D) and anisotropy (B,E,C,F) decay of d-[GCGC2APC]$_2$
(40 μM) in 0.01 M Tris, 0.1 mM EDTA with low salt (0.1 M NaCl),
(A,B,C) and high salt (5 M NaCl), (D,E,F), 20°C. Excitation at
300 nm, emission KV 380.

Wybutine in the anticodon loop of tRNA[Phe]

Like DNA, the structure and dynamics of the tRNA molecule has attached
the interest of many researchers. Kinetic processes involve, as in DNA,
base stacking and hydrogen bonding and, in addition, tertiary folding
and interaction of helix-loop regions (Rigler and Wintermeyer, 1983).
Of particular interest is the anticodon loop containing the anticodon
triplet. In the case of tRNA nature has evolved a variety of modified
nucleotides with spectroscopic properties different from ordinary purines
and pyrimidines such as wybutine (Y-base) adjacent to the anticodon triplet
at position 37 in tRNA[Phe] of several species and 7-methyl-guanosine at
position 46 in a variety of tRNAs.

These nucleotides placed in strategic positions in the tRNA structure
can be used as local structural probes. Because of its higher quantum
yield investigations have been centered on wybutine (Eisinger et al.,
1970; Beardsley et al., 1970) in studying codon-anticodon interactions

(Eisinger, 1971: Labuda and Pörschke. 1982). Investigations of the lifetimes of wybutine and tumbling rates could not be performed prior the availability of ps laser pulses and informations on conformations and dynamics of the anticodon have become available only recently (Claesens and Rigler, 1986).

EMISSION SPECTRA AND LIFETIMES OF EXCITED STATES

When incorporated in the anticodon loop of tRNA[Phe] (Fig. 6) the excitation and emission spectra show peaks at 315 and 450 nm respectively. For wybutine isolated from yeast tRNA[Phe] excitation and emission peaks are redshifted by 10 and 20 nm respectively (F.Claesens and R.Thiebe, unpublished data). Furthermore no pronounced excitation maximum below 300 nm as in tRNA is observed. For m^7G another fluorescent base present in tRNA we have measured excitation and emission spectra with peaks at 285 and 385 nm respectively. In the tRNA[Phe] emission and excitation spectra extra peaks exist (Fig.6) which are close to the spectral peaks of m^7G, however a conclusive identification has not yet been possible.

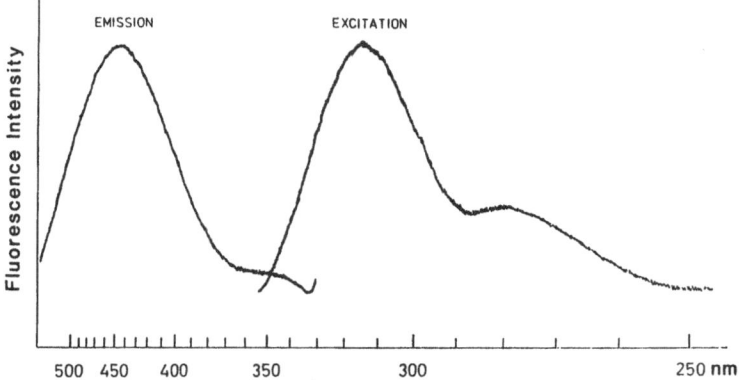

Fig. 6. Excitation and emission spectrum of wybutine (Y-base) in yeast tRNA[Phe] (3 μM) in 0.01 M tris, 0.1 M KCl, 20 mM $MgCl_2$, 0.1 mM EDTA, pH 7.4, 20°C.

The lifetimes of wybutine are complex and bear some similarity to those of ATP (Table 1) with similar weighting factors. Their distribution is concentration dependent and suggests the presence of stacking equilibria in the concentration range 1-100 μM in aqueous solvents. For m^7GMP a single lifetime of 260 ps was determined (Table 1).

When incorporated in the anticodon wybutine decays with 3 lifetimes (Table 1), the weighting factors depending on the presence and absence of the codon sequence UUC (Fig. 7A and B). Detailed analysis shows that these lifetimes correspond to individual conformational states of wybutine and the anticodon loop with wybutine firmly stacked ($\tau=6.2$ ns) or freely mobile ($\tau=100$ ps) as extremes (Claesens and Rigler, 1986).

ANISOTROPY DECAY

Local motion of wybutine in the presence of a codon is demonstrated by the initial decay of the anisotropy (Fig. 7D) with a τ_r of 370 ps (Table 2). This initial decay due to local motions is then followed by further decay (not shown) due to overall tumbling of the whole tRNA molecule (16 ns). In the absence of a codon wybutine is immobile (Fig. 7C), the anisotropy decay is only determined by the tumbling of the whole tRNA molecule.

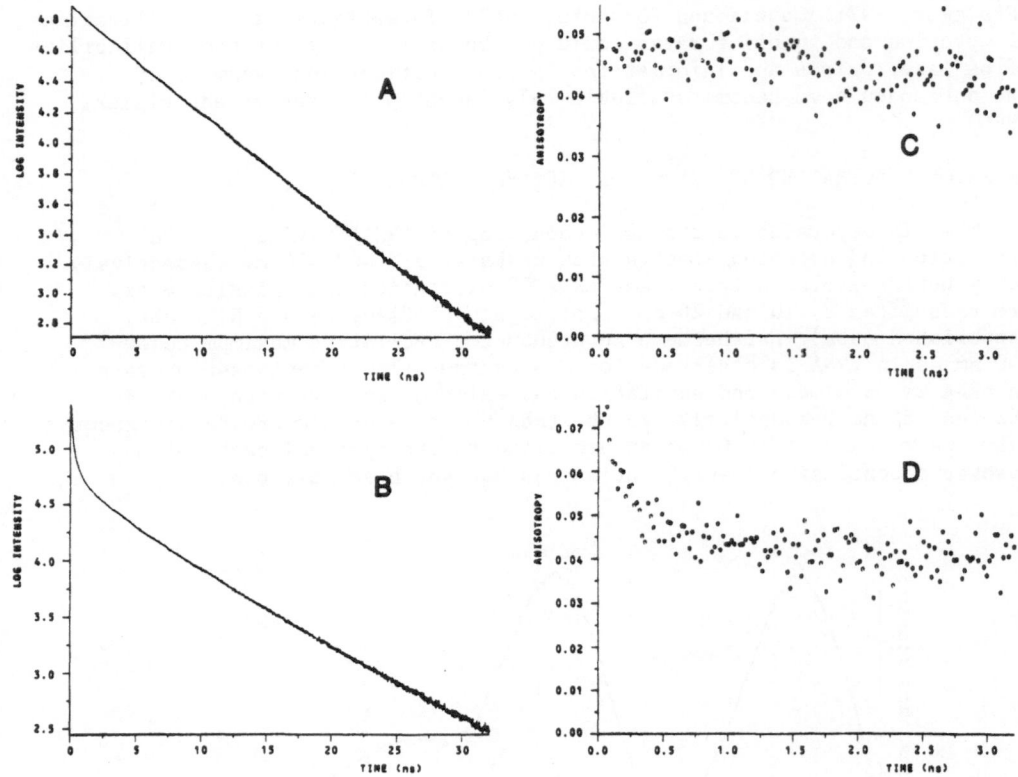

Fig. 7. Fluorescence (A,B) and anisotropy (C,D) decay cf yeast tRNA[Phe]
(30 μM) with (B,D) and without (A,C) E.coli-tRNA[Glu] codon (30 μM)
in 0.01 M Tris, 0.1 M KCl, 20 mM $MgCl_2$, 0.1 mM EDTA, pH 7.4,
20°C. Excitation 300 nm, emission KV 418.

DISCUSSION

The usefulness of modified purines like d-2AP, wybutine or m^7G as
local probes has to be discussed in relation to the emission properties
of unmodified purines and pyrimidines.

Emission spectra, fluorescence, phosphorescence and energy transfer

Similar spectra as shown in Fig. 2 were found for ATP and GDP
(Börresen, 1963) and for poly A (Ballini et al., 1982) and ApA (Morgan
and Daniels 1979) at room temperature as well as for AMP, GMP and GpC
at 77°K (Hélène and Michelson, 1966).

The existence of monomer and excimer fluorescence and phosphorescence
transition has been demonstrated and discussed previously (Eisinger and
Shulman, 1968; Hélène, 1973). When comparing our results at room temperature
with those obtained at low temperature we find the low wavelength emission
(∿330 nm) at the position of fluorescence transitions while the high
wavelength emission (>400 nm) coincides with phosphorescence transitions.
The notion of different decay mechanisms for the excited states of nucleic
acids at room temperature is also supported by the finding that the appear
ence of the short wavelength emission is related to short lived excited-
states while the long wavelength emission is related to a long lived excited
state. Similar conclusions have also been reached by Ballini and coworkers
(1984).

When comparing the excitation and emission spectra of d-2AP a strong
overlap between the 330 nm purine emission and the excitation spectrum

of d-2AP is found. For the 400 nm emission excitation bands appear to exist around 360 nm (Rigler et al., 1985) which overlap with the emission of d-2AP. Thus various pathways for energy transfer from purines (and probably also pyrimidines) to 2AP as well as from 2AP to purine must exist.

For m^7G the situation is similar to 2AP, except the fact that compared to 2AP the lifetime is a factor 40 shorter. While for wybutine energy transfer from purine is possible, the reversed case is less likely because of the large redshift of the emission spectrum of wybutine.

Rotational relaxation and decay time of 2-aminopurine in the DNA-helix

For the rotational motion of 2AP within the DNA helix a value of 400 ps in low salt conditions (0.1 M NaCl) was evaluated which increased almost a factor 2 when the measurements were performed in high salt conditions (5 M NaCl). For d-2AP alone only a 30 percent increase of τ_r is observed.

When calculating the rotational tumbling of the roughly cubic DNA-hexamer from atomic dimensions (cylinder of 10 Å radius and 10.4 Å height) or the molecular volume and partial specific volume (0.7 ml/g) τ_r values of 1.6 and 1.1 ns are calculated. In the latter situation no water of hydration has been included. This value is at least a factor 3 larger than that observed for the initial part of the anisotropy decay (Fig. 5C, Table 2) and we conclude that the rotational motion of 2AP must be due to internal motions within the DNA helix. This motion appears to be slowed down under conditions which promote the Z-form of DNA indicating a higher rigidity of the DNA helix.

Motion of the 2AP base relative to neighbouring bases with varying positions of absorption and emission moments will lead also to various efficiencies of energy transfer. If rotational motion occurs in the time scale where emission takes place coupling of rotation and energy transfer should lead to a whole spectrum of decay times, which evidently is observed (Table 1).

Lifetimes and rotational relaxation of wybutine in tRNA[Phe]

In the case of yeast tRNA[Phe] where wybutine (Y-base) has been used as conformational probe for the conformation of the anticodon loop (Claesens and Rigler, 1986) 3 different decay times of wybutine have been found (Table 1), each of which is characterized by its own rotational mode. The fastest one (370 ps) (Table 2) corresponds to internal motions of wybutine in the presence of a codon while the slowest one (16 to 19 ns) corresponds to the overall tumbling of tRNA. The change in mobility of wybutine and its lifetime spectrum is to be explained by a different conformation of the anticodon loop and the anticodon triplet stack as a result of codon-anticodon interaction.

With the development of ps and fs laser spectroscopy it has become possible to investigate the dynamics of nucleic acids and their constituents at temperatures of biological significance. We have demonstrated the applicability of modified nucleosides as local probes for investigating conformations states and motions in nucleic acid structures: deoxy-2-aminopurine for the base paired double stranded helix and wybutine for the single stranded loop.

ACKNOWLEDGEMENTS

The studies reported were supported by grants from the K.& A. Wallenberg foundation, the Swedish Natural Science Research Council and from the Karolinska Institute.

REFERENCES

Auston, D. H., and Eisenthal, K. B., 1984, Ultrafast Phenomena
    IV, Springer Series in Chemical Physics, Springer-Verlag, 38.
Ballini, J. P., Daniels, M., and Vigny, P., 1982, J. Luminescence, 27:389.
Ballini, J. P., Vigny, P., and Daniels, M., 1983, Biophys.
    Chem., 19:61.
Ballini, J. P., Daniels, M., and Vigny, P., 1984, In "Abstracts
    8th International Biophysics Congress", Bristol, England, p. 228.
Beardsley, K., Tao, T., and Cantor, C. R., 1970, Biochemistry,
    9:3524.
Bäumert, H., Kaun, E., Claesens, F., and Rigler, R., 1985, in
    preparation.
Börresen, H. C., 1963, Acta Chem. Scand., 17:921.
Claesens, F., and Rigler, R., 1986, Eur. Biophys. J. in press.
Daniels, M., 1973, In "Physico-chemical properties of nucleic
    acids", Ed. I.Duchesne, Academic Press, London-New York, p. 99.
Eisinger, J., and Shulman, R. G., 1968, Science, 161:1311.
Eisinger, J., Feuer, B., and Yamane, T., 1970, Proc. Natl. Acad.
    Sci. USA, 65:638.
Eisinger, J., 1971, Biochem. Biophys. Res. Comm., 43:854.
Hélène, C., and Michelson, A. M., 1966, BBA, 142:12.
Hélène, C., 1973, In "Physico-chemical properties of nucleic
    acids, Ed. I.Duchesne,, Academic Press, London-New York, p. 119.
Karplus, M., and McCammon, J. A., 1983, Ann. Rev. Biochem., 52:263.
Labuda, D., and Pörschke, D., 1982, Biochemistry, 21:49.
McCammon, J.A., 1983, In "Reports of Progress in Physics", Institute
    of Physics, London.
Millar, D. P., Robbins, R. J., and Zewail, A. H., 1982, J. Chem.
    Phys., 76:2080.
Morgan, J. P., and Daniels, M., 1979, Photochem. Photobiol.,
    31:101.
Pohl, F. M., and Jovin, T. M., 1972, J. Mol. Biol., 67:375.
Pohl, F. M., 1983, In "Cold Spring Harbor Symp. Quant. Biol.",
    XLVII:113.
Pörschke, D., 1977, In "Chemical Relaxation in Molecular
    Biology, Eds. I.Pecht and R.Rigler, Springer-Verlag,
    Berlin-Heidelberg-New York, p. 191.
Rigler, R., and Wintermeyer, W., 1983, Dynamics of tRNA, Ann.
    Rev. Biophys. Bioeng., 12:475.
Rigler, R., Claesens, F., and Lomakka, G., 1984, In "Ultrafast
    Phenomena IV", Eds. D.H.Auston and K.B.Eisenthal, Springer
    Series in Chemical Physics, Springer-Verlag, 38:472.
Rigler, R., Claesens, F., Kaun, E., and Bäumert, H., 1985, in
    preparation.
Tidor, B., Irikura, K. K., Brooks, B. R., and Karplus, M., 1983,
    J. Biomol. Struct. Dyn., 1:231.

Z-RNA: A LEFT-HANDED DOUBLE HELIX

Ignacio Tinoco, Jr., Phillip Cruz, Peter Davis,
Kathleen Hall, Charles C. Hardin, Richard A.
Mathies, Joseph D. Puglisi, and Mark A. Trulson

Department of Chemistry
and Laboratory of Chemical Biodynamics
University of California, Berkeley, CA 94720

W. Curtis Johnson, Jr.
Department of Biochemistry and Biophysics
Oregon State University
Corvallis, OR 97331

Thomas Neilson
Department of Biochemistry
McMaster University
Hamilton, Ontario, Canada

INTRODUCTION

The synthetic polyribonucleotide, poly [r(G-C)], has been
shown to undergo a transition from the right handed A-form to a
left-handed Z-conformation.[1] This is the first major conformational change found in double-stranded RNA.[2] Therefore it is
important to corroborate this observation, to characterize the
conditions that produce Z-RNA, and to determine some of the
properties of this novel left-handed structure. This article is a
summary of our results in this area.

SPECTROSCOPY

Four different spectroscopic methods have been used to detect
and to characterize the new conformation of RNA: nuclear magnetic
resonance, circular dichroism, ultraviolet absorbance and Raman

scattering. These methods have all been applied to poly [r(G-C)], which was synthesized using RNA polymerase on a poly [d(I-C)] template.[3] The RNA polynucleotide was purified and studied as a function of temperature in a wide range of solvents.

Nuclear Magnetic Resonance

The original evidence for a conformation in RNA similar to that of Z-DNA came from proton and phosphorous NMR.[1] Poly [r(G-C)] in 6 M NaClO$_4$ at a temperature of 45°C was found to have two phosphorous resonances separated by 1.33 ppm; one is at -2.74 ppm and one at -4.07 ppm relative to trimethylphosphate. These values are very similar to the phosphorous chemical shifts found for Z-DNA.[4] By analogy with Z-DNA the -4.07 ppm phosphorous resonance is assigned to CpG and the -2.74 ppm resonance to GpC[5]. At lower temperatures in 6 M NaClO$_4$ or at any temperature in 3 M NaClO$_4$, the two phosphorous chemical shifts occur near 4 ppm and correspond to A form RNA as seen in dilute buffer. Proton nuclear Overhauser effects (NOE) were used to establish that the conformation corresponding to the wide separation between phosphorous resonances contained guanosines in the syn conformation, as found in Z-DNA. Poly [r(G-C)] in 6 M NaClO$_4$ at 45°C showed a negative NOE between the proton resonances of the 8 position of guanine (GH-8) and the 1' position of its ribose (GH-1'). The magnitude of the NOE was about the same as that seen between the protons at the 5 (CH-5) and 6 (CH-6) positions of cytosine. The distance between CH-5 and CH-6 is 2.3 Å, therefore the distance between GH-8 and GH-1' must be similar. This can only occur if guanosine is syn. In lower salt concentrations or at lower temperatures, the much smaller NOE between GH-8 and GH-1' indicates guanosine is anti as expected for A-RNA. The general conclusion is that in 6 M NaClO$_4$ above 35°C, poly [r(G-C)) has a conformation called Z-RNA which is similar to Z-DNA. The chemical shifts for the aromatic protons are very similar for Z-RNA and Z-DNA; the sugar H-1' protons are different as expected for ribose vs. deoxyribose (see Table 1).

Circular Dichroism

The reason the NMR studies were done in 6 M NaClO$_4$ at elevated temperatures was that the circular dichroism (CD) of poly

Table 1. Comparison of the Change in Proton Chemical Shifts in the A to Z Transition of RNA with those in the B to Z Transition of DNA.

	CHEMICAL SHIFT			
PROTON	poly[r(G-C)]		poly[dG-C)]	
	A[a]	Z[b]	B[c]	Z[c]
GH8	7.51	7.82	7.82	7.81
CH6	7.63	7.16	7.32	7.25
CH5	5.11	5.17	5.28	5.14
GH1'	(5.65)	5.87	(5.82)	6.21
	or		or	
CH1'	(5.43)	5.71	(5.64)	5.65

a) In 3 M $NaClO_4$, 45°C

b) In 6 M $NaClO_4$, 45°C

c) From ref. 4.

[r(G-C)] changes from an A-RNA spectrum to a different shape as the temperature is raised in this solvent.[1] Furthermore, the UV absorbance spectrum acquires a shoulder near 290 nm as seen in Z-DNA. Although the high salt-high temperature CD spectrum is different from A-RNA, it is also different from Z-DNA in the wavelength region usually studied (220 nm - 340 nm). However, if the wavelength region is extended to 170 nm, a much more definitive conclusion can be reached.[6]

Fig. 1 (top) shows the CD spectrum from 170 nm to 340 nm for poly [r(G-C)] in 6 M $NaClO_4$ at two temperatures. The major change in the spectrum takes place between 170 nm - 220 nm. Furthermore, nearly the same change occurs in this region (Fig. 2) when the two right-handed forms of DNA (A and B) change to the left-handed Z-DNA. This is very compelling further evidence that poly [r(G-C)] goes from right-handed A-RNA to left-handed Z-RNA in high salt and high temperature. However, there are obvious differences in the CD between Z-RNA and Z-DNA (particularly above 260 nm). This may indicate differences in the relative orientation of the base pairs in the two forms, or it may be caused by differences in base-phosphate or base-counterion orientation in the two forms.

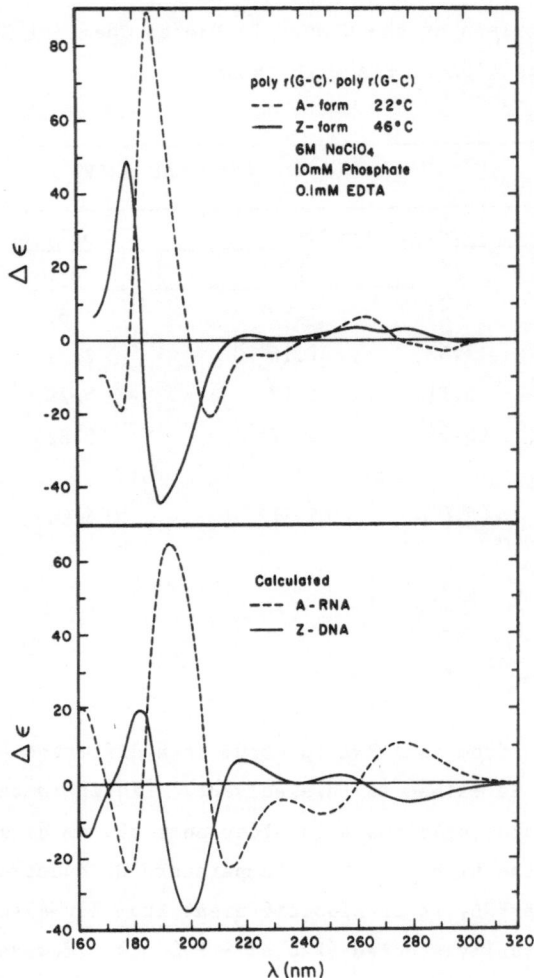

Fig. 1. (Top) Measured circular dichroism for poly [r(G-
C)] in 6 M NaClO$_4$ in A-form and Z-form. Reprinted with
permission from Riazance et al., Nucleic Acids Research,
13, 4983 (1985). (Bottom) Calculated circular dichroism
for poly (G-C) in right-handed A-RNA geometry and left-
handed Z-DNA geometry. The qualitative agreement below
220 nm is excellent.

The CD has been calculated for the various conformations of
both RNA and DNA alternating G-C polynucleotides using the unpublished X-ray coordinates.[7] Polynucleotide rotational strengths
and transition frequencies are calculated from the transition
moment magnitudes, directions, and frequencies of the nucleic acid
bases.[8,9] Interaction energies are calculated using transition
monopoles adjusted to be consistent with experimental transition
dipoles.[10] Calculations were done for a helix of twelve base

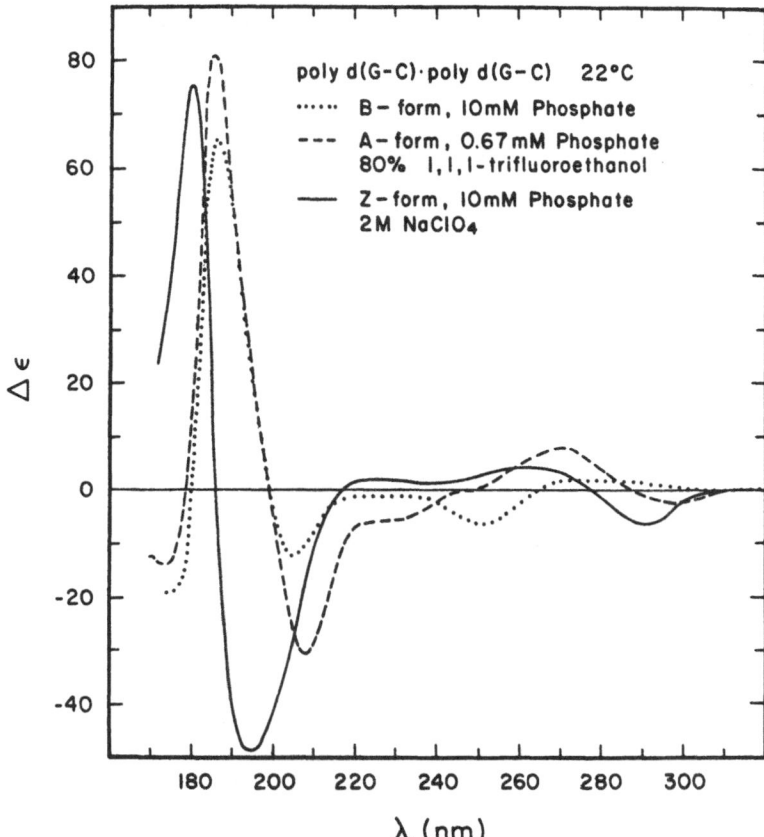

Fig. 2 Measured circular dichroism for poly [d(G-C)] in right-handed A- and B-forms and left-handed Z-form. The difference in CD below 220 nm for the right and left-handed forms is nearly the same as found in RNA. Reprinted with permission from Riazance et al., Nucleic Acids Research, 13, 4983 (1985).

pairs. The calculated results in Fig. 1 (bottom) for A-RNA and Z-DNA are in very good agreement with experiment below 220 nm.

Both the experimental and calculated CD spectra below 220 nm are clearly charateristic of the sense of the polynucleotide helix. The right-handed forms (A-RNA, A-DNA and B-DNA) have positive CD from 180 nm to 200 nm. The left-handed forms (Z-RNA, Z-DNA) have a negative CD from 185 nm to 200 nm. Measurement of the CD in this wavelength region may be the easiest method to identify left-handed polynucleotides. Commerical CD instruments can be used and the amount of material needed is less than required for other spectroscopic methods.

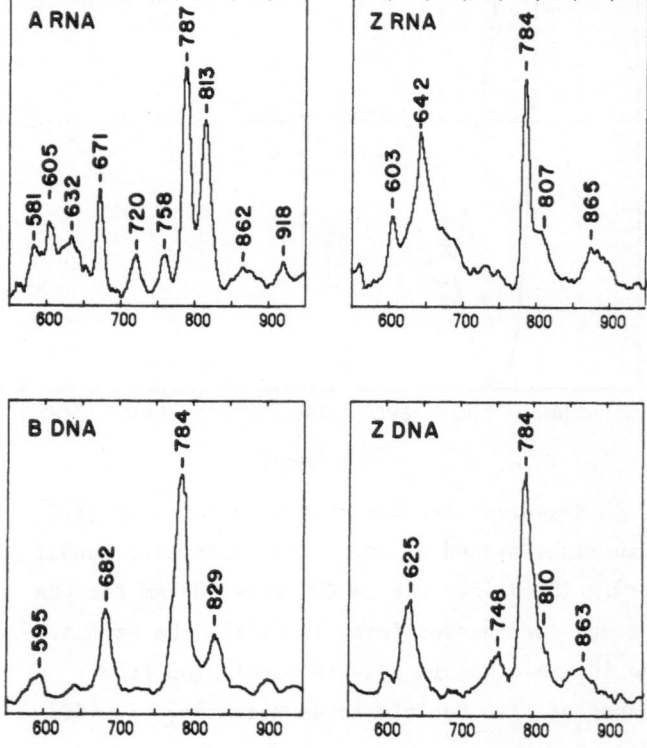

Fig. 3 The Raman spectrum of poly [r(G-C)] measured in 5.5 M NaBr at 22°C (A-RNA) and 40°C (Z-RNA). Very similar changes in the Raman spectra of DNA occur when it goes from B-DNA to Z-DNA. The DNA data are from J. M. Benevides and G. J. Thomas, Jr., Nucleic Acids Research, 11, 5747-5761 (1983).

Raman Scattering

When poly [r(G–C)] undergoes an A to Z transition in 5.5 M NaBr as shown by CD, its Raman spectrum shows characteristic changes (Fig. 3). The changes are similar to those seen in the B to Z transition in DNA.[11] The phosphodiester antisymmetric stretch at 813 cm^{-1} in the A-RNA spectrum nearly disappears in the Z-form. The guanine ring breathing mode at 671 cm^{-1} in A-RNA shifts to 642 cm^{-1} in Z-RNA. In DNA the shift is from 682 cm^{-1} in B-DNA to 625 cm^{-1} in Z-DNA. The difference in guanine frequencies between Z-DNA and Z-RNA may indicate a difference in conformation between the two forms. The long wavelength CD spectra also indicate differences in conformation between Z-DNA and Z-RNA.

Z-RNA ENVIRONMENTS

The spectroscopic techniques provide ample methods to identify left-handed RNA: proton NMR, phosphorous NMR, CD and Raman scattering. We have mainly used long wavelength CD measurements, because they require the least amount of material. However, for solvents which absorb below 300 nm, phosphorous NMR is convenient. We have surveyed many different salt solutions to see if they favored Z-RNA; we tried many of the conditions that favor Z-DNA. Table 2 summarizes our results. The general conclusion is that it is more difficult to induce Z-RNA than Z-DNA; higher salt concentrations are required. Ions such as $Co(NH_3)_6^{3+}$, which induces Z-DNA in micromolar concentrations[12,13] by site specific binding have no effect on poly [r(G–C)].

The transition temperatures shown in the Table are approximate. The kinetics of the transition are slow; the transition may take minutes or hours to occur. However, for all the solvents studied, higher temperature favors the Z-conformation. This means that the A to Z transition requires absorption of heat. With increasing temperature the equilibrium is:

$$A\text{-RNA} \rightleftharpoons Z\text{-RNA} \rightleftharpoons \text{Single strands}$$

As guanosine is in a syn conformation in Z-form, substituents on its 8 position which block the anti conformation should favor Z-RNA or Z-DNA. Bromination of poly [d(G–C)] in a Z-forming solvent does indeed stabilize Z-DNA.[14] We have found poly [r(G–C)] is also stabilized by bromination by Br_2 which produces 8-

61

Table 2. The A to Z Transition Temperature for Poly [r(G–C)] in Different Environments

Solvent	A → Z Transition Temperature
6.8 M NaBr	<25°C
6.0 M NaClO$_4$	~35°C
4.8 M NaClO$_4$, 20% ETOH	<25°C
Brominated poly[r(G–C)] in 2 M NaClO$_4$	<25°C
4.0 M MgCl$_2$	<25°C
2.9 M MgBr$_2$	<25°C
Brominated poly[r(G–C)] in pure H$_2$O	<25°C

NaCl, LiCl, CsCl, Na trifluoroacetate, Na trichloroacetate, MnCl$_2$, NiCl$_2$, Co(NH$_3$)$_6$Cl$_3$, urea, guanidinium Cl and spermidine did not show evidence of Z–RNA.

bromoguanine and 5–bromocytosine. The CD spectra showed that the brominated polynucleotide was mainly Z in 2 M NaClO$_4$ at room temperature; it was partly Z even in 0.3 M NaCl at room temperature. This is encouraging because it means that brominated poly [r(G–C)] could be used to induce anti Z–RNA antibodies.

When poly [d(G–m^{5}C)] is dialyzed exhaustively against pure water, it undergoes the transition to Z–DNA.[15] Poly [r(G–C)] does not go to Z–RNA in pure water, but circular dichroism and Raman data indicate that brominated poly [r(G–C)] does.

It should also be emphasized that there is more than one type of Z–RNA conformation. In 4 M MgCl$_2$ the long wavelength CD spectrum for poly [r(G–C)] is like the Z–DNA spectrum instead of the Z–RNA CD spectrum seen in NaClO$_4$ and NaBr. Raman spectra in MgCl$_2$ show that the poly [r(G–C)] is Z–form, and there is evidence that the transition from a conformation with one CD spectrum to another CD spectrum for poly [r(G–C)] in MgCl$_2$ is slow. Circular dichroism spectra of the brominated polynucleotide at low ionic strength are also of the Z–DNA type. NMR spectra may allow a better characterization of the different left–handed RNA forms.

TETRANUCLEOTIDE, rCpGpCpG

Although the discovery and characterization of a left-handed form of poly [r(G-C)] is of interest to physical chemists, unless Z-RNA can also occur in shorter sequences it is unlikely to have biological significance. Therefore it was pleasing to find that the tetra-ribonucleoside triphosphate rCpGpCpG (called the tetranucleotide) also undergoes an A to Z transition. A bonus from this oligonucleotide is the ability to compare the A to single strands transition with the Z to single strands transition.

CD measurements on rCpGpCpG show that at 0°C the double helix is in A-form in 1 M $NaClO_4$ and Z-form in 6 M $NaClO_4$.[15] The CD spectra cover the range from 200 nm to 340 nm; they are similar, but not identical to the corresponding polynucleotide spectra. As the temperature is raised, the A-duplex at millimolar concentrations in 1 M $NaClO_4$ melts to single strands near 25°C; the Z-duplex in 6 M $NaClO_4$ melts to single strands near 5°C. At intermediate salt concentrations there seems to be a mixture of A and Z double-stranded forms at 0°C; at -15°C both A- and Z-forms exist in 6M $NaClO_4$. We thus find that Z-RNA is more stable at lower temperatures with an oligonucleotide than with a polynucleotide. The oligonucleotide has a lower melting temperature than the polynucleotide; this seems to shift the A to Z transition to room temperature and below.

NMR studies corroborate that rCpGpCpG is in Z-form in high salt, and also show that there is slow exchange between single strands and the Z-duplex. The A-duplex and the single strands are in fast exchange as expected for either B-DNA or A-RNA dissociation kinetics for this length oligonucleotide.

The phosphorous NMR of single stranded rCpGpCpG shows three resonances near -4 ppm relative to trimethylphosphate. In 1 M $NaClO_4$ as the solution is cooled, the peaks shift (and broaden) to produce the three unique resonances corresponding to the symmetrical duplex. This shows that the A-duplex and the single strands are in fast exchange. All three phosphorous resonances in the single strands or the A-duplex are within 0.5 ppm of -4 ppm. In contrast, as the 6 M $NaClO_4$ solution is cooled, three new peaks appear in addition to those of the single strands (Fig.4). This clearly shows that the single strand to Z-duplex transition is

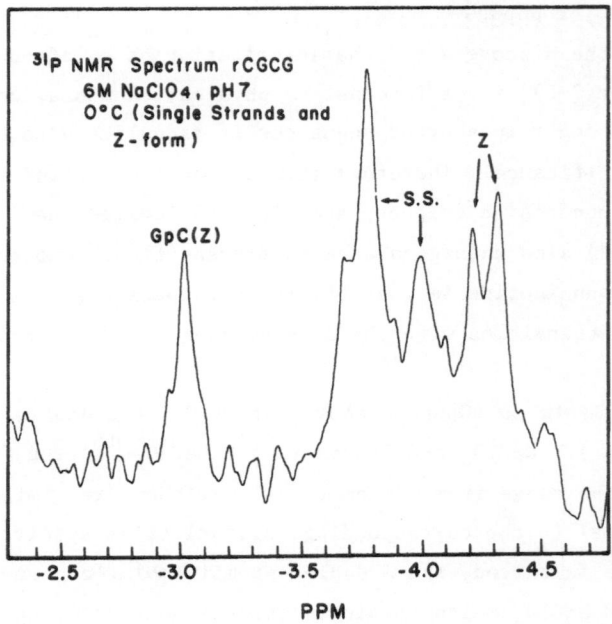

Fig. 4. The phosphorous NMR spectrum of rCpGpCpG shows that in 6 M NaClO$_4$ at 0°C, the duplex in a Z-conformation is in slow exchange with single strands.

slow on the NMR time scale (>5-10 ms). One of the new phosphorous peaks is shifted downfield and two remain near 4 ± 0.5 ppm. This is expected for a Z-duplex with the peak at -3.0 ppm corresponding to the GpC phosphorous.

Proton NMR is consistent with our interpretations. There is slow exchange between a Z-duplex and single strands as seen in 6 M NaClO$_4$ spectra at 0°C; aromatic proton peaks from both forms are observed. In low salt the A-duplex and single strand equilibrium shows only one set of aromatic protons. NOE experiments in 6 M NaClO$_4$ at 0°C provide further information about the structure and kinetics of the Z-duplex. There is the expected proton NOE from the 8 position of guanine to its 1' ribose. This is the dipolar NOE which shows that the guanosines are in a syn conformation in the Z-duplex. There is of course an NOE between H-6 and H-5 on the cytosines, but there is also an NOE from H-6 on cytosine in the Z-duplex to H-6 on single-stranded cytosine. This is an apparent NOE due to chemical exchange; it can be used to study the transition kinetics.

Much more can be learned from the NMR of rCpGpCpG (and dCpGpCpG). Complete 2D NMR spectra can provide detailed knowledge of the Z-RNA duplex conformation and a comparison with the Z-DNA duplex conformation.

BIOLOGICAL FUNCTION

There is no evidence yet that left-handed RNA exists in a naturally occurring double-stranded RNA, and there is no evidence yet for any biological function. Possible methods to search for (presumably) small amounts of a left-handed conformation in a natural RNA include use of specific anti-Z-RNA antibodies, or use of secondary structure specific reagents. Until a rigorous search is made we can only speculate about possible sites of left-handed RNA and possible functions.

An obvious site where RNA structure and dynamics are important is the ribosome. The ribosomal RNA molecules are involved in all aspects of translation of the messenger RNA.[17] The secondary structures of the 5S, 16S and 23S ribosomal RNA's of the E. coli ribosome are well established.[18] Fig. 5 shows double-stranded regions in E. coli 16S rRNA where there are sequences of four or more alternating purine-pyrimidines containing at least one G·C. The longest double-stranded sequence of alternating purine-pyrimidine contains seven base pairs (at bases 947-953, paired with 1228-1234). Four of the regions have three alternating G·C base pairs. Any of the alternating sequences could possibly form left-handed structures under the influence of an appropriate effector. The effector could be a specific protein or small molecule.

If a right- to left-handed transition in the ribosome is to have any biological function it must occur not only in E. coli, but in other bacteria as well. The sequences in Fig. 5 labelled with an asterisk are conserved in nearly all of 22 eubacteria whose 16S rRNA's have been sequenced.[19] For some regions the exact sequence is conserved; for other regions the sequence changes, but the alternation of purine-pyrimidine is conserved. The region A-C-G-C:G-C-G-U (33-36, 548-551 in E. coli) is

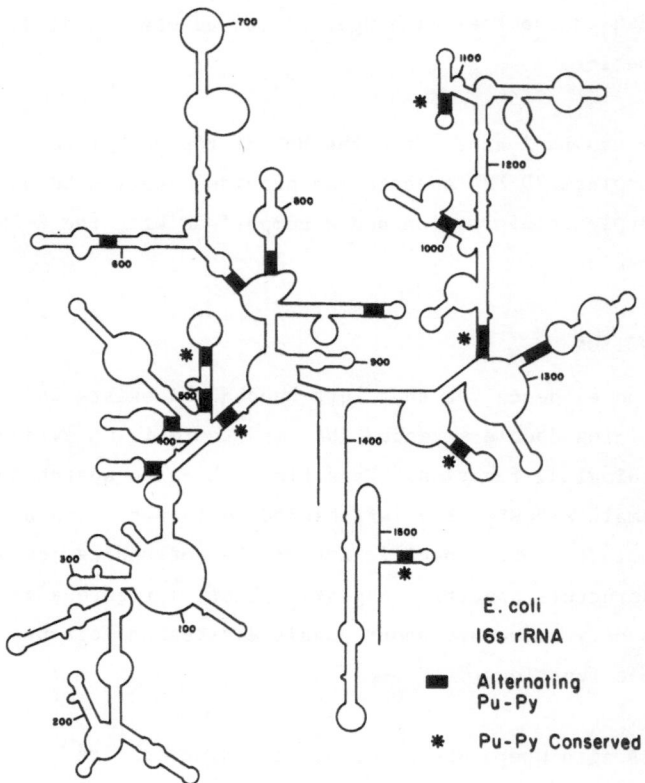

Fig. 5. The secondary structure of E. coli 16S
ribosomal RNA with regions of four or more
alternating purine-pyrimidine sequence shown. The
* regions are conserved throughout the eubacteria.

conserved exactly in all 22 eubacteria; it is known to be a
binding site for S4 ribosomal protein. The 947-953, 1228-1234 E.
coli sequence of alternating purine-pyridimines is conserved in 19
of 22 eubacteria, although the sequence is not. There is a
conserved region of four alternating purine-pyrimidines in the
colicin fragment of 16S rRNA (1510-1513, 1522-1525 in E. coli).
Some of these same regions are also conserved in other classes of
organisms such as archaebacteria, eukaryotes and mitochondria.[20]
None is conserved throughout all the different kingdoms.

    Regions of alternating purine-pyrimidine sequences are also
found in double-stranded portions of 5S and 23S rRNA.[18] Viral
RNA's which are completely double stranded contain alternating C·G
base pairs.[21] For a completely random sequence of bases with
equal amounts of each, we expect to find a C-G-C-G (or G-C-G-C)

with a probability of $4^{-4} = 3.9 \times 10^{-3}$. This means that a random sequence of 1000 nucleotides would contain four runs of C-G-C-G (or G-C-G-C).

What role could an A to Z transition have? RNA in a right-handed A conformation has approximately 11 base pairs per turn. This corresponds to a winding angle of +33°. Left-handed Z-DNA has 12 base pairs per turn and a winding angle of -30°. If Z-RNA is similar, each base pair which changes from a right to left conformation causes a rotation of -63°. This will move one part of an RNA relative to the other, unless some compensating change in another double-stranded region occurs. If covalently closed, double-stranded RNA circles exist, superhelical turns or cruciform formation must compensate for the right to left transition. But for any RNA, a right to left transition will cause motion. The A to Z transition could thus be a biochemical switch.

CONCLUSION

We have shown that both poly [r(G-C)] and rC-G-C-G can adopt a left-handed Z conformation in a variety of concentrated salt solutions. The Z-form is favored by increasing temperature and by bromination at the 8 position of guanine. Z-RNA is similar in structure, but not identical to Z-DNA. In fact CD measurements suggest that there are at least two left-handed forms; one has a CD spectrum very much like Z-DNA, the other does not. Brominated poly [r(G-C)] is at least partly Z in physiological conditions and can be used to elicit antibodies. Once antibodies are available, a serious search for naturally occuring Z-RNA can be made. Until then it is entertaining to speculate on possible locations and biological functions for Z-RNA.

ACKNOWLEDGEMENT

This research was supported in part by National Institute of Health grant GM10840, and by the U. S. Department of Energy, Office of Energy Research under contract 82ER60090.

REFERENCES

1. K. Hall, P. Cruz, I. Tinoco, Jr., T. M. Jovin, and J. H. van de Sande, Nature 311, 584-586 (1984).

2. W. Saenger, Principles of Nucleic Acid Structure, Springer-Verlag, New York, Chapter 9, 1984.

3. K. Hall, P. Cruz, and M. J. Chamberlin, Arch. Biochem. Biophys. $\underline{236}$, 47-51 (1985).

4. D. J. Patel, S. A. Kozlowski, A. Nordheim and A. Rich, Proc. Natl. Acad. Sci., U. S. A. $\underline{79}$, 1413-1417 (1982).

5. T. M. Jovin, J. H. van de Sande, D. A. Zarling, D. J. Arndt-Jovin, F. Eckstein, H. H. Fuldner, C. Greider, I. Grieger, R. Hamori, B. Kalisch, L. P. McIntosh and M. Robert-Nicoud and L. B. Clark, Cold Spring Harbor Symp. Quant Biol. $\underline{47}$, 143-154 (1983).

6. J. H. Riazance, W. A. Baase, W. C. Johnson, Jr., K. Hall, P. Cruz, I. Tinoco, Jr., Nucleic Acids Res., $\underline{13}$, 4983-4989 (1985).

7. A. Williams, Jr., C. Cheong, I. Tinoco, Jr., unpublished results.

8. I. Tinoco, Jr., C. Bustamante and M. F. Maestre, Ann. Rev. Biophys. Bioeng. $\underline{9}$, 107- 141 (1980).

9. V. Rizzo and J. A. Schellman, Biopolymers, $\underline{23}$, 435-470 (1984).

10. L. B. Clark, J. Am. Chem. Soc. $\underline{99}$, 3934-3938 (1977).

11. T. J. Thamann, R. C. Lord, A. H.-J. Wang and A. Rich, Nucleic Acids Res. $\underline{9}$, 5443-5457 (1981).

12. M. Behe and G. Felsenfeld, Proc. Natl. Acad. Sci., U. S. A. $\underline{78}$, 1619-1623 (1981).

13. R. V. Gessner, G. J. Quigley, A. H.-J. Wang, G. A. van de Marel, J. H. van Boom and A. Rich, Biochemistry $\underline{24}$, 237-240 (1985).

14. A. Moller, A. Nordheim, S. A. Kozlowski, D. J. Patel and A. Rich, Biochemistry $\underline{23}$, 54-62 (1984).

15. B. G. Feurstein, L. J. Marton, M. A. Keniry, D. L. Wade and R. H. Shafer, Nucleic Acids Res. $\underline{13}$, 4133-4141 (1985).

16. K. Hall, Ph.D. Thesis, U. of California, 1984.

17. H. F. Noller and J. A. Lake, Chapter 5 in Membrane Structure and Function, Vol. 6, E. E. Bittar, editor, John Wiley, New York, 1984.

18. H. F. Noller, Ann. Rev. Biochem. $\underline{53}$, 119- 162 (1984).

19. R. R. Guttell, personal communication.

20. C. R. Woese, R. R. Guttell, R. Gupta and H. F. Noller, Microbiol. Revs. $\underline{47}$, 621-669 (1983).

21. M. J. Leibowitz, D. J. Thiele and E. M. Hannig, pp. 457-462 in Double Stranded RNA Viruses, R. W. Compan and D. H. L. Bishop, editors, Elsevier Science, Amsterdam, 1983.

A BETTER WAY TO MAKE RNA FOR PHYSICAL STUDIES

P. Lowary, J. Sampson, J. Milligan, D. Groebe, and
O. C. Uhlenbeck

Department of Biochemistry
University of Illinois
Urbana, IL 61801

Studies on the structure and function of RNA have often been hampered by the difficulty in obtaining material in sufficient amounts for many physical methods. In addition, the variety of RNA sequences easily available for study is somewhat limited. Ideally, one would like a simple method to rapidly obtain multimilligram amounts of any RNA sequence desired. Although methods for the chemical synthesis of RNA have progressed rapidly in the past several years (1), they remain technically quite complex and are generally limited to fragments of less than 20 nucleotides. The enzymatic methods used by this laboratory to synthesize RNA by joining shorter fragments with T4 RNA ligase have also succeeded in preparing fragments up to 20 nucleotides long, but yield only very small amounts of material (2).

In this paper, we will review the methodology that has been developed in several labs to use RNA polymerase to make RNA. By combining this method with automated solid support DNA synthesis technology, it is possible to obtain multimilligram amounts of virtually any RNA sequence desired with precisely defined 3' and 5' termini. An example of such a large scale synthesis will be given.

RUNOFF TRANSCRIPTION WITH T7 RNA POLYMERASE

A typical _in vitro_ runoff transcription reaction involves incubating RNA polymerase, ribonucleoside triphosphates and a linear DNA template containing a promoter. The RNA transcript starts at the promotor and terminates at the end of the DNA. For most RNA polymerases the yields of _in vitro_ transcription reactions are less than one mole of RNA per mole of DNA template. However, perhaps because they are single polypeptide chains without dissociatable initiation factors, certain bacteriophage encoded RNA polymerases are much more active, making up to several hundred transcripts per mole of template (3).

The two phage polymerases most extensively characterized come from _Salmonella_ bacteriophage SP6 (4,5) and _E. coli_ bacteriophage T7 (6). Not only are these two enzymes structurally quite similar, but they appear to have quite similar enzymological properties. However, their promoter sequences are sufficiently different that they do not cross react. The T7 system is greatly preferred for preparing RNA for physical studies because it is possible to obtain large amounts of very pure enzyme from a recombinant bacterial strain (7). In a recent purification in our lab, we were able to obtain 35 mg of homogeneous enzyme from approximately 3 g of induced cells obtained from one liter of culture. This enzyme was assayed to be 4.5×10^5 units/mg which is close to the value of 6×10^5 units/mg reported for enzyme purified from phage infected cells (8). Scaling up this purification 10 or 20-fold should be straightforward.

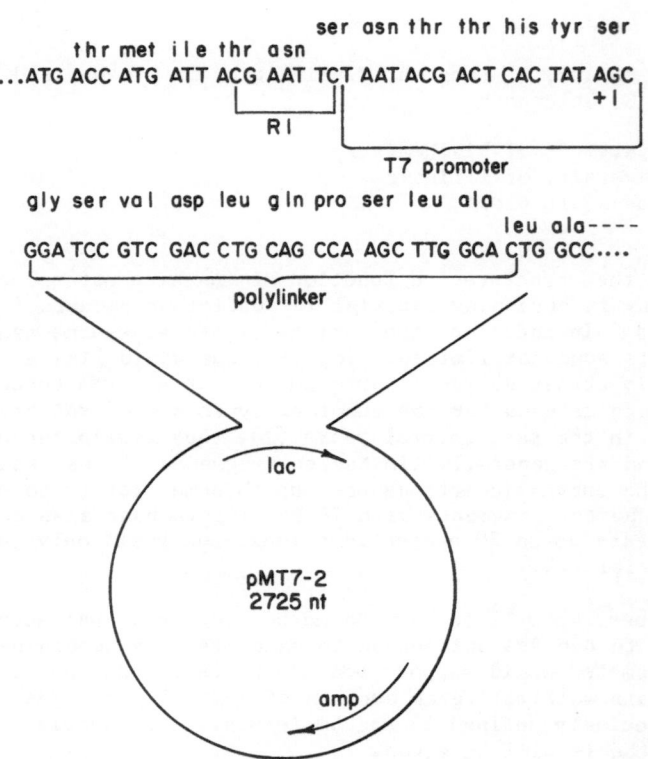

Fig. 1. pMT7-2, a T7 RNA polymerase _in vitro_ transcription vector. The reading frame of β-galactosidase is given above the DNA sequence.

If a DNA copy of the RNA sequence desired is available and one is not concerned about extra nucleotides at the 3' and 5' ends of the RNA, a number of cloning vectors containing the T7 promoter upstream of a poly-linker cloning site are available. An example of such a vector prepared in our lab (pMT7-2) is shown in Figure 1. An advantage of this vector is that when the DNA sequence of interest is cloned into the polylinker, the reading frame of β-galactosidase will be disrupted, and the bacterial colony will be white on X-Gal plates. Similar selection vectors with more complete polylinkers have recently been prepared (D. Mead, personal communication) and several commercial vectors without such a selection method are also available.

By cleaving the DNA of pMT7-2 (or a related vector) at the Hind III site in the polylinker, a substrate for T7 runoff transcription is prepared that gives a relatively short RNA product. By using this substrate to optimize T7 transcription reaction conditions, the transcription initiation step was emphasized. The optimal reaction conditions were found to be: 20 nM T7 promotor, 0.7 mM of each NTP, 6 mM $MgCl_2$, 1 mM spermidine, 5 mM DTT, 40 mM Tris-Cl pH 8.1, 100 µg/ml serum albumin and 10 units/µl T7 polymerase. These conditions are quite similar to those previously determined for the T7 (3) and SP6 (5) enzymes. The reaction rate slowed considerably by 1 h and incorporation was complete by 2 h at 37°C. From 20 to 30% of the NTP's were incorporated into RNA which corresponds to 500 moles of RNA per DNA template. It is not entirely clear why the reaction stops when it does. The addition of more enzyme, DNA or NTPs either before the reaction starts or after it is completed does not give substantially more RNA product. Since the addition of 25 units/ml pyrophosphatase increases the yield somewhat, accumulation of the reaction product pyrophosphate may be inhibiting.

THE 3' AND 5' TERMINI

A variety of different restriction enzymes were used to cleave pMT7-2 and several other vectors in the polylinker to generate T7 transcription templates with different types of 3' termini. The transcription reactions were analyzed on high resolution polyacrylamide gels to determine the length, homogeneity and yield of the transcript. In several cases, the product of the reaction was cut out of the gel, eluted and 3' end labeled using RNA ligase (9). Hydrolysis of the labeled RNA allowed determination of the 3' terminal nucleotide. It was clear from preliminary experiments that templates produced using restriction enzymes which gave 3' overhanging ends were not very active in transcription reactions. Thus, Table 1 only summarizes results using templates with 5' overhanging ends. In each case, a transcript is obtained that has the length and 3' terminus that would be expected if the very last nucleotide of the template strand is copied. However in some cases, transcription appears to terminate one nucleotide earlier as well, resulting in a doublet on the sequencing gel. In two cases, a transcription product one nucleotide longer than expected is obtained, where the enzyme apparently inserts a randomly selected

Table 1. 3' Termini of Runoff Transcripts

Enzyme	DNA Terminus	RNA Transcript	3' Terminal Nucleotide
Eco Rl	G CTTAA	one	U
Hind III	A TTCGA	one	U
Xba I	T AGATC	doublet	A + C (upper) G
Acc l	GT CAGC	doublet	A (upper) G (lower)
Bst Nl	CC GGT	one	A
Sal l	G CAGCT	one	A

71

residue before terminating. Since not enough termini have been tested, it
is not yet possible to predict the homogeneity of the 3' terminus of the
transcript. It is clear, however, that at least some of the product will
be a copy of the entire template strand.

Transcription with T7 polymerase has been shown to initiate at a
unique site with respect to the T7 promotor (10). This site, labeled +1
in Table 2, is a G residue for nearly all T7 promoters, but can be an A
residue. Since pMT7-2 and all the other vectors have a G at +1, it is not
surprising that when several of the transcripts were dephosphorylated and
5' ^{32}P labeled, the 5' terminal nucleotide was shown to be a G residue.
This is consistent with the observation that transcripts are labeled when
[γ -^{32}P] GTP, but not [γ -^{32}P] ATP, is included in the reaction mixture.
It is not yet known whether T7 polymerase would initiate with A, U, or C
if the proper promoter were prepared. SP6 polymerase has been reported to
be able to initiate with A (4).

A comparison of the 17 T7 promoter sequences suggests a consensus
sequence which extends from -17 to +5 (11). Although it is unclear
whether T7 RNA polymerase interacts with all these residues, it seems
possible that part of the promoter may lie downstream of +1. If this is
the case, it may not be possible to change the DNA sequence in this region
without effecting the transcription reaction. This in turn might place
restrictions on the sequence of the transcript near the 5' terminus. In
order to test this possibility, we have prepared several templates with T7
promoters that have the same sequence from -17 to +1 and different nucleotides from +2 to +5. It appears that although the expected transcript is
always present, the yield of transcript can be greatly reduced when +2 to
+5 is changed. In the case of these "weak" promoters, a much greater
proportion of product is in the form of short oligonucleotides that run
faster than the bromophenol blue dye marker on a 20% acrylamide gel.
Although the sequence of these oligomers remains to be determined, they
are labeled when [γ - ^{32}P] GTP is present in the reaction. This suggests
that they may be abortive initiation products similar to those found for
certain E. coli promoters (12). Attempts to stimulate these weak
promoters by altering the reaction conditions or including transcription
primers has not been successful. It therefore appears that there are some
restrictions on the sequences that can be made with T7 polymerase. Future
experiments will determine which residues in the +2 to +5 region can be
tolerated.

Table 2. The T7 Promoter Extends Past +1

```
       -17                   +1   +5
    ...TAATACGACTCACTATAGGGAG....
    ....ATTATGCTGAGTGATATCCCTC...
                                |
                                ↓

                    pppGGGAG....
```

+1 to +5 Sequence	Runoff Transcript	Short(<8) Oligomers
GGGAG	+++	+
GCGGA	++	++
GAGCT	+/-	+++
GAACA	+/-	+++

72

Modern automated DNA synthesizers can be used to make substantial amounts of defined DNA fragments 50 nucleotides or longer. For example in several recent 1 µmole scale syntheses of 40mers at our university, we obtained from 3 to 5 mg of purified product. Since the T7 promotor is only about 20 nucleotides, it should be possible to directly synthesize templates for T7 polymerase that will produce RNAs 30 residues or longer. This strategy avoids the labor of cloning and the necessity of using a restriction enzyme to define the 3' end of the template. Figure 2 shows two templates designed to produce a 24 nucleotide fragment of RNA (also shown) which is a variant of the R17 replicase translational operator that is expected to bind R17 coat protein (13). Template A is made of two synthetic DNA fragments with the wild type T7 promoter adjacent to the sequence desired. Template B uses the same bottom (template) strand as A, but only has the T7 promoter (residues -17 to +1) as the top strand. The construction of template B was motivated by the success of transcription of 5' overhanging ends generated by restriction enzymes. As can be seen in the gel lanes in Fig. 2, both templates make a transcript of the proper length. The RNA was eluted from the gel, shown to have the expected U at the 3' terminus and found to bind R17 coat protein in a millipore filter binding assay.

We have studied the transcription reaction using synthetic templates such as A and B in some detail and find that while the optimal buffer conditions are the same as determined for plasmid DNA, the promoter concentration can be raised to 200 nM and the enzyme concentration to 60 units/µl. Under these conditions the number of turnovers is somewhat less than plasmid DNA even when the oligomers are highly purified. Templates of the A type generally give from 5 to 10 moles of RNA per mole of DNA. Surprisingly, templates of the B type give 4 to 8 times more turnovers than the A type. The reason for the increased activity of B type templates is unclear. Since the bottom strand shows no activity by itself,

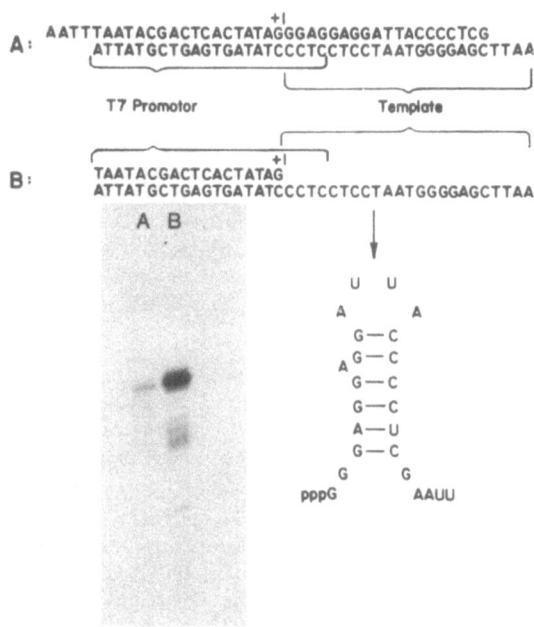

Fig. 2. *In vitro* transcription from synthetic DNA.

the double stranded promoter is presumably needed to bind T7 polymerase. Perhaps transcription can initiate or propagate more rapidly when the top strand does not have to be displaced. The fact that both the transcript and the template have substantial secondary structure may be important as well. Whatever the reason for their increased activity, the fact that B type templates work at all is extremely convenient since it is only necessary to synthesize a single DNA fragment in order to make a new RNA.

The somewhat lower activity of synthetic DNA as a template does not prevent this method from being an excellent way to make large amounts of short RNAs. For example, if 100 µg of a synthetic 40mer template strand were annealed to 50 µg of promoter top strand, this would provide enough B type template for a 37 ml transcription reaction. By adding 2.2×10^6 units (4.5 mg) T7 polymerase to ths reaction, it is reasonable to expect at least 75 turnovers which would produce 4 mg of a RNA fragment 22 residues long. We have not yet run such large scale reactions using B type templates, but several reactions at the 10-15 ml scale using A type templates were successful although the yields were 4 to 8 fold lower. By purifying the products of the reaction on HPLC using a Fractogel-DEAE resin, it is possible to recover both the product and active template in good yield. Thus, the way appears clear to prepare nearly any RNA sequence of less than 30 or 40 nucleotides in amounts suitable for NMR and other physical studies.

A YEAST tRNA^{Phe} TRANSCRIPT

Yeast phenylalanine tRNA is probably the most extensively studied RNA molecule to date. A high resolution X-ray crystal structure is available. It has been the subject of study by nearly every known physical technique that can be applied to macromolecules, including an extensive analysis by NMR (14-15). The biochemical properties of yeast tRNA^{Phe} have also been well studied, including its interaction with its tRNA synthetase and the protein synthesis apparatus. In analogy to similar studies that have been carried out with proteins, it would be very useful to be able to specifically alter individual nucleotides at predefined positions in the molecule. The physical and biochemical properties of these mutant tRNAs could aid our understanding how the molecule "works".

It has, however, been rather difficult to obtain such specific yeast tRNA^{Phe} variants. Site directed mutagenesis of the gene has been complicated by the fact that the promoter of yeast tRNAs is within the coding sequence so that many mutants are not transcribed (16). In addition, even when a mutant tRNA gene is transcribed, it is often not processed properly (17) so it is not possible to recover the mutant RNA for physical or biochemical studies. Methods developed to directly modify the sequence of the mature tRNA^{Phe} by cleavage and repair strategies have been moderately successful (18), but are quite complex and are limited to making changes in the anticodon loop or the CCA terminus.

Figure 3 shows the construction of PT7YP-0, a plasmid designed such that T7 polymerase can be used to make a 76 nucleotide runoff transcript with a sequence identical to yeast tRNA^{Phe}. Six oligonucleotides were synthesized and ligated together to make a 105 nucleotide fragment containing the tRNA^{Phe} sequence with the T7 promoter upstream and a Bst N1 site downstream. This fragment was then cloned into a high copy number plasmid (pUC13) to make pT7YP-0. It was important to confirm the sequence of the plasmid since mutants occured at a low but significant frequency. One interesting mutant was pT7YP-1 where the G at position +4 was changed to an A, thereby introducing an A · U pair in place of a G · U pair in the acceptor stem.

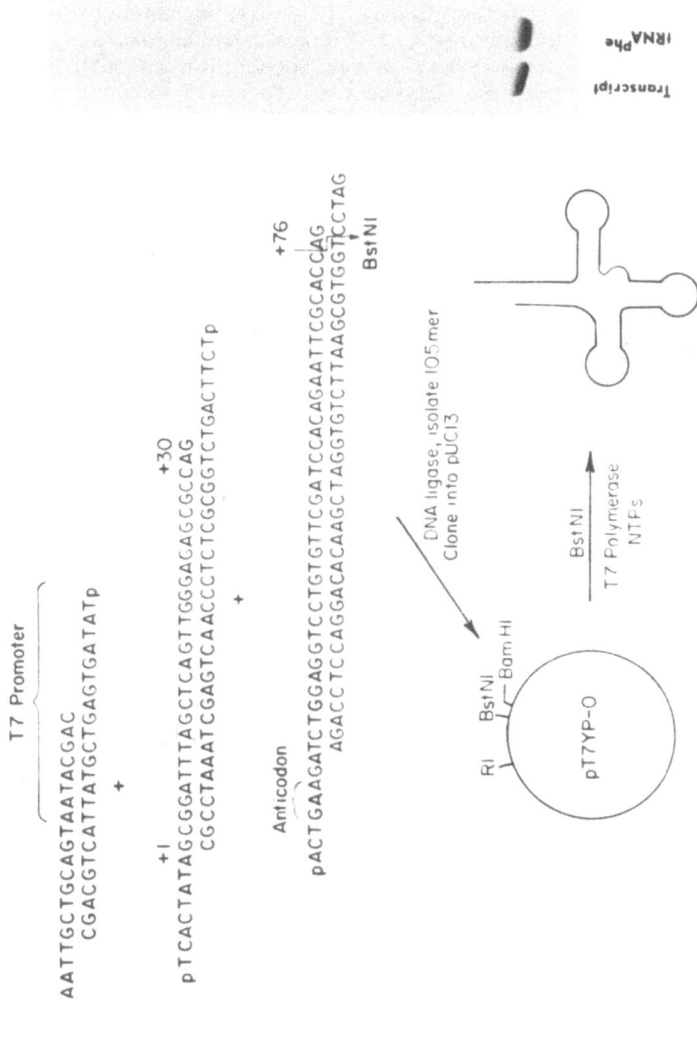

Fig. 3. Construction of pT7YP-0 and in vitro transcription of Bst N1 cut plasmid. pT7YP-1 is identical except that position +4 is an A instead of a G.

As shown on the gel in Figure 3, transcription from Bst N1 cut pT7YP-0 gives a unique transcript with a length identical to authentic yeast tRNA^{Phe}. The transcript was shown to have the expected nucleotide sequence and terminate with an A residue. Preliminary aminoacylation studies with purified yeast phenylalanyl synthetase indicate that the transcript can be aminoacylated to the expected level of 1300 pmoles/A_{260} unit, but the rate of aminoacylation is somewhat reduced. Presumably the absence of the methyl group on G-10 (19) and perhaps other modifications are responsible for this lower rate of aminoacylation.

In order to prepare sufficient amounts of transcript for NMR studies, a large scale transcription reaction was carried out. The 15 ml reaction contained the same buffer described above, 15 μmoles of each NTP, 1.5 mg of pT7YP-1 cut by Bst N1 (0.8 nmoles), 250 units pyrophosphatase and 390,000 units (0.8 mg) T7 polymerase. After incubation at 40°C for 2 h, the reaction was phenol extracted and the nucleic acids were ethanol precipitated. The tRNA transcript was purified from short oligonucleotides, NTPs and DNA on a HPLC column of Fractogel-DEAE. The recovered yield of transcript was 160 A_{260} units (256 nmoles) which corresponds to 320 moles of RNA per mole of DNA. Approximately 80% of the DNA was recovered for reuse.

It therefore appears that either by repeating the cloning with different oligonucleotides or by site directed mutagenesis of pT7YP-0, it should be possible to prepare transcripts with defined nucleotide changes at different positions in the molecule. These altered transcripts can be assayed for biochemical function and their structure can be examined by NMR. Since this methodology is not restricted to this particular RNA, it appears that the difficulty of obtaining enough RNA is solved.

REFERENCES

1. J. H. van Boom, and C. T. J. Wreesman, in: Oligonucleotide Synthesis: A Practical Approach, Chapter 7, pp. 153 M. J. Gait, ed., IRL Press, Oxford (1984).
2. D. Beckett and O. C. Uhlenbeck, ibid, Chapter 8, pp. 185.
3. M. Chamberlin and T. Ryan, in: The Enzymes, Chapter 4, pp. 87 P. Boyer ed., 3rd Edition, Academic Press (1982).
4. E. Butler and M. Chamberlin, J. Biol. Chem. 257, 5772 (1982).
5. D. A. Melton, P. A. Krieg, M. R. Rebagliati, T. Maniatis, R. Zinn, and M. R. Green, Nuc. Acids Res. 12, 7035 (1984).
6. Osterman, H. L. and Coleman, J. E., Biochemistry 20, 4884 (1981).
7. P. Davanloo, A. Rosenberg, J. Dunn, and F. W. Studier, Proc. Natl. Acad. Sci. USA 81, 2035 (1984).
8. G. Kassavetis, and M. Chamberlin, J. Virol. 29, 196 (1979).
9. T. E. England, A. G. Bruce, and O. C. Uhlenbeck, Methods in Enzymology 65, 65 (1980).
10. M. Rosa, Cell 16, 815 (1979).
11. J. J. Dunn, and F. W. Studier, J. Mol. Biol. 166, 477 (1983).
12. J. D. Gralla, A. J. Carpousis, and J. E. Stefano, Biochemistry 19, 5864 (1980).
13. O. C. Uhlenbeck, J. Carey, P. Romaniuk, P. Lowary, and D. Beckett, J. Biomol. Struct. and Dynam. 1, 539 (1983).
14. D. Roy, and A. Redfield, Biochemistry 22, 1386 (1983).
15. A. Heerschap, C. A. G. Haasnoot, and C. W. Hilbers, Nuc. Acids Res. 11, 4501 (1983).
16. R. Koshi, S. Clarkson, J. Kurjan, B. Hall, and M. Smith, Cell 22, 415 (1980).
17. J. D. Smith, Prog. Nuc. Acids. Res. 16, 25 (1976).
18. A. G. Bruce and O. C. Uhlenbeck, Biochemistry 21, 855 (1982).
19. B. Roe, M. Michael, and B. Dudock, Nature New Biology 233, 274 (1971).

Mg^{2+} - INNER SPHERE COMPLEXES AT "ENDS" AND "BENDS" OF POLYNUCLEOTIDES AND THEIR POTENTIAL ROLE AS LONG RANGE INDUCERS OF CONFORMATION CHANGES

Dietmar Porschke

Max Planck Institut für biophysikalische Chemie
3400 Göttingen, FRG

ABSTRACT

A short review is given on Mg^{2+}-binding to oligo- and polynucleotides with special emphasis on the formation of inner sphere complexes. Due to the rather slow exchange of ligands in the inner sphere of Mg^{2+} these complexes are localised at a given site with a relatively long lifetime, whereas outer sphere complexes are much more mobile (ion atmosphere binding). Measurements of the binding kinetics demonstrate that Mg^{2+} binds to polymers like poly(A) mainly in the form of outer sphere complexes, whereas short oligoriboadenylates are shown to form inner sphere complexes. Evidence is presented that the measured increase of the inner sphere binding constant with decreasing chain length results from preferential inner sphere complexation at chain ends. Inner sphere complexes are also formed at "bends" of polynucleotide chains. An example is given for the anticodon loop of tRNA^{Phe}, where Mg^{2+} preferentially forms an inner sphere complex with one of two loop conformations. The special properties of Mg^{2+}-ligands may be used for transmission of a "signal" along a polynucleotide chain.

INTRODUCTION

Magnesium ions Mg^{2+} are required for virtually all biological functions of nucleic acids. Replacement of Mg^{2+} by other ions usually results in loss of biological activity or in a reduction of this activity to a relatively low level. Apparently the molecular properties of Mg^{2+} are unique. Examination of these properties reveals an individual character

77

of Mg^{2+} mainly with respect to its dynamics of ligand binding. Ligands attached directly to the inner sphere of Mg^{2+}-ions are bound rather tightly and thus their exchange against other ligands requires a relatively high activation energy.[1,2] The kinetics of ligand exchange has been characterised for a large number of different ligands. These investigations demonstrated an S_N1 mechanism for all the exchange reactions with a typical rate constant of $10^5 s^{-1}$ for the substitution of H_2O-molecules by any ligand. It could be demonstrated that a closely corresponding rate constant is valid for the exchange of a H_2O-molecule in the inner hydration sphere by another H_2O-molecule.[3]

In many respects the thermodynamics of complex formation by Ca^{2+}-ions is quite similar to that of Mg^{2+}-ions, whereas the dynamics of ligand binding proved to be completely different.[1,2] Due to the larger ion radius of Ca^{2+}, ligands in its inner hydration sphere are not attached as tightly as in the case of Mg^{2+} and thus the activation barrier for dissociation is not nearly as high. Measurements with various ligands demonstrated that the substitution of water molecules in the inner hydration sphere of Ca^{2+}-ions is associated with a rate constant of about $3 \cdot 10^8 s^{-1}$. Thus the substitution rates of Mg^{2+} and Ca^{2+} are different by more than 3 orders of magnitude.

The special ligand binding dynamics of Mg^{2+} may provide at least a partial explanation for their unique biological function. The dynamic properties can also be used to characterise the nature of Mg^{2+}-complexes with various ligands including polynucleotides.

OUTER AND INNER SPHERE COMPLEXES IN SINGLE STRANDED OLIGO- AND POLYNUCLEOTIDES

Since the inner sphere of $Mg(H_2O)_6^{2+}$ in aqueous solutions is substituted with a rate constant of about $10^5 s^{-1}$, this reaction can be conveniently characterised by chemical relaxation measurements. Among the various relaxation methods, the electric field jump procedure proved to be particularly useful, since electric field pulses selectively perturb reactions involving charge compensation. In a first investigation, the rate constants of Mg^{2+}-binding to single stranded polynucleotides[4] were found to be between 1 and $2 \times 10^{10} M^{-1} s^{-1}$. These high rate constants indicate a diffusion controlled reaction and the formation of "outer sphere" complexes, where the inner sphere remains completely filled with H_2O-molecules and $Mg2+$ does not form any direct contact to the ligand.

A separate relaxation effect reflecting a conversion from "outer" to "inner" sphere complexes has not been detected. This result is surprising at a first glance, but may be explained by the molecular properties of the reactants. The ion radius of Mg^{2+} is 0.7Å, whereas the distance between adjacent phosphates in a stacked polynucleotide chain is about 7Å. Thus Mg^{2+}-ions cannot compensate their two charges by binding at two phosphates simultaneously. When a Mg^{2+}-ion is attached at a single internucleotide phosphate, a positive charge remains which appears to be unfavorable. The experimental data indicate that Mg^{2+}-ions bind to polynucleotides in the form of a mobile ion cloud, which is usually described by the term "ion atmosphere" binding.[5]

During a later investigation[6] it was found by accident that oligoriboadenylates in the presence of Mg^{2+}-ions clearly exhibit an inner sphere relaxation process. A quantitative analysis of this process for a given oligomer as a function of the Mg^{2+}-concentration provides the equilibrium and rate constants for inner sphere complexation. The results of these measurements performed for various oligomer chain lengths are compiled in Table I. The equilibrium constant K_2 observed for the transition from an "outer" to an "inner" sphere complex decreases from a relatively high value of 5.4 for the shortest chain length (4 phosphate residues) to virtually zero for the polymer. Thus the chain length dependence found for oligoriboadenylates is consistent with the earlier observation of preferential "ion atmosphere" binding of Mg^{2+}-ions to polynucleotides. The decrease of the inner sphere binding constant with increasing chain length is reflected in the kinetic parameters mainly by the forward rate constant k_{12}, whereas k_{21} is almost independent of chain length. The rate constant $k_{12} = 6.5 \cdot 10^4 s^{-1}$ observed for $A(pA)_4$ is rather close to the value $10^5 s^{-1}$ usually observed for inner sphere complexation, whereas $k_{12} = 10^3 s^{-1}$ found for $A(pA)_{17}$ is unusually low.

The formation of inner sphere complexes by short oligoadenylates and their absence in long chains may be explained in general terms by a higher flexibility of short oligomer chains. For example, short oligomers may be bend to a loop structure with a sufficiently close approach of phosphate residues for a connection by a Mg^{2+}-inner sphere complex, providing a full compensation of the Mg^{2+} and phosphate charges. A loop model for inner sphere complexation may also explain the decrease of k_{12} with increasing chain length by an increase of steric hindrance effects.

Loop structures formed in the presence of Mg^{2+}-ions should be much more compact than single stranded helices of the same chain length. Thus the formation of loop structures from single stranded helices may be tested by hydrodynamic measurements. Recently it has been possible to extend the time resolution of electric dichroism measurements[7] to the range required for the characterisation of the rotation diffusion of relatively short oligomers. The overall rotation time constants of oligoriboadenylates have been measured as a function of the chain length both in the presence and the absence of Mg^{2+}-ions.[7] The data have been analysed by a weakly bending rod model providing a "persistence length" 1 as a measure for the stiffness of the nucleotide chain. The 1-value measured in the presence of Mg^{2+} (88Å) was clearly higher than that obtained in its absence (53Å). These results indicate that the riboadenylate chain is more rigid in the presence of Mg^{2+} and do not provide any evidence for bending by Mg^{2+}-ions.

According to the rotation diffusion measurements the formation of loop structures appears to be unlikely and thus conformation changes required for inner sphere complexation are probably restricted to some local rearrangement of the structure at the ends of the chain. The decrease

Table I

Parameters obtained by UV titration and electric field jump relaxation for the binding of Mg^{2+} to $A(pA)_n$. $K_1(K_2+1)$ overall binding constant, K_2 inner sphere equilibrium constant, $k_{12}(k_{21})$ rate constant for the formation (dissociation) of inner sphere complexes.[6]

	$K_1*(1+K_2)$	K_2	k_{12}	k_{21}
$A(pA)_4$	$3.2*10^3$	5.4	$6.5*10^4$	$1.2*10^4$
$A(pA)_5$	$14.3*10^3$	5.0	$4.9*10^4$	$1.0*10^4$
$A(pA)_6$	$19.*10^3$	3.8	$5.0*10^4$	$1.3*10^4$
$A(pA)_7$	$32.*10^3$	0.56	$1.0*10^4$	$1.8*10^4$
$A(pA)_9$	$31.*10^3$	0.27	$0.5*10^4$	$2.0*10^4$
$A(pA)_{17}$	$190.*10^3$	0.04	$0.1*10^4$	$2.7*10^4$
poly(A)	$3800.*10^3$	–	–	–

of the inner sphere binding constant with increasing chain length may then be explained by a competition effect between binding sites at the end of the chain and all the other "interior" sites. The relative number of interior sites increases with chain length. Furthermore the affinity of these interior sites is expected to increase with chain length due to an overlap of the electrostatic potentials resulting from the phosphate residues. Thus the probability of ion binding at the ends of the chain decreases with chain length. Obviously these considerations are only valid for the first ion(s) bound to the oligonucleotide chain, which correspond to the one(s) identified by the relaxation measurements owing to the optical detection procedure.[6]

The inner sphere relaxation has been detected so far only for oligoriboadenylates, whereas oligomers formed from deoxyriboadenylates, ribocytidylates, ribouridylates and riboinosinic acids did not show any corresponding effect.[6] These results indicate that both specific base and sugar residues are required for inner sphere complexation and suggest the possibility that these complexes may be used as markers for recognition.

Mg^{2+} BINDING TO A LOOP STRUCTURE

As discussed above the formation of inner sphere complexes may be supported by bending and folding of the polynucleotide chain to complex structures providing binding sites with an increased number of potential contacts. A structure with a relatively complex folding of the polynucleotide chain has been found for tRNA molecules.[8] These molecules are known to bind a large number of Mg^{2+}-ions over a relatively wide affinity range. The process of Mg^{2+}-binding to one of these sites has been characterised recently by equilibrium titrations and relaxation measurements.[9] The most simple mechanism consistent with the experimental data is as follows:

$$tRNA_A \rightleftharpoons tRNA_B$$

$$tRNA_B + Mg^{2+} \rightleftharpoons C_o \rightleftharpoons C_i$$

Mg^{2+}-ions are bound preferentially to one of two conformations of the tRNA, which are coupled to each other by a reaction with a time constant of about 100μs. The outer sphere complex C_o is converted to an inner sphere complex with a relatively low rate constant of about $10^3 s^{-1}$, indicating some conformational barrier. These details of the binding mechanism could be characterised by measurements of the fluorescence intensity of the Wye base, a modified residue located in the anticodon loop of tRNA[Phe]. The large fluorescence change upon addition of Mg^{2+} indicates that the binding site is very close to the Wye base in the anticodon loop. A corresponding site has been identified in the crystal structure by x-ray analysis [10,11] it is found almost at the centre of the anticodon loop with one direct contact to a phosphate residue and several indirect contacts via water molecules to various base residues. Among these residues is the Wye base, providing an explanation for the large fluorescence change. Another contribution to this change comes from the rearrangement of the loop conformation. The structure found in

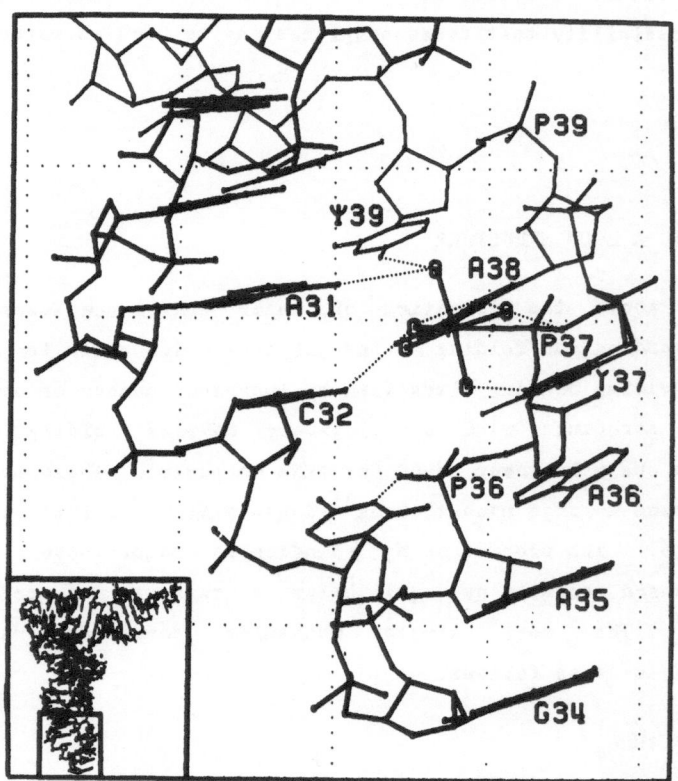

Fig. 1. View of the Mg^{2+}-ion in the anticodon loop of tRNA[Phe] (yeast). The Mg^{2+} is directly coordinated to a phosphate oxygen of residue 37. The waters of hydration form many hydrogen bonds; one of them is directed to the fluorescent Wye base (Y). The insert shows the location of the diagram relative to the whole molecule (from ref. 12 with permission).

82

the crystal, corresponding to the state B, is the "3'-stacked" form of the anticodon loop. In the absence of Mg^{2+} another form of the anticodon loop A is found to be in equilibrium with B at an almost equal ratio. Probably the state A is equivalent to the "5'-stacked" form of the loop, where the anticodon bases are stacked on the other side of the helix stem. Preferential binding of Mg^{2+} to the "3'-stacked" form can be easily explained by many contacts available for the Mg^{2+}-ion in this conformation. According to a model proposed by Woese[13] the conformation change between 3'- and 5'-stacked forms is involved in the ribosomal translation process.

SIGNAL TRANSMISSION ALONG A POLYNUCLEOTIDE BY Mg^{2+}-COMPLEXES

The special properties of Mg^{2+}-complexes may be used to construct a simple model for the transmission of a signal along a polynucleotide chain.[14] According to the data discussed above it is possible that a Mg^{2+}-inner sphere complex is formed at an end of a polynucleotide chain. Since Mg^{2+}-ions at interior sites preferentially form outer sphere complexes, these ions cannot be localised exactly. Nevertheless the probability to find a Mg^{2+}-ion close to every second phosphate residue will be enhanced due to electrostatic interactions. The spacing of these Mg^{2+}-ions will then be determined by a Mg^{2+}-inner sphere complex located at one end of the chain and thus may induce a relative decrease of the Mg^{2+}-ion concentration at the other end of the chain

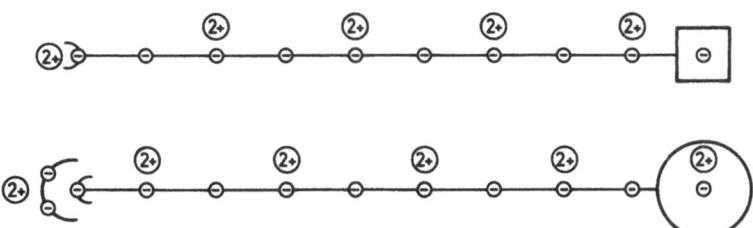

Fig. 2. Transmission of a "signal" along a polynucleotide chain by Mg^{2+}. Addition of a ligand at one end of the polymer changes the Mg^{2+}-binding site, induces a shift in the spacing and thus increases the probability of a conformation change at the other end (from ref. 14).

(cf. Fig. 2). Binding of another ligand, e.g. a complementary oligomer, at the end with the inner sphere complex may shift the location of this complex by a distance corresponding to one nucleotide residue. Due to electrostatic interactions this shift may be transmitted along the polymer chain by the spacing of Mg^{2+}-ligands. The change in the spacing by one residue induces a relative increase of the Mg^{2+}-concentration at the remote end and may support, for example, formation of an inner sphere complex at this end together with a conformation change. According to this mechanism a "signal" received at one end of the chain by binding of a ligand may be transmitted to a remote end owing to long range electrostatic interactions.

At present this mechanism is merely hypothetical. However, it may provide an explanation for some results indicating a long range transmission of conformation changes. For example, various observations indicate that codon binding induces a conformation change of tRNA molecules.[14,15] This effect is strongly dependent upon the presence of Mg^{2+}-ions. Since the interaction of a codon triplet with the anticodon involves only a relatively small part of the tRNA molecule, the codon binding apparently induces a change transmitted to remote parts of the tRNA.

A rather large change of the tRNA conformation induced by a relatively minor substitution at one end of the molecule has also been reported for the case of aminoacylation.[16] Since oligoriboadenylates form inner sphere complexes at their end(s) [cf. above], the CCA-end of tRNA molecules appears to be a good candidate for inner sphere complexation of Mg^{2+}-ions. Aminoacylation at the CCA-end is clearly expected to reduce the affinity for Mg^{2+}-binding at this site and thus may induce a change of the charge pattern transmitted to remote parts of the tRNA molecule.

Acknowledgement

The comments of Dr. H. Diebler on the manuscript are gratefully acknowledged.

REFERENCES

1. H. Diebler, M. Eigen and G.G. Hammes, Z. Naturf. 15b, 554 (1960)

2. M. Eigen, Pure Appl. Chem. 6, 97 (1963)

3. J. Neely and R. Connick, J. Amer. Chem. Soc. 92, 3476 (1970)

4. D. Porschke, Biophys. Chem. 4, 383 (1976)

5. G. Manning, Quart. Rev. Biophysics 11, 179 (1978)

6. D. Porschke, Nucleic Acids Research 6, 883 (1979)

7. D. Porschke and M. Jung, J. Biomol. Structure & Dynamics 2, 1173 (1985)

8. G.J. Quigley and A. Rich, Science 194, 791 (1976)

9. D. Labuda and D. Porschke, Biochemistry 21, 49 (1982)

10. A. Jack, J.E. Ladner, D. Rhodes and A. Klug, J. Mol. Biol. 111, 315 (1977)

11. G.J. Quigley, M.M. Teeter and A. Rich, Proc. Natl. Acad. Sci. USA 70, 2683 (1978)

12. M.M. Teeter, G.J. Quigley and A. Rich in Nucleic Acid–Metal Ion Interactions, Ed. T.G. Spiro, John Wiley & Sons, New York (1980)

13. C. Woese, Nature 226, 817 (1970)

14. D. Labuda, G. Striker and D. Porschke, J. Mol. Biol. 174, 587 (1984)

15. H.G. Gassen, Progress in Nucleic Acid Research and Mol. Biology 24, 57 (1980)

16. R. Potts, M.J. Fournier and N.C. Ford Jr., Nature 268, 563 (1977)

PSEUDOKNOTS IN RNA: A NOVEL FOLDING PRINCIPLE

Cornelis W.A. Pleij, Alex van Belkum, Krijn Rietveld and
L. Bosch

Department of Biochemistry
University of Leiden
The Netherlands

I INTRODUCTION

Models of the secondary structure of RNA usually contain a number of
characteristic structural elements like base paired stem regions and various
kinds of single stranded regions like hairpin, bulge, interior and bifurcation loops (Zuker and Stiegler, 1981). In such models mostly about one third
or more of the nucleotide residues remains unpaired, which in some cases is
confirmed by experimental data or computer-aided predictions. Although stem
regions unquestionably are an important feature in the structure of RNA
molecules, the final three-dimensional structure will be determined mainly
by the interactions of the residues left in the single stranded regions.
This is clearly illustrated in the case of tRNA, where the T-, D- and
variable loop are largely responsible for maintaining the typical L shape
of the molecule (Kim et al., 1974; Robertus et al., 1974). The tertiary
interactions in the native conformation of tRNA often involve non standard
base pairs or base triplets, while only in a few cases normal Watson-Crick
base pairs are found. In fact Watson-Crick base pairing between complementary
sequences might be considered an obvious possibility for tertiary interactions. Such interactions have been proposed indeed for the ribosomal 5S RNA
(Pieler and Erdmann, 1982; Trifonov and Bolshoi, 1983) and were previously
proposed on theoretical grounds (Studnicka et al., 1978).
Tertiary interactions of this kind were called knotted or pseudoknotted
structures depending on whether or not they could give rise to real knots
in the RNA chain, especially when the resulting stem regions are in the
range of one turn of an RNA double helix (Cantor, 1980). However, the chance
that real knots do occur in RNA will be very low, so that we prefer the term
of pseudoknot coined originally by Studnicka et al. (1978), if a tertiary
interaction with Watson-Crick base pairing occurs.
    We here describe a pseudoknot of a special type involving hairpin loops
and which gives rise to a compact folding of RNA molecules and formation of
extended double helical stem regions. This type of pseudoknot plays an important role in the three-dimensional folding of the 3' termini of some
plant viral RNAs, but appears to be present in ribosomal and messenger RNAs
as well.

II PSEUDOKNOTS IN tRNA-LIKE STRUCTURES AT THE 3' TERMINI OF PLANT VIRAL RNAs

    The genomic RNAs of a number of plant viruses like turnip yellow mosaic
virus (TYMV), brome mosaic virus (BMV) and tobacco mosaic virus (TMV) can be

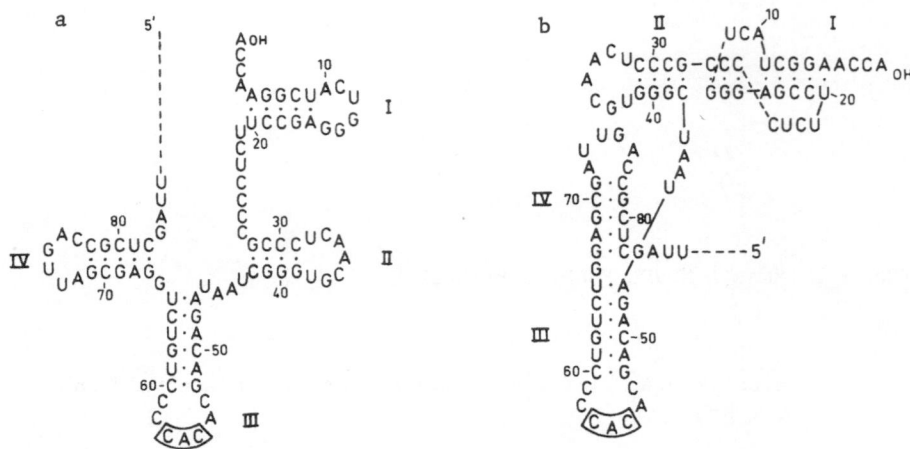

Fig. 1. The structure at the 3' end of TYMV RNA.
Secondary structure (a) and L arrangement (b) were
proposed on the basis of chemical modification and
enzymatic digestion data (for details see Rietveld
et al. (1982)). The boxed region indicates the valine specific anticodon.

esterified at their 3' end to a specific amino acid in a manner analogous to
tRNA (Hall 1979, Haenni et al., 1982). It was therefore concluded that the
3' termini of these RNAs should have a three-dimensional structure resembling that of canonical tRNA. One naively would expect a cloverleaf
structure to be present at these 3' termini. However, studies in our
laboratory revealed that in all three viral RNAs mentioned above, a special
folding of the RNA deviating from the cloverleaf structure, had to be
proposed in order to understand their interaction with tRNA specific enzyme
(Rietveld et al., 1984).
We first examined a 3' terminal fragment of TYMV RNA by structure mapping
using chemical modification according to Peattie and Gilbert (1980) and
enzymatic digestion with single or double strand specific RNases. Our
results and those of Florentz et al. (1982) pointed to the existence of a
secondary structure as depicted in Fig. 1a. It consists of 4 regular stem
and loop structures in the last 86 nucleotides, being the minimal size of
a 3' terminal fragment still fully active in aminoacylation (Joshi et al.,
1982). No indications were obtained for an aminoacyl stem, in which the 3'
terminus base pairs with the 5' end. Looking at the lower three quarters of
the proposed structure a close resemblance with a cloverleaf structure can
be seen. Especially hairpin III seems to be a good candidate for the anticodon stem and loop because of the presence of the anticodon CAC for valine
at the right position, as already noted earlier (Silberklang et al., 1977;
Briand et al., 1977). The problem rather was the presence of hairpin I,
which excludes the formation of a conventional aminoacyl acceptor stem.
The solution to this seemingly anomalous feature in the secondary structure
was based on the following three considerations: 1. The presence of an ACC(
sequence right after hairpin I, suggesting that the stem region of the latt
might function as part of the aminoacyl acceptor arm. 2. Some experimental
data obtained in the presence of 10 mM Mg^{2+} did not agree with the proposed
model of Fig. 1a. Cobra venom RNase digestion and dimethyl sulphate (DMS)
modification of the cytidine residues under these conditions showed that
the triple G sequence in the loop of hairpin I and the triple C sequence
adjacent to the stem of hairpin II were involved in base pairing interactions. 3. The conserved complementarity between the two sequences mentioned
under (2) in other plant viral RNA termini (Rietveld et al., 1982).

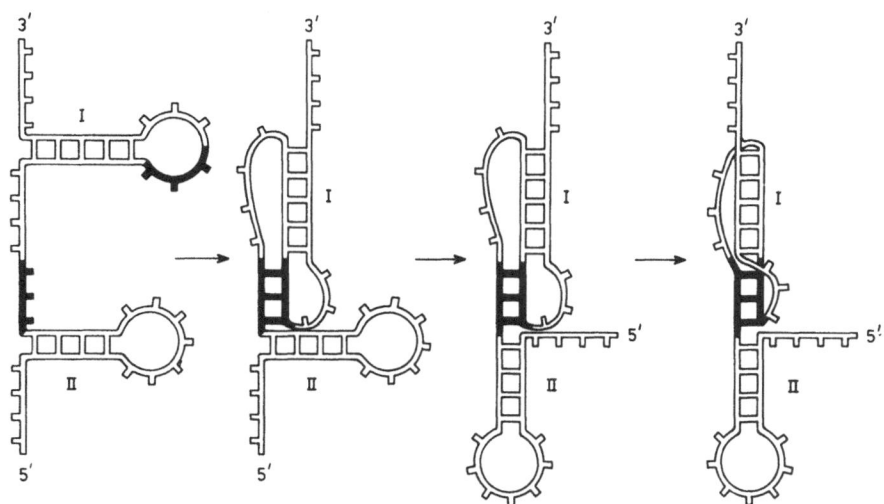

Fig. 2. Schematic presentation of the stepwise folding of the 3'
terminal 46 nucleotides of TYMV RNA into the aminoacyl
acceptor arm. Hairpin I and II correspond to those given
in Fig. 1a. The black regions represent the complementary
triple G and triple C sequences.

These data played a crucial role in our proposal for the aminoacyl acceptor
arm in the tRNA-like structure of TYMV RNA. Fig. 2 shows in a schematic way
how the folding of the last 42 nucleotides into an aminoacyl acceptor arm
is achieved. Base pairing of the triple G and triple C sequences (in black)
can lead to a coaxial stacking of the three double helical segments, giving
rise to a structure which strongly resembles the aminoacyl acceptor arm of
canonical tRNA. Both in tRNA and in this tRNA-like structure this domain is
formed by a stack of 12 base pairs in a double helix of the RNA-A type with
on either side the ACC(A) sequence and a sevenmembered hairpin loop (see
also Fig. 1b).
A major difference with tRNA is the number of three double helical segments
in the aminoacyl acceptor arm of TYMV RNA, whereas two are found in canonical tRNA. Moreover, two single stranded pieces of RNA are running outside
the double helical domain in the case of TYMV RNA, connecting the three
stem segments.
    Assuming that hairpin III and IV were part of the anticodon arm (Fig. 1b)
we were able to propose a model of the global three-dimensional folding of
the tRNA-like structure of TYMV RNA. For further details about the striking
resemblance with tRNA see Rietveld et al. (1982, 1983).
We recently determined the sequence at the 3' terminus of a number of other
viral RNAs from the tymovirus group of which TYMV is the type member. They
all could be folded in an identical secondary and tertiary structure,
thereby fully supporting the proposed model of TYMV RNA (van Belkum et al.,
to be published).
    Structure mapping of the tyrosine accepting 3' terminus of BMV RNA and
that of the histidine accepting TMV RNA also led to proposals for the spatial
folding of the tRNA-like structure (Rietveld et al., 1983, 1984). These
models comprising the last 134 and 95 nucleotides, respectively, (Joshi et
al., 1983; Joshi et al., 1985) were also strongly supported by the sequences
of related RNAs from the same virus group (Rietveld et al., 1984; Joshi et
al., 1983). Although the three tRNA-like structures examined to date differ
in many details, they all share the same type of tertiary interaction needed
for building the aminoacyl acceptor arm. The question we will now further
concentrate on is whether the peculiar foldings in the aminoacyl acceptor

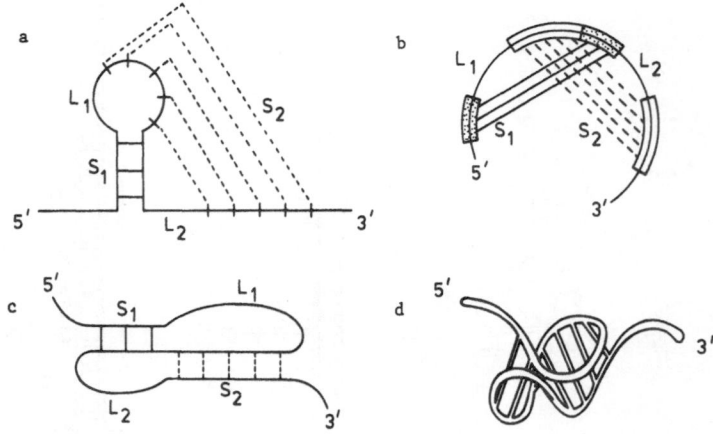

Fig. 3. The formation of a pseudoknot of the special kind.
S_1 and S_2 are stem regions arising from normal Watson-Crick
base pairing. L_1 and L_2 are single stranded regions of arbitrary size, connecting the double helical segments S_1 and S_2.
a) Conventional presentation. The dashed lines indicate the
tertiary interaction leading to pseudoknotting. b) Circular
presentation of the RNA chain. The boxed regions are engaged
in base pairing interactions as indicated. c) Schematic
folding. d) Three-dimensional folding showing the quasi-
continuous double stranded helix formed by stem S_1 and S_2.

arms in the three viral RNAs are sterically acceptable and whether they
are present outside these tRNA-like structures in other RNA molecules as
well. For this reason we analyzed this type of pseudoknotting in more detail
to derive some of its general properties (Pley et al., 1985).

III PROPERTIES OF A SPECIAL TYPE OF PSEUDOKNOT

     The principle underlying the construction of the plant viral aminoacyl
acceptor arms described above can be generalised as illustrated in Fig. 3a.
It shows the formation of a pseudoknotted structure by base pairing of a
few nucleotides in a hairpin loop with a complementary region elsewhere in
the RNA chain. We have found that thus far the sequence in the hairpin loop
taking part in the tertiary interaction always borders immediately to the
stem region of the hairpin. Consequently, two of the four base pairing
stretches of RNA involved, are adjacent with no single nucleotide in between.
This is clarified in the circular presentation of the RNA chain as shown in
Fig. 3b.
Generally speaking, each time that two separate stem regions give rise to
intersecting chords as in Fig. 3b, one has a pseudoknotted structure. This
means that for instance interior and bulge loops can also take part in the
formation of pseudoknots (see also section IV and Pley et al., 1985). Note
also that base pairing interactions which give rise to these intersecting
chords are not admissible in current algorithms for the prediction of
secondary structures of RNA (Studnicka et al., 1978; Zuker and Stiegler,
1981).
Here we only consider those pseudoknots in which stem adjacent nucleotides
from a hairpin loop are involved. From here on we define such a tertiary
interaction as the pseudoknot of the special kind. This pseudoknot,

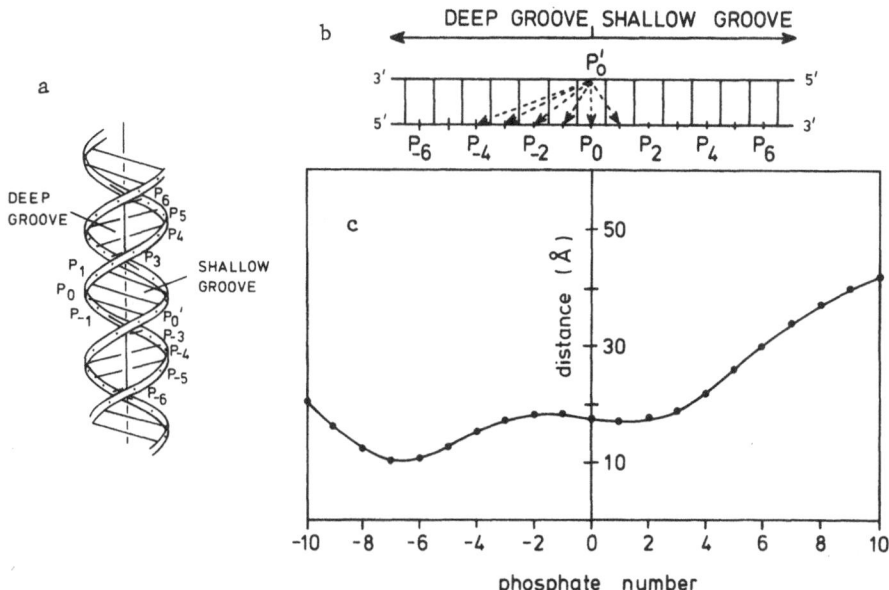

Fig. 4. Interphosphate distances in a regular RNA double helix of
the A-type. The distance between one fixed phosphate on
one strand (P_0') and the phosphates on the opposite strand
was calculated, using the coordinates reported by Arnott
et al. (1972) a) Schematic three-dimensional representation
showing the deep and shallow grooves of A-RNA and the position in space of the phosphates used in the calculations.
b) Two-dimensional representation of the RNA double helix.
c) Graphical representation of the distance from P_0' on one
strand to all other phosphate residues on the opposite strand.

illustrated in Fig. 3, favours the formation of an elongated double stranded
helix by coaxial stacking of the two separate double helical segments S_1
and S_2 (Fig. 3c and 3d). These two stem segments are stacked on top of each
other in such a way that a quasi-continuous right-handed double helix is
formed. It must be stressed here that this in fact is an important assumption because no direct proof for such a stacking is available as yet,
although our models of the aminoacyl acceptor arms in the plant viral
tRNA-like structures strongly suggest that a stacking of this kind will be
a more general property of this type of pseudoknot. Fig. 3c and 3d also
clearly show that each pseudoknot is characterised by the presence of two
single stranded pieces of RNA or so-called connecting loops which cross
the quasi-continuous double helix either over the deep groove (L_1) or over
the shallow one (L_2). They both have the same function, that is connecting
two opposite strands of the double helix. On the other hand they have
different characteristics due to the geometric properties of a regular RNA-A
duplex as will be explained below.
     From an analysis of all tRna-like structures in plant viral RNAs, known
to possess pseudoknots of this special type, we learned that in fact no
upper limit for the length of connecting loops L_1 and L_2 exists (Pley et
al., 1985). They may contain hundreds of nucleotides, which of course in
their turn will fold into a secondary and tertiary structure themselves. The
lower limit for the length of L_1 and L_2 appeared to be the two nucleotides.
A closer look at the geometry of the RNA-A helix, based on the data of
Arnott et al. (1972) (Fig. 4a) made it clear that these two nucleotides were
indeed the shortest connecting loops possible, at least in the case of L_1.
In Fig. 4b and 4c it is shown that the shortest distance connecting the two

opposite strands over the deep groove is about 10 Å. Surprisingly, in spanning this distance, 6-7 base pairs are crossed. Such a distance of 10 Å might be relatively easily bridged by a dinucleotide. This is what actually is observed in the case of the pseudoknot present in BMV RNA (Rietveld et al., 1983).

For connecting loop L_2 these figures are 17 Å and 2-4 base pairs respectively (Fig. 4c). Note that the nucleotides needed to span the narrow groove is comparable to the formation of the shortest hairpin loop possible. In the latter case this corresponds to the connection of phosphate P_0' with P_0 (Fig. 4b). Hairpin loops of two nucleotides only are considered not to occur (Gralla and Crothers, 1973).

One has to keep in mind that the various distances might vary somewhat depending on whether the quasi-continuous double helix is distorted in some way and thereby will deviate from the regular A helix (see also Sundaralingam, 1980). That irregularities are likely to occur will be shown below for TMV RNA, where even one single nucleotide residue appears to be sufficient for bridging the deep groove (section IV). Consequently, the two opposite strands across the deep groove have to come into closer proximity so that a distance of 7 Å or less is reached. The latter is the length of one single nucleotide when stretched out maximally (Saenger, 1984).

Model building studies of the aminoacyl acceptor arm of TYMV RNA (see Fig. 1b and 2) showed that bridging the narrow groove by A_{10}-U_{12} appeared to be just possible without adopting unacceptable torsion angles or van der Waals interactions (unpublished observations). An important feature of this trinucleotide connecting loop was that the base residues could not be turned towards the base pairs of the double helix to enable multiresidue interactions. This observation is in good agreement with our result that A_{10} is the only adenine residue in the tRNA-like structure of TYMV RNA which is highly accessible to DEP modification over the entire temperature range of 0-90°C (to be published). Bridging the deep groove with U_{21}-C_{24} (Fig. 1b) posed no problems at all as was expected from our analysis of the RNA duplex geometry. If L_1 (as well as L_2) has no other function than to make the connections discussed above, then it would be interesting to see whether a reduction in the size of this particular connecting loop to three or two nucleotides will affect the overall structure and function of this tRNA-like structure.

Finally, it is worth mentioning that in most cases, the ribose phosphate backbone in these pseudoknots are forced to make sharp turns at the junction of connecting loop to double helix. Interestingly such sharp turns were observed in the crystal structure of tRNA[Phe] from yeast in the anticodon and T-loop, respectively (Quigley and Rich, 1976; Saenger, 1984).

IV PSEUDOKNOTS IN OTHER RNAs

So far our discussion of the pseudoknot of the special kind was based on its occurrence in the tRNA-like structures of some plant viral RNAs. The question can therefore be raised how widespread this novel feature in RNA folding is in other natural RNA molecules. We will show here that this special folding is also present outside the tRNA-like structures and in strategic regions of other RNA molecules.

a. The 3' noncoding region of TMV RNA

An interesting example of the importance of pseudoknots in the three-dimensional structure of RNA is given by the 200 nucleotides of the 3' non-coding region of TMV RNA. We have reported earlier (Rietveld et al., 1984) that the tRNA-like structure of TMV RNA, comprising the last 95 nucleotides at the 3' end (Joshi et al., 1985) contains two pseudoknots. The first one is present in the aminoacyl acceptor arm, analogous to the situation in TYMV RNA, while the second pseudoknot was found to involve a large bulge loop in the anticodon arm. The latter is therefore not a pseudoknot of the

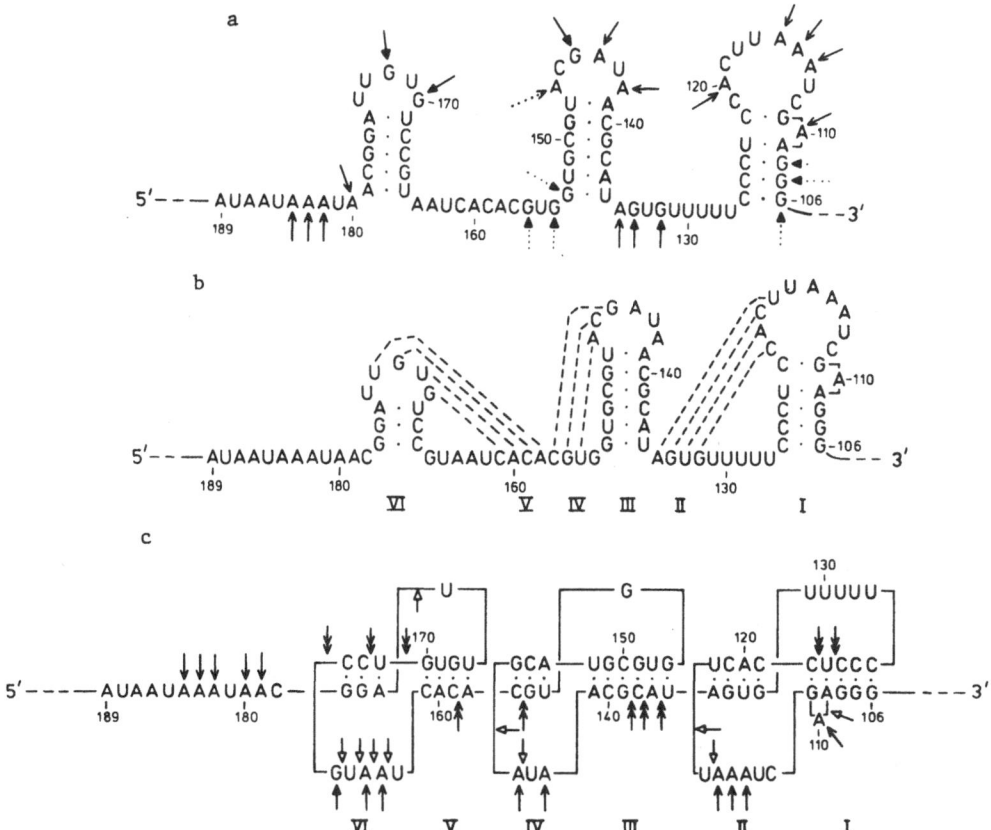

Fig. 5. Models of the structure of a part of the 3' noncoding region
of TMV RNA located just upstream of the tRNA-like structure
a) Secondary structure under semidenaturing conditions (1
mM EDTA). b) Secondary structure showing the tertiary interactions (dashed lines) which give rise to pseudoknot formation upon addition of Mg^{2+}. Roman numbers correspond to the
six double helical segments involved. c) Structure under
native conditions (10 mM Mg Cl_2). The spatial folding is
obtained by coaxial stacking of helices I-VI. Arrows
indicate base residues or phosphodiester bonds accessible
to the various probes used ($\longrightarrow$ RNase T_1; $\longrightarrow$ nuclease S_1;
$\longrightarrow$ cobra venom RNase; $\longrightarrow$ diethyl pyrocarbonate). The dotted
arrows correspond to weak bands in the autoradiographs.

special type as described in section III. For a discussion of the tRNA-like
structure of TMV RNA, the reader is referred to Rietveld et al. (1984).

To our surprise, analysis of the sequence upstream of the tRNA-like
structure revealed the presence of another highly structured domain consisting of three more pseudoknots arranged in tandem, and located just after
the second pseudoknot of the tRNA-like structure. Evidence for the existence
of these three consecutive pseudoknots was obtained in two ways (van Belkum
et al., 1985).

Structure mapping by chemical modification of A residues with diethyl
pyrocarbonate (DEP) and by enzymatic digestions with S_1 nuclease, RNase
T_1 and cobra venom RNase led to the results as summarised in Fig. 5. The
results obtained under semidenaturing conditions (Fig. 5a) and under native
conditions (Fig. 5c) could be explained best by proposing the three pseudoknotted structures as indicated in Fig. 5b. Especially revealing in this
respect was a number of G and A residues becoming inaccessible under native
conditions (10 mM Mg^{2+}) towards RNase T_1 and DEP, respectively (Fig. 5a and
5c). Moreover, the cuts in five out of the six proposed double helical
segments, observed with the double strand specific cobra venom RNase, can
also be considered as a strong argument in favour of this model.
Note that the three pseudoknots all meet the definition of the pseudoknot
of the special kind as described above. Furthermore their arrangement in
tandem yields a perfect quasi-continuous double helix of 25 base pairs, in
which the six separate stem regions are each crossed by one connecting loop,
alternately over the deep and the shallow groove.
Further support for these three pseudoknots forming this "stalk-like" domain
was derived from sequence comparisons among the tobamoviral RNAs, whose 3'
terminal sequences are known (Takamatsu et al., 1983; Meshi et al., 1983).
Both the RNA from the tomato, watermelon and the cowpea strain could be
folded in exactly the same way as in the common strain of TMV, despite large
differences in the primary structure. Many conservative base pair changes
occur in all stem regions, which usually is seen as a strong support for a
proposed secondary structure of RNA (Fox and Woese, 1975).
All this together means that in the 3' noncoding region of the tobamoviral
RNAs evidence was obtained for the existence of 5 different pseudoknots, 4
of which are of the special type discussed in this paper (see also van
Belkum et al., 1985).

Two aspects of the structure presented in Fig. 5c are worth mentioning
here in view of the properties of pseudoknots in general. First, the
nucleotides in the connecting loops crossing the deep groove (the upper
three single stranded regions) appear to be more protected against cutting
by RNase T_1 and nuclease S_1 than the residues bridging the narrow groove
(the lower three single stranded stretches of RNA). This might be explained
by the fact that the residues crossing the deep groove are more or less
located within the outer envelope of the double helical cylinder, which
makes them less vulnerable to enzymatic attack for steric reasons. The
better accessibility of the residues bridging the narrow groove correlates
well with our model building studies on the pseudoknot in the aminoacyl
acceptor arm of TYMV RNA described in section III.
Secondly, it is interesting to note that two out of three connecting loops
bridging the deep groove apparently consist of one nucleotide only (G_{154}
and U_{174}, Fig. 5c). As outlined above this would demand a distortion of
the quasi-continuous double helix in order to bring together the phosphates
across the helical groove. This might be achieved by a special base sequence
in the double stranded regions involved, analogous to DNA (Dickerson,
1983), by binding multivalent cations like polyamines (Quigley et al.,
1978; Sundaralingam, 1980) or by partial destacking at the border of two
double helical segments. NMR or X-ray diffraction of pseudoknot containing
RNA fragments will be able to shed light on this question.

Analysis of the 3' noncoding regions of all aminoacylatable plant viral
RNAs brought to light that especially the middle pseudoknot of TMV RNA as
illustrated in Fig. 5b was also very well conserved in RNAs outside the
tobamovirus group (Fig. 6). This particular pseudoknot with the single
G residue bridging the deep groove was found at similar positions in the
RNAs from the bromovirus group (CCMV) and the hordeivirus group (BSMV)
(to be published). This strongly conserved feature immediately raises the
question of its biological significance, if any. Up till now we can only
speculate about a possible role of the stalk-like domain in the replication
cycle of these plant viruses.

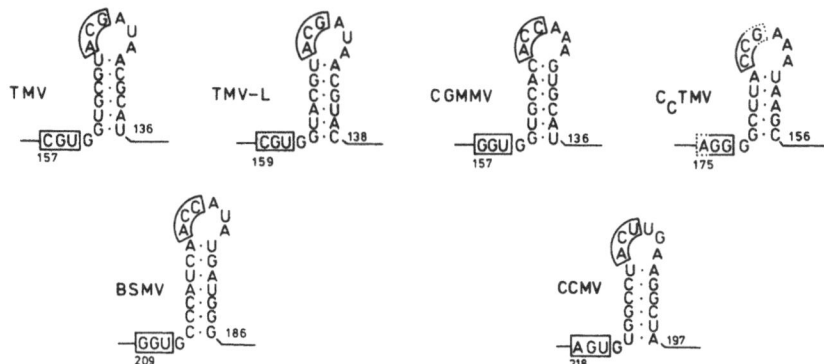

Fig. 6. Pseudoknot conservation in aminoacylatable plant viral RNAs.
The pseudoknot of TMV vulgare corresponds to the middle one
of Fig. 5 (stem regions III and IV). The upper row shows the
conservation in the tobamoviral RNAs (for further details
see Van Belkum et al., 1985). The bottom row shows the same
pseudoknot as found in RNA2 of barley stripe mosaic virus
(BSMV) which belongs to the hordeiviruses (see also Kozlov
et al., 1984) and RNA3 and 4 of the bromovirus cowpea chloro-
tic mottle virus (CCMV, for the primary structure see Ahlquist
et al., 1981). Numbering of the nucleotides is from the 3'
end. The boxed regions indicate the complementarities needed
for pseudoknot formation.

b. Pseudoknots in ribosomal RNA and autocatalytic selfsplicing of RNA

        In the course of our work on the aminoacylatable plant viral RNAs
we noted that at least in two other RNA systems pseudoknots may play an
important role. The first example is found at the 5' side of 16S rRNA
(Fig. 7a). Here, three double helical stem regions (1-3) were proposed,
mainly on the basis of phylogenetic comparisons. The coexistence of helix
1 and 2 was puzzling however because these two helices appeared to be
mutually exclusive (Maly and Brimacombe, 1983). In the light of the properties of pseudoknots as mentioned above it will be clear that these helices
can stack coaxially so that they form an extended double helix of 9 base
pairs (Fig. 7b), which eventually might be extended further by including
stem region 3. This implies that the 5' terminus of the ribosomal RNA from
the small subunit can undergo a considerable conformational change on
adopting the pseudoknot structure. Such a conformational transition may
play an important role as a switch in ribosome functioning. Another example
stems from the field of the autocatalytic self-splicing of RNA precursor
molecules (Cech, 1983). Phylogenetic comparisons of the intron and exon
sequences in fungal mitochondrial precursor RNAs led to a proposal for a
generalized secondary structure model (Davies et al., 1982, Waring and
Davies, 1984). In this model (Fig. 7c) the interaction between a few
nucleotides in the hairpin loop, closing stem region P_1, and a complementary
region of the 3' exon is especially noteworthy. This interaction, which
again fully meets the definition of a pseudoknot of the special type, will
bring the splice junctions in close proximity (Fig. 7d), thus creating
a geometrically favourable conformation for the cleavage and ligation
reactions to take place in one concerted reaction.
Furthermore a coaxial stacking of stem P_2 on top of stem P_1 and P_{10} and
the stacking of stem E-E' on top of P_8 are worth mentioning, being further
examples of the formation of extended quasi-continuous double helices. It
remains to be seen whether these structural elements will adopt these
conformations during the autocatalytic self-splicing reaction.

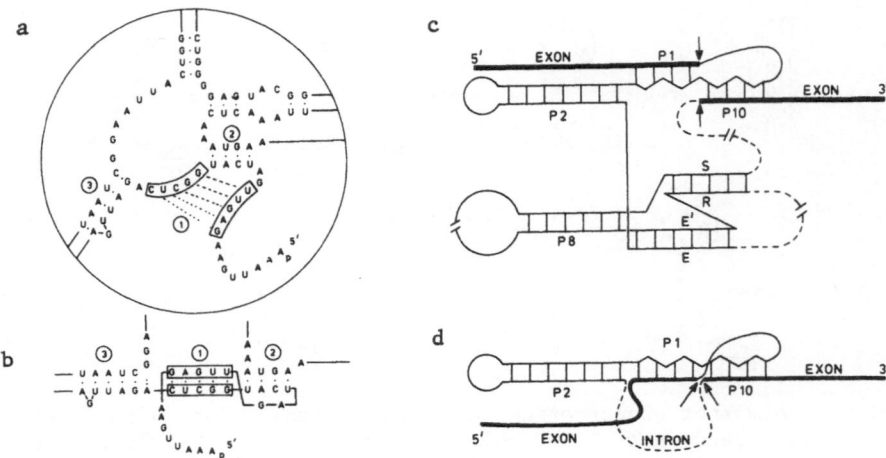

Fig. 7. Models of RNA secondary structures showing the possible role
of pseudoknots in ribosome functioning and RNA self-splicing.
a) Part of the secondary structure of 16S rRNA from E. coli
(see Maly and Brimacombe, 1983). b) Pseudoknot formation in
16S rRNA by coaxially stacking of helix 1 and 2. c) Generalised secondary structure model of fungal mitochondrial
introns. For a full description of this model and the symbols
used see Davies et al. (1982). The arrows correspond to the
splice sites. d) Coaxial stacking of helical segments P_1,
P_{10} and P_2, which brings the two splice sites into closer
proximity.

V CONCLUDING REMARKS

Our studies on the 3' noncoding regions of the aminoacylatable plant
viral RNAs have revealed the important role of pseudoknotting in the three-
dimensional folding of RNA. In the case of the tRNA-like structures it is
clear how the viral RNA takes advantage of this folding principle to adopt
a functional three-dimensional structure, although the exact role of the
tRNA-like structure in the replication cycle of the virus still remains
elusive (Florentz et al. (1984) and references therein). Taken together
with the other examples mentioned in this paper one now can say that
pseudoknots in RNA are no longer a merely theoretical possibility
(Studnicka et al., 1978), although it cannot be excluded that the pseudoknots of the special type will appear to be mainly a peculiar property of
some plant viral RNAs. Thus far we could not find these pseudoknots in the
3' termini of other plus-stranded plant or animal viral RNAs (unpublished
results).

In principle it is relatively easy to find possible pseudoknots of the
special kind in an arbitrary RNA chain of sufficient length. We developed a
computer program which searches for this type of structure according to the
principle illustrated in Fig. 3b (Rietveld, 1984). If one allows short stem
regions (e.g. 3 base pairs) and long connecting loops to be formed, one ends
up with a very large number of possible pseudoknots, more or less equivalent
to the number of all possible stem regions of 6-8 base pairs in the same RNA
chain. For instance we have searched in the genomic RNA of TMV (about 6500
nucleotides, Goelet et al., 1982) for other pseudoknots while reducing the
maximal connecting loop size to 10 nucleotides. We found some interesting
combinations of possible pseudoknots, some of them arranged in tandem like
in the 3' noncoding region. Unfortunately, no other complete tobamoviral
RNA sequences are available so that the powerful method of sequence comparisons could not be applied here to get the support that was available for all
the examples mentioned in this paper.

96

A big problem is how the pseudoknot formation can be incorporated in existing or new algorithms for the prediction of secondary and tertiary structures of RNA. Apparently, they were not excluded without reason from currently used algoritms (Zuker and Stiegler, 1981; Studnicka et al., 1978). It is not justified to rely simply on existing models of secondary structures to look for possible pseudoknots, because in some cases parts of this structure have to be melted first before the more stable pseudoknot can be formed (see also Rietveld et al., 1983). In using the Zuker and Stiegler algorithm on the 3' terminal 200 nucleotides of TMV RNA we obtained a secondary structure which differed to a large extent from the one we propose in fig. 5. Some of the simple hairpins, the loops of which are involved in the pseudoknot formation were not predicted at all. This illustrates again that we are still far away from the prediction of correct secondary and tertiary structures of RNA. On the other hand the principle of pseudoknotting in RNA has to be considered seriously in any new algorithm to be developed in the future.

ACKNOWLEDGEMENTS

We thank Jan Pieter Abrahams, Jiang Bing Kun, Kees Linschooten and Paul Verlaan for their contributions to the work described.

REFERENCES

Ahlquist, P., Dasgupta, R. and Kaesberg, P. (1981) Cell 23, 183-189.
Arnott, S., Hukins, D.W.L. and Dover, S.D. (1972) Biochem. Biophys. Res. Commun. 48, 1392-1399.
Briand, J.P., Jonard, G., Guilley, H., Richards, K. and Hirth, L. (1977) Eur. J. Biochem. 72, 453-463.
Cantor, C.R. (1980) in "Ribosomes (Chambliss, G., Craven, G., Davies, J., Davis, K., Kahan, L. and Nomura, M, eds.) University Park Press, Baltimore pp. 23-49.
Cech, T.R. (1983) Cell 34, 713-716.
Davies, R.W., Waring, R.B., Ray, J.A., Brown, T.A. and Scazzocchio, C. (1982) Nature 300, 719-724.
Dickerson, R.E. (1983) J. Mol. Biol. 166, 419-441.
Florentz, C., Briand, J.P., Romby, P., Hirth, L., Ebel, J.P. and Giegé, R. (1982) EMBO J. 1, 269-276.
Florentz, C., Briand, J.P. and Giegé, R. (1984) FEBS Lett. 176, 295-300.
Fox, G.E. and Woese, C.R. (1975) Nature 256, 505-507.
Goelet, P., Lomonossoff, G.P., Butler, P.J.G., Akam, M.E., Gait, M.J. and Karn, J. (1982) Proc. Natl. Acad. Sci. USA 79, 5818-5822.
Gralla, J. and Crothers, D.M. (1973) J. Mol. Biol. 73, 497-511.
Haenni, A.-L., Joshi, S. and Chapeville, F. (1982) in "Progress in Nucleic Acids Research and Molecular Biology" (Cohn, W.E., ed) Academic Press, New York, 27, pp. 85-102.
Hall, T.C. (1979) Int. Rev. Cytol. 610, 1-26.
Joshi, R.L., Joshi, S., Chapeville, F. and Haenni, A.-L. (1983) EMBO J. 2, 1123-1127.
Joshi, R.L., Chapeville, F. and Haenni, A.-L. (1985) Nucleic Acids Res. 13, 347-354.
Joshi, S., Chapeville, F. and Haenni, A.-L. (1982) Nucleic Acids Res. 10, 1947-1962.
Kim, S.-H., Suddath, F.L., Quigley, G.J., McPherson, A., Sussman, J.L., Wang, A.H.J., Seeman, N.C. and Rich, A. (1974) Science 185, 435-440.
Kozlov, Y.V., Rupasov, V.V., Adyshev, D.M., Belgelarskaya, S.N., Agranovsky, A.A., Mankin, A.S., Mozorov, S,Y., Dolja, V.V. and Atabekov, J.G. (1984) Nucleic Acids Res. 12, 4001-4009.
Maly, P. and Brimacombe, R. (1983) Nucleic Acids Res. 11, 7263-7268.
Meshi, T., Kiyama, R., Ohno, T. and Okada, Y. (1983) Virology 127, 54-64.
Nussinov, R. and Tinoco, I. (1981) J. Mol. Biol. 151, 519-533.

Peattie, D.A. and Gilbert, W. (1980) Proc. Natl. Acad. Sci. USA 77, 4679-4682.

Pieler, T. and Erdmann, V.A. (1982) Proc. Natl. Acad. Sci. USA 79, 4599-4603.

Pley, C.W.A., Rietveld, K. and Bosch, L. (1985) Nucleic Acids Res. 13, 1717-1731.

Quigley, G.J. and Rich, A. (1976) Science 194, 796-806.

Quigley, G.J., Teeter, M.M. and Rich, A. (1978) Proc. Natl. Acad. Sci. USA 75, 64-68.

Rietveld, K. (1984) Ph.D. Thesis, University of Leiden.

Rietveld, K., van Poelgeest, R., Pley, C.W.A., van Boom, J.H. and Bosch, L. (1982) Nucleic Acids Res. 10, 1929-1946.

Rietveld, K., Pley, C.W.A. and Bosch, L. (1983) EMBO J. 2, 1079-1085.

Rietveld, K., Linschooten, K., Pley, C.W.A. and Bosch, L. (1984) EMBO J. 3, 2613-2619.

Robertus, J.D., Ladner, J.E., Finch, J.T., Rhodes, D., Brown, R.S., Clark, B.F.C. and Klug, A. (1974) Nature 250, 546-551.

Saenger, W. (1984) in "Principles of Nucleic Acid Structure", Chapter 15, Springer Verlag, New York.

Silberklang, M., Prochiantz, A., Haenni, A.-L. and RajBhandary, U.L. (1977) Eur. J. Biochem. 72, 465-478.

Studnicka, G.M., Rahm, G.M., Cummings, I.W. and Salser, W.A. (1978) Nucleic Acids Res. 5, 3365-3387.

Sundaralingam, M. (1980) in "Biomolecular Structure, Conformation, Function and Evolution" Vol I (Srinivasan, R. ed.) Pergamon Press, Oxford.

Takamatsu, N., Ohno, T., Meshi, T. and Okada, Y. (1983) Nucleic Acids Res. 11, 3767-3778.

Trifonov, E.N. and Bolshoi, G. (1983) J. Mol. Biol. 169, 1-13.

Van Belkum, A., Abrahams, J.P., Pley, C.W.A. and Bosch, L. (1985) Nucl. Acids Res. 13, 7673-7686.

Waring, R.B. and Davies, R.W. (1984) Gene 28, 277-291.

Zuker, M. and Stiegler, T. (1981) Nucleic Acids Res. 9, 133-148.

PROTON NMR STUDIES OF RNA'S AND RELATED ENZYMES USING ISOTOPE LABELS

A.G. Redfield, B.-S. Choi, R.H. Griffey, M. Jarema,
P. Rosevear, P. Hoben*, R. Swanson*, and D. Soll*

Department of Biochemistry   *Department of Molecular
Brandeis University   Biology and Biophysics
Waltham, MA 02554   Yale University
    New Haven, CT 06511

INTRODUCTION AND HISTORY

This article gives a brief history of proton NMR in RNA's, mainly
tRNA, followed by a descriptions of recent work in our lab based primarily
on proton NMR aided by deuteron and nitrogen 15 labeling, directed more and
more towards studies of protein-tRNA interaction. The field is young in
that the first promising spectra of tRNA were reported by Kearns et al[1]
roughly fifteen years ago; but tRNA is now one of the older areas of biochemical NMR, being overtaken, in amount of research, by studies of smaller
RNA fragments and small enzymes. Transfer RNA, and especially tRNA
synthetases, remain one of the larger molecules studied by NMR. Over the
years tRNA has provided a testing ground for some types of NMR methodologies, and we hope that this will continue to be true. Transfer RNA is
too large, probably, for us to expect to solve complete structures as
appears possible for smaller proteins and nucleic acids, but on the other
hand it is a small molecule by biological standards. It has well-studied
interactions with enzymes but much remains to be learned, such as details
of recognition by aminoacyl tRNA synthetases. Thus it is of interest to
continue research on NMR of this important molecule.

Early research on tRNA established the feasibility of obtaining
spectra of continually improving quality (reviewed by Reid[2]). The field
was dominated by study of the downfield resonances of ring NH protons,
namely the N1H protons of guanosine and the N3H protons of uridine, because
their resonances are relatively well spread-out in the spectrum, shifted
downfield away from other resonances, and potentially provide dynamic
information. Successful identifications of NMR lines with specific protons
began with that of S^4U8 by Mildred Cohn and coworkers and of several methyl
protons by Kan, Sprinzl and others. Pulsed Fourier transform (FT) methods
had to be developed to obtain spectra in nearly 100% H_2O solvent which is
necessary for observation of these labile protons, and were combined with
use of also the nuclear Overhauser effect (NOE)[3]. NOE was first used for
identifying resonances GU pairs (thereby showing for the first time that
they truly existed in the form proposed long ago by Crick), and to
distinguish AU from GC resonances. Studies of proton solvent exchange rates
used line broadening, saturation-recovery, and "real-time" exchange
(observation of NMR after a rapid change of solvent from H_2O to D_2O), to
get a rough idea about tRNA dynamics.[4,5]

Thus, about five years ago good samples and spectra were available, as well as methods for learning about dynamics, and a good X-ray structure was available for one tRNA. Nevertheless the spectra remained to be convincingly assigned. While full assignment of the ring NH protons in a tRNA remains difficult even now, and is only possible to achieve nearly completely in a few favorable cases, there were at this point several developments that led to identification of large parts of the spectra. These were: 1) Deuteron labeling was used to test NOE assignments, thereby distinguishing, for example, resonances of reverse Hoogsteen base-pair protons from those of Watson-Crick base-pairs[6]. 2) Siddhartha Roy observed NOE's between, rather than within, base pairs[7]. By correlating NMR connectivity with base-pair sequence, he performed the first massive set of NOE assignments in any nucleic acid, of the D-stem in yeast tRNA^{Asp}. 3) Hare and Reid demonstrated that such sequential assignments could be made rapidly in a 500 MHz NMR instrument, because of its sensitivity and resolution[8]. 4) NOE and isotopes were used to study unusual parts of tRNA such as the TΨC loop[9], unusual base pairs and triples[10,11] and surroundings of methyl groups[9,12,13]. 5) Samples of ^{15}N labeled tRNA were produced and studied, immediately providing a fool-proof new identification of many NH resonances using difference decoupling and other methods (below)[14,15].

Added to these developments might be 2D NMR; however, this important technology has been of only minor importance so far for tRNA[14,15,16]. There are relatively few NOE's that are useful in a larger molecule like tRNA; many potentially interesting NOE peaks are so overlapping as to be useless. 2D is most useful for cases where many resolvable cross peaks are found; this is not the case for tRNA. Also, except for ^{15}N or ^{13}C labeled samples, spin-spin couplings are not resolved in tRNA, which renders many 2D methods useless.

There has been only one recent published comprehensive NMR study of an RNA smaller than tRNA, on the anticodon stem of yeast tRNA^{Phe} which showed stacking similar to the same stem in the X-ray structure[17]. 5S RNA is the only larger RNA other than tRNA that has been studied by NMR; some of this work is described in another article in this volume by Moore and Jarema. Transferred NOE has been used by Clore and Gronnenborn to infer the structure of the complex of a tRNA with its complementary codon triplet RNA sequence[18]. Proton NOE has been used together with phosphorous NMR to find a transition to left-handed Z conformation in an RNA duplex[19].

A comprehensive proton NMR study starts with purification of at least 100 units of pure RNA and observation of a good spectrum. An NOE survey follows, of stronger NOE's between neighboring base pairs, from methyl resonances, and within base pairs. This distinguishes classes of resonances (AU, GC, GU, methylated bases) and establishes some connectivities between them. This survey can be one-or two-D NOE. Generally the connectivities can be mapped to the sequence at least in part, but often runs of more than two adjacent GC pairs cannot be followed. This survey is generally supplemented by selected one-D NOE runs with a long (> 0.5 sec) irradiation to identify spins further apart, using NOE connectivity by multiple flip processes, to make a few further identifications. Sometimes it is most desirable (and not so hard) to produce a sample deuterated at the purine C8 position, to identify reverse-Hoogsteen pairs whose uracil N3H to adenine C8H NOE's will vanish in such a sample. Deuteration of adenine at C2 is also useful[20]. Finally, with greater investment of money and labor, ^{15}N labeled tRNA can be produced to provide a few further positive identifications and with them potential geometric information from NOE's.

Structural information that comes from such studies include: 1) identity of base pairs of reasonable stability; NMR shows that these generally include GU pairs, for example. 2) Local geometrical information about areas such as the T loop and T stem. 3) In one case, confirmation that the acceptor stem and T stem stack stably on each other[21]. Unfortunately NMR does not yet tell us the angle between the T loop and the D loop helix axes, or show whether the invariant tertiary base pair G19-C56, at the corner of the molecule, is always formed, although it does confirm general proximity of the TψC and D loops through a series of NOE's.

INFORMATION ON DYNAMICS

We discuss this field only very briefly partly because limited information previously available may be augmented soon with higher field measurements. Inferences concerning dynamics have been drawn largely from studies of rates of ring NH proton exchange with solvent, assuming that this exchange rate will be greater for more flexible parts of the molecule[4,22]. Most studies have been performed on samples containing low magnesium. It is interesting to note that even a few Mg^{++} ions per tRNA affect rates of exchange of many protons, and that NMR shifts vary continously with Mg^{++} concentration, indicating rapid motion of the ion along the surface, and rapid exchange of it between tRNA molecules[23]. As expected, Mg^{++} decreases the solvent exchange rate of most protons substantially, into a region where it is hard to measure by NMR at low temperature. Spectroscopy at high temperature generally confirms the expectation that melting is cooperative in the presence of Mg^{++}.

At low temperature in zero Mg^{++} and moderate (~0.1M) ionic strength the results of various studies indicate the following: 1) The T stem is highly stable, and the TψC loop is fairly so. 2) Tertiary structure does not disappear before secondary structure, and individual helixes do not open as a unit[4,13,22]: Individual base pair protons on the same helix exchange at quite different rates and in one case, the tertiary pair U8-A14 of yeast tRNA[Asp] is the most stable one by NMR kinetics[24]. 3) Although the details of proton exchange kinetics remain poorly understood, it has recently been established that the opening rate of a base pair is considerably greater than the exchange rate of its internal NH proton[25]. The rate shows buffer catalysis by tris or imidazole and a study of rate vs. catalyst indicated opening rates of order 0.1 minute^{-1} for the T stem, and 100 sec^{-1} or greater for some bases of the acceptor. The nature of the "open" state is not clear but it seems to involve unpairing of only one or a few bases.

A study of a tRNA from a thermophilic bacteria was recently completed, and the expected greater stability was observed[26].

Much more precise and unambiguous measurements of solvent exchange rates could be made using, for example, ^{15}N labels to edit spectral lines (see below). However, the measurements are tedious and the results do not have obvious significance at least in the case of tRNA.

ISOTOPE LABELS

As already indicated, ^{15}N and deuteron labels are quite useful for identification of proton lines, and will continue to be so. Carbon 13 has also been used for methyl proton resonance identification[27], but may be less generally useful than ^{15}N because of the greater expense and difficulty of introducing it into tRNA. Potentially all the methods described below for ^{15}N labels could be applied to ^{13}C labels, and they probably soon will be.

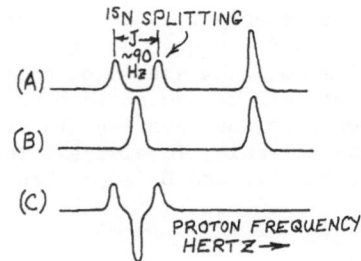

Figure 1. A) Proton NMR spectrum without decoupling, of a
sample having one proton donated by a ^{15}N labeled nitrogen
(resonance to the left) and another proton on an unlabeled
nitrogen or carbon. (B) Spectrum of the same sample with
decoupling applied at exact resonance frequency of the ^{15}N spin.
This frequency is very far from the proton resonance frequency,
and affects the proton spectrum only indirectly, by collapsing
the splitting of the ^{15}N labeled proton. (C) Difference between
the first two spectra. Unfortunately the pattern of the
difference spectrum becomes even more complicated if the [15]N
irradiation is slightly off resonance, often making it difficult
to interpret surveys of spectra versus ^{15}N irradiation
frequency.

However, the primary motivation in developing ^{15}N label proton-observe
NMR has been to try to observe specific protons in tRNA-protein complexes
and draw inferences from these specific observations. Proton NMR at 500
MHz is sensitive enough to allow us to observe single protons in a
molecular weight ~100,000 complex within a few hours, but an isotope like
[15]N on the proton's donor tags the proton resonance and permits us to see
it above the background of hundreds of other protons. This approach will
work until the proton-nitrogen spin-spin coupling is exceeded by the proton
line width, probably for complexes greater than 100,000 molecular weight.

METHODS

Isotopes have been used in biopolymer NMR for many years, and the
advantages of proton-observe nitrogen-perturb methods have also been
demonstrated long ago. However, there was a considerable resurgence in
general interest in proton-observe and ^{13}C or ^{15}N perturb methods as a
result of recent work by Griffey, Poulter, Bax and their collaborators who
first identified a number of resonances using ^{15}N labeled tRNA and also
demonstrated the first 2D ^{15}N-proton correlation methods using proton
observation, in a tRNA-sized molecule[14]. Dr. Roy and other members of our
group have also studied a number of ^{15}N labeled tRNA's[15,26]. In the
course of these studies, we have investigated a large number of methods of
^{15}N-proton NMR observation. Of these we will describe three which seem
most useful, all developed by others. We call them: difference
decoupling; difference echo, and 2D forbidden echo. All three give, to
some degree, the proton resonance shift of every proton attached to a
label, and the nitrogen shift of the nitrogen associated with a given
proton resonance. In other words, they are "two dimensional" in the sense
that every ^{15}N-H group gives a resonance with both a proton and a nitrogen
shift associated with it.

Unfortunately the most sensitive method, difference decoupling, gives the most confusing results, while the highest resolution method, 2D forbidden echo (2DFE) is the least sensitive and does not work for systems larger than 5S RNA. All these methods work by taking difference spectra between experiments which perturb ^{15}N different ways. Protons remote from ^{15}N give identical signals, and therefore their resonances cancel. The trick is to invent ways to extract the most useful information from the remaining ^{15}N labeled resonances, at high sensitivity.

Difference-decoupling relies on the simple fact that the proton splitting, of about 90 Hertz, produced by the donor ^{15}N of an ^{15}NH group, can be collapsed to a singlet by applying strong RF power at the nitrogen frequency (Figure 1). This effect is sensitive to the value of the frequency applied to the nitrogen spins. By varying this frequency, one can determine the nitrogen resonance frequency and thus the ^{15}N shift. Since, for example, the GN1 nitrogens of GC pairs resonate in a different range than those of other kinds of base pairs, this allows us to tell their proton resonances apart in the few cases where proton NMR alone does not discriminate between them.

These experiments are generally done as differences between a spectrum with irradiation very far from ^{15}N resonance, as a control, and a spectrum with irradiation near to ^{15}N resonance. By taking a series of difference spectra with the latter ^{15}N resonance frequency systematically shifted in steps through the expected region of resonance, one can in principle look for the nitrogen frequency that gives the largest and sharpest effect, thereby associating a nitrogen shift with each proton resonance. Unfortunately in practice if there are more than a few ^{15}NH groups in the molecule with similar N and H shifts, the patterns become very confused. Thus difference decoupling is useful if there are only one or a few non-overlapping ^{15}NH groups in specific known places in a molecule. Then it is the most sensitive method, requiring only a few tens of minutes for 1 mM tRNA in 0.2 ml of solvent[14,15]. Unfortunately tRNA cannot now be labeled in this way except perhaps in a few sites.

The opposite technique[28] in terms of complication, poor sensitivity, and theoretical complexity, is described in Figure 2. We call it 2D forbidden echo (2DFE) because the proton part appears to be observation of a "spin echo" (a phenomenon discovered years ago by E.L. Hahn), in which the latter half of the spin-echo which occurs starting at time t_1+2t_a is recorded as a function of the time t_2 after the echo. (For an explanation of spin echoes see, for example, reference 29). This signal, in a molecule like tRNA where proton spin-spin interactions are swamped by the proton line width, would be similar to that right after the first 90° pulse except for relaxation (below) and the effect of nitrogen irradiation on ^{15}NH groups. The nitrogen "noise decoupling" during this part of the sequence is also relatively simple in its result: it eliminates all ^{15}N splittings in the proton dimension of the 2D map. The function of the two ^{15}N 90° pulses is much more complicated: if the equal times t_a are both approximately equal to $(2J)^{-1}$, where J is the ^{15}NH splitting of the proton resonance (Figure 1a), the first pulse transfers the proton coherence from an observable magnetic signal, just after the first 90° pulse, to a hidden, not directly observable, coherence between the nitrogen and proton spins, after the first nitrogen pulse. The second nitrogen pulse and waiting period t_a reverse this. Between the two ^{15}N pulses this coherence evolves according to the nitrogen chemical shift. This sequence is repeated for many values of t_1, and data is recorded for each value of t_1, and stored as a data set of signal versus t_1 and t_2. Then a somewhat specialized 2D Fourier transform is performed which produces a map of peaks, one for each ^{15}NH group, whose coordinates are the proton and nitrogen resonance frequencies of the proton and its attached nitrogen. The detailed

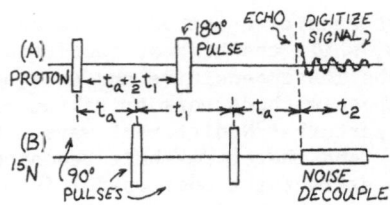

Figure 2. Pulse sequence used for 2 dimensional forbidden echo and
          difference echo experiments. A 90° pulse is one that flips spin
          magnetization exactly 90° from its equilibrium direction along
          the z-axis, and a 180° pulse is twice as long and reverses the
          magnetization. (The interested reader should refer to a book on
          NMR for a complete description of their effect). (A) Proton
          sequence. The proton signal is not observed until the middle of
          the spin echo (for $t_2 > 0$). (B) Nitrogen 15 sequence. "Noise
          decoupling" is broadband (non-monochromatic) power applied
          continuously to flip all or a group of ^{15}N spins, to decouple
          their protons. The experiment is repeated with all possible
          combinations of 90° radio frequency phase shifts of the three
          proton pulses, and the two nitrogen pulses, with respect to each
          other. For the one-dimensional difference echo, the time t_1 is
          zero and the nitrogen pulses are either in phase, forming a 180°
          pulse, or effectively omitted.

theoretical and experimental aspects of this method are far beyond the
scope of this paper; as simple a theoretical explanation as is possible is
given in reference 28.

This technique gives proton and ^{15}N shifts with the greatest possible
resolution, about 20 Hertz, or less than 0.5 ppm for ^{15}N. It has the same
advantage over one-D NMR that 2D chromatography has over one D
chromatography, namely, individual signals can be separated and followed.
Unfortunately, resonances of GC pairs are only slightly separated from each
other in the ^{15}N dimension, as well as being more numerous and harder to
label than those of AU pairs. Also, the method seriously fails in terms of
signal to noise ratio for larger complexes, because the signals in NMR
always decay with a first order rate constant T_2^{-1} which becomes greater
for larger molecules. When the corresponding time T_2 becomes less than the
time $2t_a$, or 5 to 10 millisec for the typical ^{15}NH splitting of ~100 Hz,
the 2D signals are badly degraded. For tRNA alone, T_2 is around 15
millisec for many protons but for the smallest tRNA-synthetase complexes it
is around 5 msec, and 2D runs do not give useful signals in 24 hours.

The final method we use[30] is called difference echo and uses the same
sequence as 2DFE except that t_1 is set equal to zero. This in effect
collapses the two ^{15}N 90° pulses so that they form a single 180° pulse,
that flips over all ^{15}N spins. This experiment can be understood
classically by viewing species having ^{15}N spin "up" as a different chemical
species from ^{15}N spin "down" and assuming that the two ^{15}N pulses in close
succession simply interchange these two species. The effect of the two ^{15}N
pulses is simply to invert the signals, during t_2, relative to what they
would have been without the ^{15}N pulses. A difference spectrum with and
without the ^{15}N pulses, and Fourier transformation of the signal with
respect to t_2, yields a normal looking spectrum showing only lines of

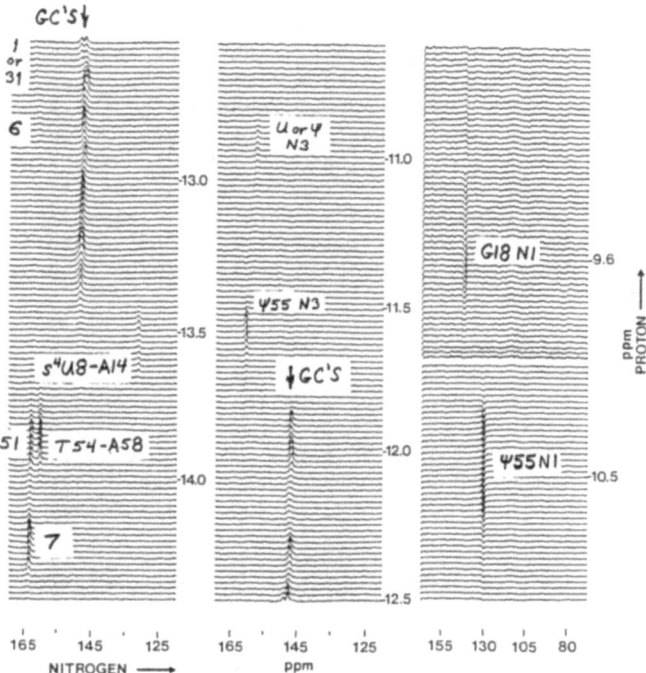

Figure 3. Two dimensional forbidden echo spectrum of fully ^{15}N labeled E. coli tRNA^{Phe}. This is the result of a run on about 200 nanomoles of tRNA taken overnight. Identified peaks labeled with numbers are secondary pairs. Identifications were made by one-D methods using NOE and isotopes as described in text.

protons donated by ^{15}N; other lines are "edited" out. The ^{15}N pulses can be made somewhat selective with respect to ^{15}N shift by making them long and weak, so that the UN3 proton resonances can be shown without the confusing lines of GN1 protons, to discriminate resonances of AU pairs from those of GC pairs, in generally ^{15}N labeled samples. This technique yields less confusing proton spectra than does difference decoupling, but no better information on ^{15}N shifts, and it suffers from the same loss of signal when T_2 is short as mentioned above. However, it is more sensitive than 2DFE in our experience because it is somewhat simpler and does not require the added time t_1. We have been able to use it for tRNA-synthetase complexes in 12 hour runs (below).

So far we have described methods which select proton resonances of labeled ^{15}NH groups. These can be combined with NOE or relaxation measurements to select subsets of proton resonances as determined by their ^{15}N shifts. In the case of NOE this will highlight "target" spins to which saturation has been transferred by preirradiation of resonances of other nearby protons. We have also developed a simple method whereby a label appears to be able to direct the "source" resonance of the NOE experiment, to a labeled proton in an overlapping group of unlabeled protons[31]. The method uses differences in NOE, with and without ^{15}N decoupling during the preirradiation time, and, in a sense, the NOE appears as if coming from irradiation of only the labeled proton. This method is called isotope-directed NOE and it has been used for identification[31] as well as for determination of protein geometry near a labeled proton (unpublished).

RESULTS

We now present a few new results from our work. In addition to these results, ^{15}N labels in tRNA have been used to confirm or establish the positions of unusual resonances, such as those of GC11 (13.6 ppm) and the reverse Hoogsteen resonance T54-m^{1}A58 (12.4 ppm) in yeast tRNA^{Phe}; to identify seperately the G and U resonances of GU pairs[14,15]; and to establish that the Ψ's in AΨ pairs are, in the several cases studied so far, always in the <u>anti</u> conformation[32].

A considerable body of information has accumulated on ^{15}N chemical shift variations in tRNA's. These are greater in ppm than proton shift variations. These shift variations must reflect hydrogen bonding variation, but are not now understood.

Figure 3 shows a 2DFE map obtained (by M. Jarema in collaboration with R. Swanson and D. Soll) of fully ^{15}N labeled <u>E. coli</u> tRNA^{Gln}. The relatively few AU and AΨ resonances are well separated but the GC peaks are only partly resolved. For comparison Figure 4 shows a one dimensional NMR spectrum of the same tRNA.

Figure 5 shows a small portion of the map of uracil ^{15}N labeled <u>E. coli</u> tRNA^{Val} (obtained by B. Choi). As the temperature is raised, more rapidly exchanging proton peaks vanish from the map. This happens when the lifetime before solvent exchange becomes less than $2t_a$ (Figure 2), or about 10 msec.

The same kind of information can be obtained from one-D difference echo methods. Figure 6 shows a series of "saturation recovery" runs on labeled yeast tRNA^{Phe} in which a band of resonances are presaturated by broadband irradiation that wipes out most of the exchangeable NH resonances. This is followed by a delay τ during which saturated spins can recover, and then an observation sequence using difference echos to select ^{15}N labeled resonances. Resonance intensity recovers exponentially

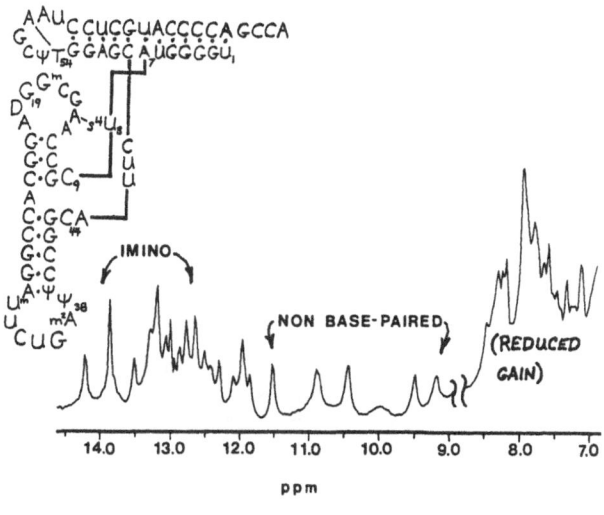

PROTON

Figure 4. One dimensional spectrum of the sample shown in Figure 3, and the sequence of the tRNA . The sequence is drawn to emulate the yeast tRNA[Phe] tertiary structure but in fact we have NMR evidence only for the two reverse Hoogsteen pairs indicated with light lines. Only peaks that are not uniquely identified with spots in Figure 3 are marked on this spectrum. It is likely to be impossible to identify the anticodon and D-stem resonances because they contain so many GC pairs.

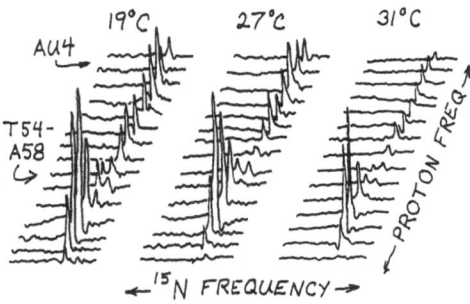

Figure 5. Portions of a 2D forbidden echo map of ^{15}N labeled E. coli tRNA[Val], at a series of temperatures. Only uracil is labeled, and only a small portion of the maps are shown. The earlier disappearance of AU4 shows that it is less stable than the tertiary base pair T54-A58.

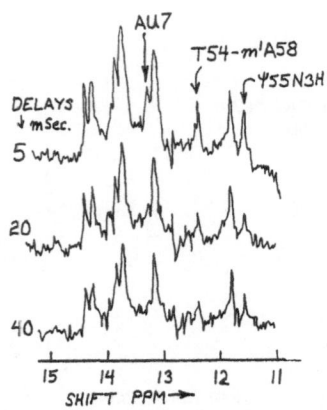

Figure 6.   Saturation-recovery run of AU and Aψ pairs in ^{15}N-uracil labeled
            yeast tRNA[Phe], at 30°C. All these proton lines are identified
            and most recover by magnetic relaxation at this temperature.
            However, the three peaks labeled are recovering more rapidly
            (that is, in the form of double difference spectra shown, these
            peaks disappear more rapidly for increasing delay time). We
            know from studies at other temperatures that this faster
            recovery results from a significant solvent exchange rate for
            these three protons. The behavior of T54-m[1]A58 in this tRNA is
            to be contrasted with the highly stable T54-A58 in E. coli
            tRNA[Val] (see Figure 5). This sample was prepared by M.
            Papastavros (see Reference 15).

toward its equilibrium value as τ is increased, but in Figure 6 the signal
for very large τ has been computer subtracted from those for small τ, to
give a series of signals that descrease exponentially with increasing τ.
The resonances that are marked disappear more rapidly than most others.
The resonance of T54-m[1]A58 is completely obscured by many GC resonances in
unlabeled tRNA[12,15]. The rapid solvent exchange of T54-m[1]A58 suggests that
the TψC loop is relatively mobile in this tRNA. In contrast, in other
tRNA's which are not methylated at A58 this proton is especially stable[26].

    We conclude by presenting the first results of our experiments to
observe useful signals from ^{15}N labeled tRNA's complexed to their cognate
aminoacyl sythetases. We have studied two systems, namely E. coli tRNA[fMet]
with a truncated methionine synthetase[33] (by P. Rosevear, R. Griffey and B.
Choi in colaboration with D. Barker), and E. Coli tRNA[Gln] with its complete
synthetase[34] (by M. Jarema in collaboration with P. Hoben, R. Swanson and
D. Soll).

    The fMet system was studied by comparing difference decoupling
spectra of the AU, GU and Aψ NH proton resonances of ^{15}N labeled tRNA[fMet],
with and without added enzyme. The strong NOE of the GU pair in unlabeled
tRNA was also compared. No striking differences other than general
broadening of spectra were found, on adding equimolar enzyme. Broadening
and small but definite shifts of the GU resonances were observed. The only
conclusion that can be drawn from these observations is that the tRNA
structure, including the tertiary interactions U8-A14 and T54-m[1]A58, are

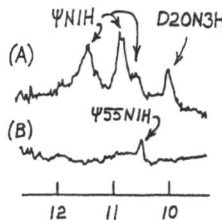

Figure 7. (A) Difference echo spectra of fully ^{15}N labeled tRNA[Gln] with
equimolar cognate aminoacyl tRNA synthetase, using nitrogen
pulses that show only UN1 and DN3 protons. (B) Spectrum
obtained in exactly the same way on labeled tRNA in the absence
of enzyme.

not much perturbed by this truncated synthetase. Conceivably this lack of
perturbation results from the fact that the enzyme is a truncated cloned
fragment. The synthetase preparation was assayed by us and found to be
fully active. This enzyme fragment is nearly the same size as the fragment
whose x-ray structure has recently been determined[35].

The tRNA[Gln] system proved to be more dramatic. Figure 7 shows a
difference-echo spectrum of ^{15}N labeled tRNA[Gln] with and without enzyme.
The experiment is performed in such a way that only dihydrouracyl N3 and
ψN1 peaks would show up even through all nitrogens are labeled. This is
achieved by making the two [15]N 90° pulses (Figure 2) long and, therefore,
partly selective; and setting their frequencies to the ^{15}N resonance of
ψN3.

The spectrum without enzyme added shows only one peak which had been
previously identified with ψ55N1H by NOE from the T54 methyl group.
However, other runs using difference decoupling show a peak at 10.9 ppm
which must be the N1 proton of either ψ38 or ψ39; and a peak at 10.0 ppm
which is the N3 proton of dihydrouracil 20 (D20). The latter peak is
assigned from its proton and ^{15}N shifts with the help of observations on a
model system (R. Griffey, unpublished). These two peaks do not appear in
Figure 7A because, probably, they have too small an exchange time compared
to the time needed to prepare the spin system before observation, as
already mentioned.

When enzyme is added (Figure 7A) three ψN1H peaks appear, as well as a
greatly strengthened D20 N3 proton peak. The most upfield ψN1H peak
appears as a shoulder at 10.5 ppm in Figure 7A, but other spectra taken
with difference decoupling have repeatedly indicated that there is a peak at
this position of intensity comparable to the other two peaks. Thus we
conclude that the D20N3 proton, and the N1 protons of, most likely, ψ38 and
39, are affected by the enzyme kinetically, by having their solvent
exchange rate decreased; and that the resonance of ψ55N1H is shifted
slightly. We can't say where it shifts since we don't know which ψN1
resonance in Figure 7A is which ψ, nor can we rule out the possibility that
the exchange rate of ψ55N1H is increased by the presence of the enzyme.

Other spectra (not shown) which were taken to optimize observation of
UN3H and ψN3H protons, and thus AU and Aψ internal NH protons, show a very
strong new peak at 13.4 ppm in the complex, which is undetectable in the
tRNA alone. This peak could be due to the tightening of either AU1, or of
Aψ31, as a result of enzyme binding.

A 2D forbidden echo spectrum, of the type shown in Figure 3 for tRNA alone, was attempted for the tRNA-enzyme complex but yielded only one definite weak peak, of T54-A58, which was unshifted. Other peaks disappeared, most probably because of the shorter T_2 they have in the tRNA-synthetase complex.

The article in this book by Moore and Jarema shows a useful map of a fragment of RNA complexed with a ribosomal protein. In that case the molecular weight of the complex is much smaller than for tRNA^{Gln} with enzyme (~30,000 Daltons compared to ~90,000).

Although these are not the first NMR experiments on nucleic acid-protein complexes, they (together with those described by Jarema and Moore in this book) are the first in which both definite, and also interpretable, changes occur. They illustrate both the strengths, and the difficulties, of the ^{15}N labeling method. In this case they show that the enzyme shields, and perhaps interacts with, the two ψ's at the anticodon as well as ψ55 in the TψC loop. It also shields the D20 proton; and it stablizes either AU1 at the acceptor end, or Aψ39 near the anticodon. Perturbation of D20 is particularly interesting since the pattern of dihydrouridine bases could be important for tRNA recognition.

CONCLUSION

This article has focused on research by us, and related research by others, on tRNA using proton NMR. The future is likely to see greater emphasis on studies of small RNA fragment models because of the greater ease of obtaining these and the pre-existing technology for studying them, already implemented for small DNA duplexes and small proteins. It is also to be expected that isotope methods emphasized in the last part of this article will continue to be applied to larger systems such as RNA-protein complexes. Recently double labeling as well as isotope-directed NOE have been applied to ~20 KD proteins to identify resonances, perform solvent exchange studies, and determine geometry (R. Griffey, R. Dahlquist, and M. Weiss, unpublished). These methods should also be applicable to RNA-protein complexes.

We thank Mary Papastavros and Sara Kunz for technical assistance. This work was supported by U.S. P.H.S. grant GM20168 to A. Redfield and N.I.H. grants to D. Soll. A. Redfield is also with the Physics Department and the Rosenstiel Center for Basic Medical Science Research at Brandeis University. This is paper No. 1562 of the Brandeis University Biochemistry Department.

REFERENCES

1. D.R. Kearns, D.J. Patel, and R.G. Shulman, Nature 229:338 (1971).

2. B.R. Reid, Ann. Rev. Biochemistry 50:96 (1981).

3. P.D. Johnston and A.G. Redfield, Nucleic Acids Res. 4:3599 (1977) and 5:3913 (1978).

4. P.D. Johnston and A.G. Redfield, Biochemistry 20:3996 (1981).

5. P.D. Johnston, N. Figueroa, and A.G. Redfield, Proc. Nat. Acad. Sci. USA 76:3130 (1979).

6.   V. Sanchez, A.G. Redfield, P.D. Johnston, and J.S. Tropp, Proc. Nat. Acad. Sci. USA 77:5659 (1980).

7.   S. Roy and A.G. Redfield, Nucleic Acids Res. 9:7073 (1981).

8.   D.R. Hare and B.R. Reid, Biochemistry 21:1835 (1982).

9.   J. Tropp and A.G. Redfield, Biochemistry 20:2133 (1981).

10. E. Schejter, S. Roy, V. Sanchez, and A.G. Redfield, Nucleic Acids Res. 10:8297 (1982).

11. B. Choi and A.G. Redfield, Nucleic Acids Res. (in press) (1985).

12. A. Heerschap, C.A.G. Haasnoot, and C.W. Hilbers, Nucleic Acids Res. 10:6981 (1982); 11:4483 and 11:4501 (1983).

13. S. Roy and A.G. Redfield, Biochemistry 22:1386 (1983).

14. R.H. Griffey, C.D. Poulter, A. Bax, B.L. Hawkins, Z. Yamaizumi, and S. Nishimura, Proc. Nat. Acad. Sci. 80:5895 (1983).

15. S. Roy, M.Z. Papastavros, V. Sanchez, and A.G. Redfield, Biochemistry 23:4395 (1984).

16. C.W. Hilbers, A. Heerschap, C.A.G. Haasnoot, and J.A.L.I. Walters, J. Biomol. Struct. Dyn. 1:183 (1983).

17. G.M. Clore, A.M. Gronenborn, E.A. Piper, L.W. McLaughlin, E. Graeser, and J. VanBoom, Biochem J. 221:737 (1984).

18. G.M. Clore, A.Gronenborn, and L.W. McLaughlin J. Mol. Biol. 174:163 (1984).

19. K. Hall, P. Cruz, I. Tinnoco, Jr., T.M. Jovin, and J.H. van de Sande, Nature 311:5986 (1984).

20. S. Roy, M.Z. Papastavros, and A.G. Redfield, Nucleic Acids Res. 10:8341 (1982).

21. D.R. Hare and B.R. Reid, Biochemistry 21:5129 (1982).

22. J.S. Tropp, Nucleic Acids Res. 11:2121 (1983).

23. M. Gueron and J.L. Leroy, Biophys. J. 38:231 (1982).

24. N. Figueroa, G. Keith, J.L. Leroy, P. Plateau, S. Roy, and M. Gueron, Proc. Nat. Acad. Sci. USA 80:4330 (1983).

25. J.L. Leroy, N. Bolo, N. Figueroa, P. Plateau, and M. Gueron, J. Biomolec. Struct. Dyn. 2:915 (1985).

26. B. Choi, Thesis, Brandeis University (1985).

27. C. Smith, P.G. Schmidt, J. Petsch, and P.F. Agris, Biochemistry 24:1434 (1985).

28. A. Bax, R. Griffey, and B.L. Hawkins, J. Magn. Reson. 55:301 (1983).

29. O. Jardetzky and G.K.C. Roberts, "NMR in Molecular Biology", Academic Press (New York, 1981).

30. M. Emshwiller, E.L. Hahn, and D. Kaplan, Phys. Rev. 188:414 (1960).

31. R.H. Griffey, MA. Jarema, S. Kunz, P.R. Rosevear, and A.G. Redfield, J. Amer. Chem. Soc. 107:711 (1985).

32. D. Davis, R.H. Griffey, C.D. Poulter, J. Biol. Chem. (in press) (1985).

33. D.G. Barker, J.-P. Ebel, R. Jakes, and C.J. Briton, Eur. J. Biochem. 127:449 (1982).

34. P. Hoben, N. Royal, A. Cheung, F. Yamao, K. Biemann, and D. Soll, J. Biol. Chem. 257:11644 (1982).

35. C. Zelwer, J.L. Risler, and S. Brunie, J. Mol. Biol. 155:63 (1982).

GU BASE PAIRS AND VARIABLE LOOP

IN YEAST tRNA[Asp]

D. Moras, P. Dumas and E. Westhof

Laboratoire de Cristallographie Biologique, IBMC du CNRS
15, rue R. Descartes, 67084 STRASBOURG cedex, France

INTRODUCTION

Yeast tRNA(Asp) is an elongator tRNA with 4 bases in its variable loop. Few other features characterize its primary structure shown in Figure 1 together with that of yeast tRNA(Phe) (1,2) ; an exceptionally high GC content, 3 GU and one Gψ base pairs in three of its four stems and an almost self- complementary anticodon (GUC).

The crystal structure of this tRNA has been elucidated (3) and is now refined to 3Å (4), a resolution close to that available for yeast tRNA(Phe), the first tRNA known at the molecular level (5,6,7). Effects of sequence variations, like the variable loop size or the composition, can therefore be analysed with enough accuracy. Moreover, the important amount of solution studies done on these two tRNAs and many others, enable good structure function correlation. In this paper, we will limit ourselves to the study of the structural environments of GU base pairs and variable loop and correlate observations to solution investigations by NMR (8) and chemical labelling (9,10).

COMPARISON WITH YEAST tRNA[Phe]

A superposition of yeast tRNA(Phe) and yeast tRNA(Asp) in their inertial systems is shown in Figure 2. The two molecules superpose well in the core and diverge most at their extremities, i.e. the acceptor and anticodon ends. As a first approximation, the influence of the different variable loop length in the opening of the arms can apparently be neglected. The larger angle between the arms of the L in tRNA(Asp) is then described as coming from an opening below the base pair C29-G41 of the AC-stem in tRNA(Asp) and to a kink between the T-stem and the AA-stem in tRNA(Phe). This can be put in evidence by monitoring the distances between a chosen phosphorus atom (for the AA-arm, phosphorus atom of residue 6 and for the anticodon AC-arm, phosphorus atom of residue 25) and the phosphorus atoms on the other "strand" of the helical arm. The distance between phosphorus atoms 25 and 39 is larger in tRNA(Asp) than in tRNA(Phe) (around 10. Å instead of 7.5 Å). The distances between phosphorus

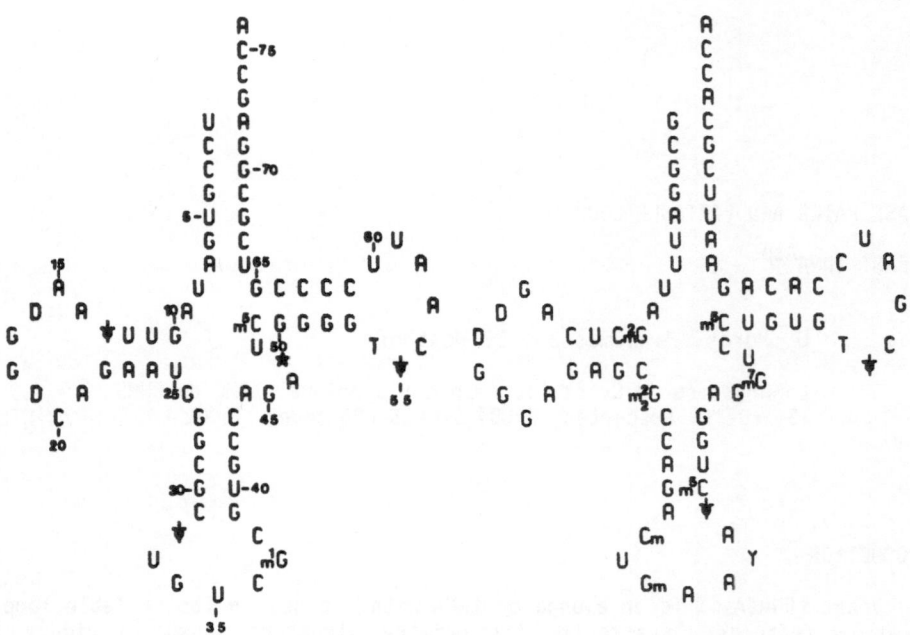

Fig. 1 : Cloverleaf structures of yeast tRNA(Asp) (left) and of yeast tRNA(Phe) (right).

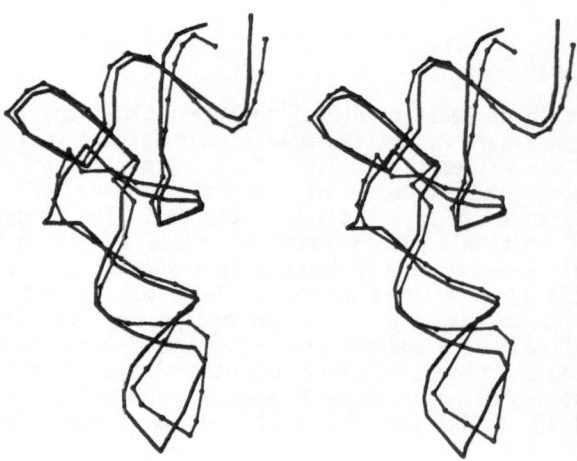

Fig. 2 : Stereoview of the superposition of yeast tRNA(Asp) and tRNA(Phe) in their inertial systems.

TABLE I : Helical parameters (a) of the stem chains in yeast tRNA-asp (first line) and yeast tRNA-phe (second line, Sussman, et al. 1978).

Chain	rms (Å)	R (Å)	Angle/res	Rise/res
2 - 7	0.617	9.09	33.7	2.57
	1.000	9.27	34.3	2.20
66 - 72	0.439	8.97	33.2	2.64
	0.921	9.38	33.0	2.81
49 - 53	1.070	8.84	36.8	1.97
	1.297	8.63	34.1	2.34
61 - 65	0.514	8.59	34.7	2.20
	1.362	9.78	32.8	2.36
22 - 25	2.251	6.06	38.1	2.90
	2.832	5.95	31.0	2.99
26 - 31	2.085	8.96	36.2	3.09
	2.380	8.83	35.1	3.00
39 - 44	1.461	10.1	31.2	1.71
	1.285	9.71	32.1	1.14
10 - 13	0.584	9.72	31.7	1.96
	1.372	8.82	35.1	1.69

(a) Those parameters were calculated in the following way. First, each arm was separately set in the inertial system of a RNA-11 helix. Then the helical parameters were obtained for each chain fragment by non-linear least-squares. A better fit would be obtained if each helical chain were fitted separately to a RNA-11 helix. The first step gives the direction cosines of each arm of the L-structure. For the AA- and T- arms, they are (-.8707, .4569, -.1819) and, for the D- and AC- arms, they are (-.5781, .0560, -.8140). Thus, in this approach, the angle between the arms is 112°.

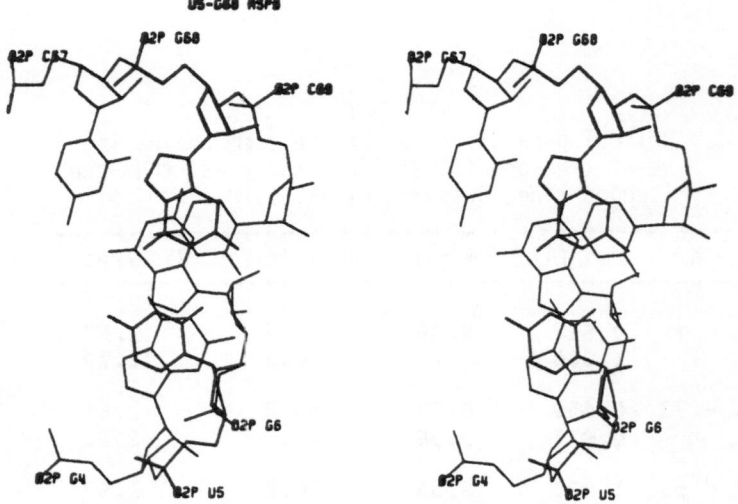

Fig. 3 : Stereoview of the stacking of U5-G68 in yeast tRNA(Asp).

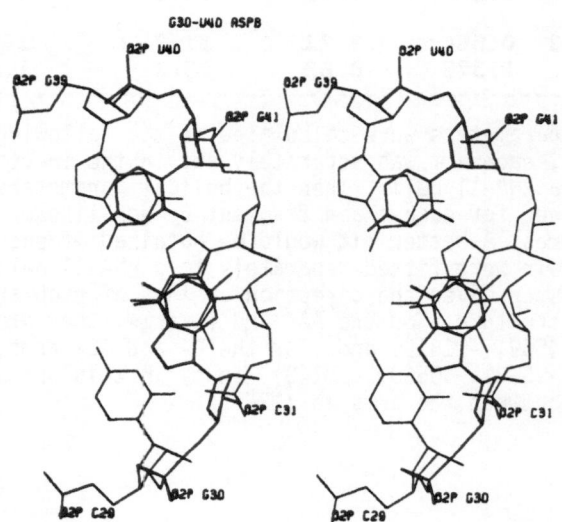

Fig. 4 : Stereoview of the stacking of G30-U40 in yeast tRNA(Asp).

atom 6 and the phosphorus atoms of residues 61 to 66 are much larger in tRNA(Phe) than in tRNA(Asp), which follows the behaviour expected for a RNA-11 helix between residues 61 and 67.

The helical parameters of the stems in yeast tRNA(Asp) shown in Table I, exhibit differences with the parameters of the RNA-11 helix (11), which are 32.7° and 2.81 Å for the angle and rise per residue, respectively. Significant differences were also noted in the structure of yeast tRNA(Phe) (12,13). In both cases the rise per residue is significantly less than in RNA-11, but when using the same procedure for the two tRNAs, the values of the rise per residue in tRNA(Asp) have a tendency to be smaller than those of tRNA(Phe). If this difference is real, it could be due to the higher ionic strength of the crystallization conditions (14). The anticodon arm deviates the most from RNA-11, especially from ψ13 to G10 and A44 to G39, from which no meaningful helical parameters could be extracted. The tertiary base pair G26-A44 at the interface between the dihydrouridine and anticodon has a pronounced propeller-twist with the angle between the normals to the base planes around 30°. This conformation is very similar in both elongator tRNAs. The angle between the helical axes of the D and anticodon stems is around 25°. A similar break at the G26-A44 pair was observed in yeast tRNA(Phe), leading to similar values for the angles between helical axes. On the other hand, the angle between the acceptor and thymine stems, which form the other arm of the L, is around 8°.

GEOMETRY AND STEREOCHEMISTRY OF GU BASE PAIRS

Among the torsion angles around the bonds of the sugar-phosphate backbone, those with the largest sigmas are the ones around P-O(5') and around C(4')-C(5'). The correlation between those two torsion angles was predicted theoretically (15), and is now described in terms of "crankshaft" bonds (16). Residues which deviate most from this strong negative correlation do not belong to helical stems, except for G10, which forms a G-U pair and is the first residue of the D-stem. ψ13, G30, G51 and G53 follow the correlation but away from the preferred domain (gauche-minus, gauche-plus) ; of those, two are G-U pairs . In the crystal structure of yeast tRNA(Phe) the conformation of the sugar-phosphate backbone in the only G-U pair of the structure, which is inside the AA-stem (G4-U69) varies from one form to another. One refinement (6) concluded that the sugar-phosphate backbone deviates only minimally from the helical conformation, while another refinement led to a trans conformation for the torsions about P-O(5') and C(4')-C(5') of G4 (5). In the structure of yeast tRNA(Asp), the U-G pair of the AA-stem displays a standard helical conformation, while the G-U pair of the AC-stem as well as the ψ-G pair of the D-stem have a double trans conformation for G30 and ψ13, respectively. In the double trans conformation about P-O(5') and C(4')-C(5'), the phosphate-phosphate distance is barely altered, but the anionic oxygens of the phosphate group point more toward the exterior than in the double gauche conformation. Since G-U pairs seem to promote this conformation, G-U pairs might play a structural role in tRNA-protein interactions. In other words it would be a way to induce a sequence dependent "non-specific" interaction. G10-U25 presents a trans, gauche-minus for the torsions about P-O(5') and C(4')-C(5') of G10. The fact that this phosphate enters the stem after a sharp turn might be responsible for this unusual conformation.

The pairing in the wobble G-U base pair involves O6(G)...N3(U) and N1(G)...O2(U). In the unusual pair G22-ψ13, the pairing involves O6(G)...N3(P) and N1(G)...O4(P). The stacking of the base pair G4-C69, U5-G68 and G6-C69 is shown on Fig. 3. The difference in stacking patterns between the consecutive base pairs is striking. The exact reverse pattern can be seen in Fig.4 which shows the stacking of C29-G41, G30-U40 and C31-G39. The former situation is that of a 5'-end U-G pair and the latter situation of a 5'-end G-U pair (17). It is of interest to note that the AC-stem of tRNA(Asp)

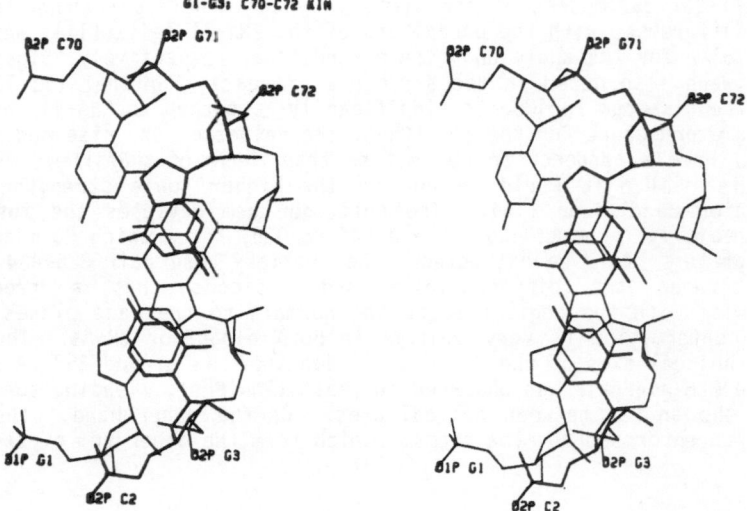

Fig. 5 : Stereoview of the stacking in the first three base pairs of the AA-stem of yeast tRNA(Phe).

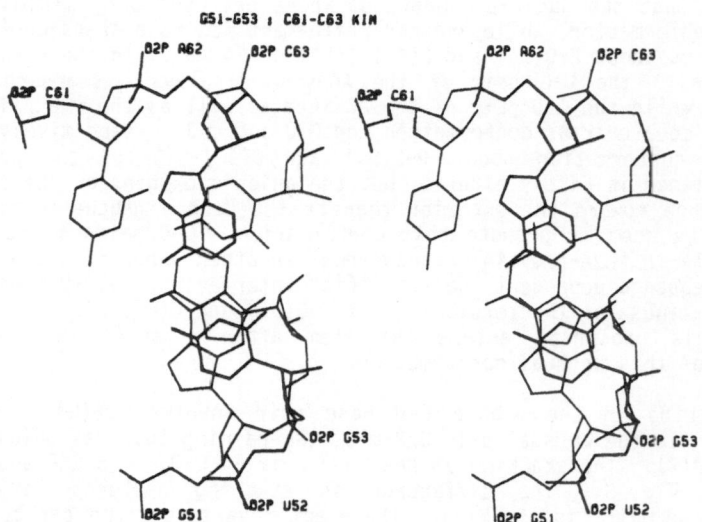

Fig. 6 : Stereoview of the stacking in the last three base pairs of the T-stem of yeast tRNA(Phe).

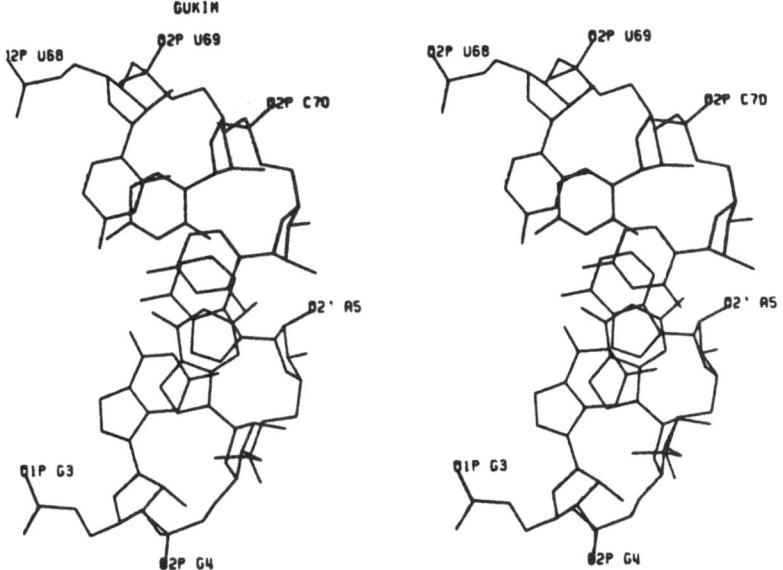

Fig. 7A : Stereoview of the stacking of G4-U69 in yeast tRNA(Phe).

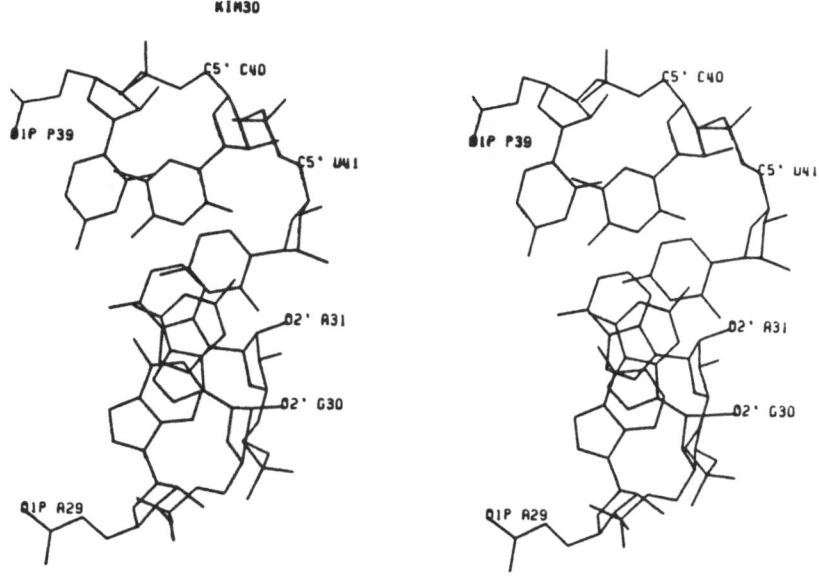

Fig. 7B : Stereoview of the stacking of G30-C40 in yeast tRNA(Phe).

deviates from that of tRNA(Phe) at G30-U40, after the weak stacking between C29-G41 and G30-U40. It is also interesting to note that base pair 29-41 is always of Watson-Crick type (18). Similarly, but in a reverse fashion, the departure of the AA-stem from a RNA-11 helix occurs after the weak stacking between U5-G68 and G6-C69.

In order to appreciate the influence of G-U base pairs on the stacking, it is of interest to compare the structural environments of U5-G68 and G30-U40 with closely related situations. Fig. 5 and 6 show sequences with alternating purine pyrimidine as observed in the structure of tRNA(Phe) (5). Fig. 5 represents G1-C2-G3 paired to C72-G71-C70, a situation quite similar to that of G4-U5-G6 and C67-G68-C69 in tRNA(Asp). The same succession of intra and inter-strand stacking is observed imposed by the alternative purine pyrimidine sequence. Fig.6, again, shows a similar situation with U-G being replaced by U-A between two G-C base pairs. By comparing three different base pairs (U-G, C-G, U-A) sandwiched between two G-C it appears that the sequence variation induces only small rearrangements of the stacking and conserves the basic features.

Fig. 7A shows the environment of the G-U base pair in tRNA(Phe) (5). The 5'-end G is surrounded by two purines. Fig. 7B represents the structure of a comparable situation (A-G-A, U-C-P). In these two cases the influence of the G-U mismatch on the stacking is also relatively minor. However, both stackings are quite different from the ones analysed above : they display very little interstrand stacking. In other words, the difference between alternating purine-pyrimidine sequences and all purines, all pyrimidines, is far more important than G-U mismatches on the stacking properties. This is again confirmed when looking at the two remaining base pairs of tRNA(Asp), those of the D-stem, G10-U25 and ψ13-G22. Although these bases are involved in tertiary interactions, the influence of these extra-bonds on the stacking of the base pairs is hardly detectable. Fig. 8 and 9 show respectively G10-U25 stacked on U11-A24 and U12-A23 on ψ13-G22.

INFLUENCE OF THE VARIABLE LOOP ON THE ENVIRONMENT OF A9 AND A21

A set of tertiary interactions analogous to that observed in yeast tRNA(Phe) is present in yeast tRNA(Asp) with two main differences : the position of A21 with respect to the reversed Hoogsteen pair U8-A14 and the absence of interactions between G18 and C56 (at least for the B-form of tRNA(Asp)). The interactions toward the broad groove of the D-stem are the following : G45...[G10...U25], A46...[G22...ψ13], and A9...[A23...U12]. The Levitt pair occurs between A15 and U48 in a trans orientation of the glycosyl bonds. The G26...A44 pair is present, as well as the second reversed Hoogsteen between T54 and A58. The only interaction between the D- and T-loops involves G17, which intercalates between A57 and A58 and makes a hydrogen bond with P55.

Fig. 10 shows the environment around A21...[U8...A14] in yeast tRNA(Asp) and in yeast tRNA(Phe) (5) seen perpendicularly to the plane of A14. While, in tRNA(Phe), A21 interacts only with the sugar of U8, the base and the sugar of A21 interact also with A14 in tRNA(Asp). A comparison of the two views shown in Fig. 10 indicates a rotation of the Levitt pair 15-48 with respect to U8-A14 in going from tRNA(Asp) to tRNA(Phe). In order to maintain stacking interactions, the base A21 followed that rotation, thereby moving away from A14. The very slow exchange of the amino protons of A14 in tRNA(Asp) has been explained on the basis of the closer position of A21 to A14 (8). Also, atom N7 of A21 is accessible to diethylpyrocarbonate in tRNA(Asp), but not in tRNA(Phe) (9). In tRNA(Phe) the rotation of A21 has moved its N7 on top of G46, which, being itself methylated on N7, contributes to block access to the N7 of A21.

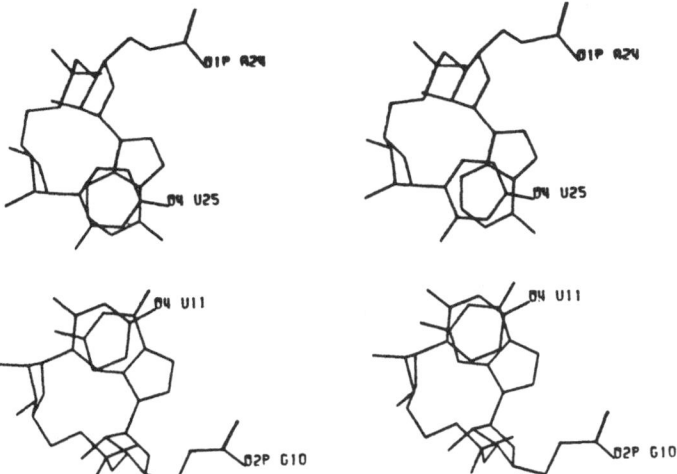

Fig. 8 : Stereoview of the stacking of G10-U25 on the A24-U11 in yeast tRNA(Asp).

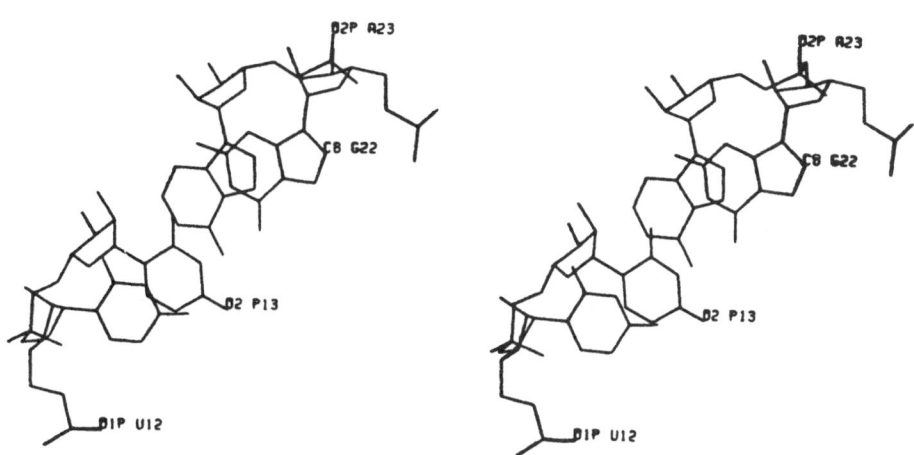

Fig. 9 : Stereoview of the stacking of U12-A23 on the ψ13-G22 in yeast tRNA(Asp).

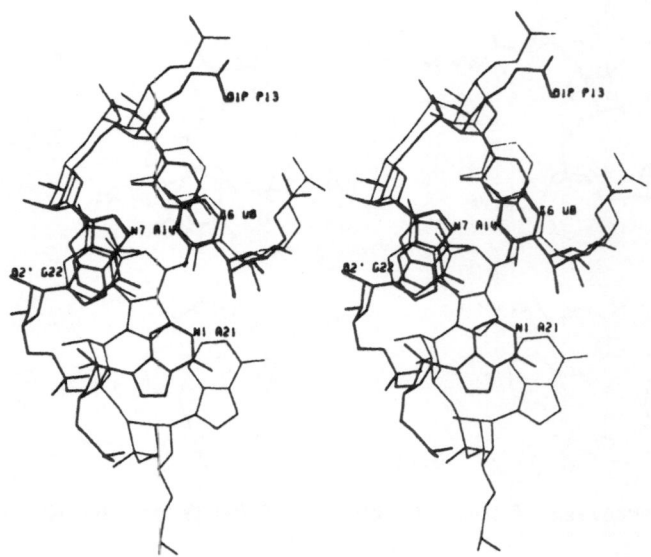

Fig. 10 : Stereoview of the environment of the U8-A14-A21 in yeast tRNA(Asp)
(heavy line) and in yeast tRNA(Phe) (light line). Notice the relative rotation
of A21 and G22.

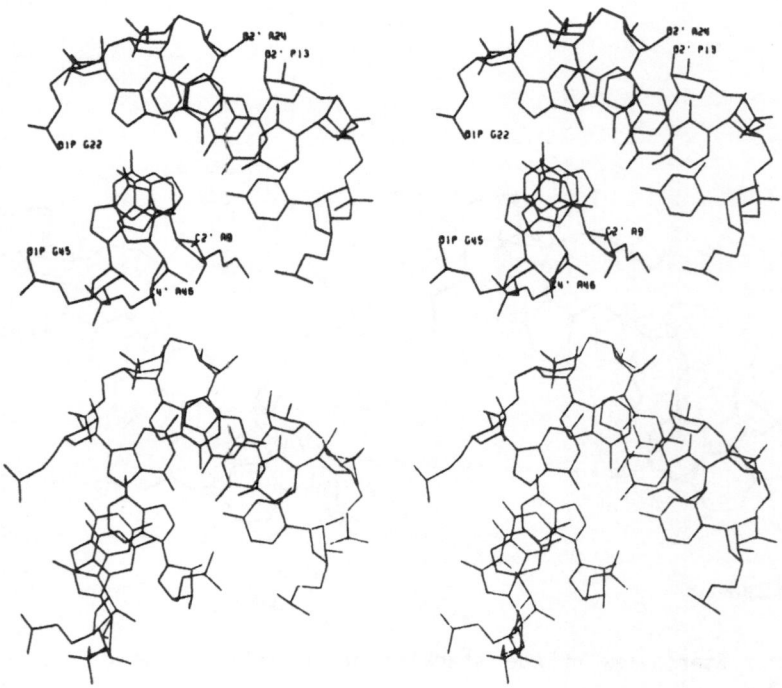

Fig. 11 : Stereoviews of the environment of A9 in yeast tRNA(Asp) (top) and in
yeast tRNA(Phe) (bottom). View down the helical axis of the D-stem. Notice the
shift of residues G45 and A46 with respect to tRNA(Phe). This shift induces a
better stacking of A9 in yeast tRNA(Asp).

The interactions made by A9 in tRNA(Asp) and tRNA(Phe) are compared in Fig. 11. In tRNA(Phe), the ribose O(2') of A9 accepts a hydrogen bond from the amino N4 of C11 and is the donor of a hydrogen bond to N7 of G10, while a phosphate anionic oxygen of A9 is the acceptor of a hydrogen bond from the amino N4 of C13. In tRNA(Asp), residues 11 and 13 are a uracil and a pseudouridine, respectively. This sequence difference apparently leads to a reorientation of the ribose of A9 (the torsion angle about C(4')-C(5') is trans, instead of gauche (+) so that O(2') still gives a hydrogen bond to N7 of G10 but receives one from N2 of G45, while the phosphate group has slightly rotated away from ψ13). The missing interactions are replaced by two C-H...O contacts : O4(U11) to C8(A9) and O2(P13) to C2(G46). Fig. 11 shows distinctly the slight movement of A9, the large slippage of A46 toward the inside of the molecule, and the slight follow-up movement of G45.

The different environments of A9 and A21 can both be correlated to the absence of residue 47 : the extra residue in the variable loop of tRNA(Phe). The movement of A46 and the rotation of U48, the two bases nearest to the missing residue, are most likely a consequence of that removal. The correlation with solution studies confirms such an interpretation. It is interesting that the N7 of G45 is not methylated by DMS in tRNA(Asp) while it is in tRNA(Phe) (10).

ACKNOWLEDGEMENTS

We thank P. Romby and R. Giegé for fruitful discussions. This work was supported by Grants from the Centre National de la Recherche Scientifique, and the Ministère de la Recherche et de la Technologie.

REFERENCES

1. Gangloff, J., Keith, G., Ebel, J.P. and Dirheimer, G. (1971). Nature New Biol. 230, 125-127.
2. Rajbhandary, U. L. and Chang, S.H. (1968), J. Biol. Chem., 243, 598-608.
3. Moras, D. , Comarmond, M. B., Fischer, J;, Weiss, R., Thierry, J.C., Ebel, J.P. and Giegé, R. (1980). Nature (London), 288, 669-674.
4. Westhof, E., Dumas, P. and Moras, D. (1985) J. Mol. Biol., 184, 119-145.
5. Sussman, J. L., Holbrook, S.R., Wade Warrant, R., Church, G.M. and Kim, S.H. (1978). J. Mol. Biol. 123, 607-630.
6. Hingerty, B., Brown, R.S. and Jack, A. (1978). J. Mol. Biol. 124, 523-534.
7. Stout, C.D., Mizuno, H., Rao, S.T., Swaminathan, P., Rubin, J., Brennan, T. and Sundaralingam, M. (1978). Acta Crystallogr. sect. B, 34, 1529-1544.
8. Figueroa, N., Keith, G., Leroy, J.L., Plateau, P., Roy, F. and Gueron, M. (1983). Proc. Nat. Acad. Sci., U.S.A. 80, 4330-4333.
9. Giegé, R., Romby, P., Florentz, C., Ebel, J.P., Dumas, P., Westhof, E. and Moras, D. (1983). In Nucleic Acids : The Vectors of Life (Pullman, B. and Jortner, J., eds), pp. 415-426, D. Reidel Publishing Company, Dordrecht.
10. Romby, P., Moras D., Bergdoll M., Dumas P., Vlassov V.V., Westhof, E., Ebel, J.P. and Giegé, R. (1985) J. Mol. Biol. 184, 455-471.
11. Arnott, S., Hukins, D. W., Dover, S.D., Fuller, W. and Hodgson, A.R. (1973). J. Mol. Biol. 81, 107-122.
12. Jack, A., Ladner, J. E. and Klug, A. (1976). J. Mol. Biol. 108, 619-649.
13. Holbrook, S.R., Sussman, J.L., Wade-Warrant, R. and Kim, S.H. (1978). J. Mol. Biol. 123, 631-660.
14. Giegé, R., Moras, D. and Thierry, J.C. (1977). J. Mol. Biol. 115, 91-95.
15. Yathindra, N. and Sundaralingam, M. (1976). Nucl. Acids Res. 3, 729-737.

16. Olson, W.K. (1981). In Biomolecular Dynamics (Sarma, R.H., ed.), pp. 327-336.
17. Mizuno, H. and Sundaralingam, M. (1978). Nucl. Acids Res. 5, 4451-4461.
18. Grosjean, H.J., Cedergreen, R.J. and McKay, W. (1982). Biochimie. 64, 387-397.

CORRELATION BETWEEN CRYSTAL AND SOLUTION STRUCTURES IN tRNA.

YEAST tRNA^Phe AND tRNA^Asp THE MODELS FOR FREE AND MESSENGER RNA BOUND tRNAs

Richard Giegé,   Anne-Catherine Dock,   Philippe Dumas,
Jean-Pierre Ebel, Pascale Romby, Eric Westhof and Dino Moras

Institut de Biologie Moléculaire et Cellulaire du CNRS
15, rue René Descartes, 67084 Strasbourg Cedex, France

MODEL MOLECULES AND OBJECTIVES

The three-dimensional structures of two elongator transfer RNAs are known in great details : first that of yeast tRNA(Phe) (1-4) and more recently that of yeast tRNA(Asp) (5-8). As seen in Figure 1, both molecules are folded in an L-shaped conformation, which is also found for initiator tRNAs (9,10). Since the conserved or semi-conserved residues (11) are involved in the tertiary interactions which stabilize this folding (1-10), the L-shaped structure represents the general structural organization of all tRNA molecules. Such structural similarity is satisfactory to explain common functions of tRNAs, namely the interaction with ribosomes, and in the case of elongator tRNAs with elongator factors, but not sufficient to account for specific functions. In that case specific structural features must be involved. Differences in the anticodon sequences account for the decoding of the genetic message on mRNA. Recognition by aminoacyl-tRNA synthetases, which leads to the specific aminoacylation of tRNAs is more complex. Because many aminoacyl-tRNA synthetases recognize isoacceptor tRNAs and catalyze tRNA mischarging (12), these features cannot be simple linear nucleotide sequences, but more likely structural domains found at the three-dimensional level.

Considering these facts it becomes important to compare thoroughly three-dimensional structures of tRNAs in order to find local conformational differences and similarities. This can be done by comparing crystal structures. But, since it is strongly suggested on the basis of biochemical and biophysical arguments that tRNAs undergo subtle conformational changes during their functioning (13,14), crystal studies must be correlated with solution data.

In the present report tRNA(Asp) and tRNA(Phe) will serve as model molecules to establish structural rules for explaining conformations of tRNAs, or more generally of RNAs for which no crystallographic data are available. Such studies are prerequisites to a structural understanding of the functional versatility of tRNAs. Following these lines, several examples will be presented indicating (i) at the structural side, that tRNAs possess a consensus T-loop structure and that base stacking in double-stranded regions is modulated by the nature of neighboring residues and by the regularity of the RNA helix ; (ii) at the functional side, that following codon-anticodon association, tRNA molecules undergo a long range conformational change in the D and T-loop region.

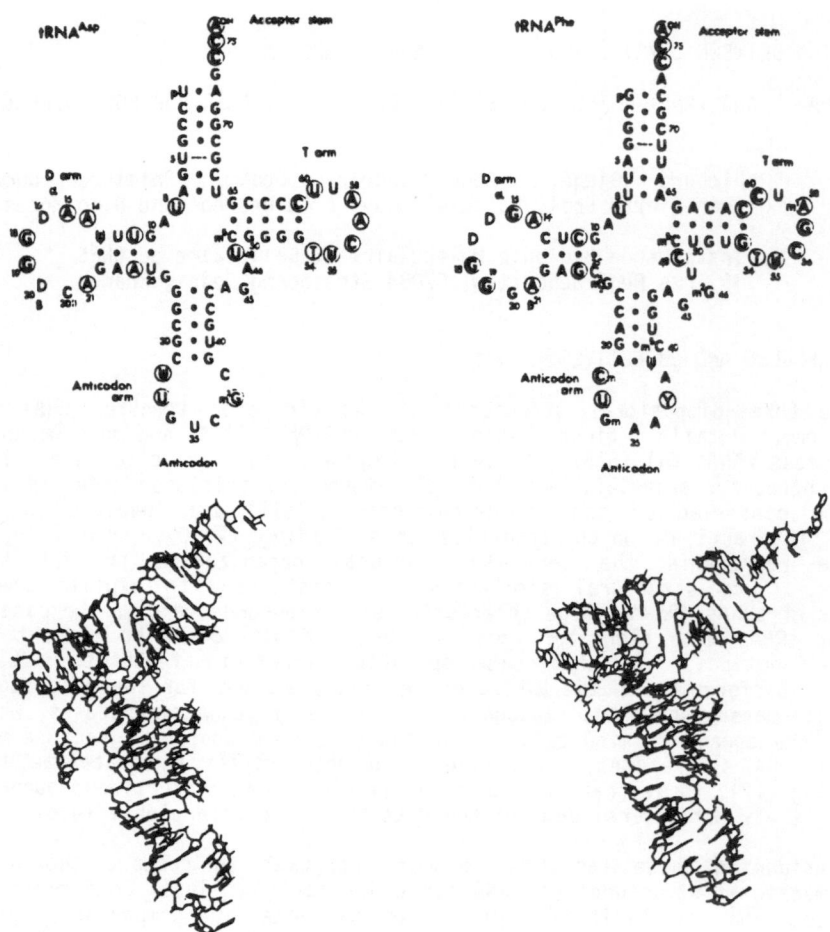

Fig. 1 :   Cloverleaf folding (11) and three-dimensional structures of yeast tRNA(Asp) (left) and tRNA(Phe) (right) (8, 22). In the cloverleaves, the numbering system of nucleotides is that of tRNA(Phe) (11); the conserved or semi-conserved residues in elongator tRNAs are circled.

EXPERIMENTAL APPROACHES

The structures of the tRNAs in solution were mapped with different enzymatic or chemical probes recognizing specific structural domains in the molecules. Here we will focus on chemical reagents because (i) these small

126

probes allow to map discrete conformational features in tRNAs and (ii) interpretation of results is less affected by steric hindrance as would be with bulky nucleases. The principle of the chemical approach derives from the chemical sequencing methodologies of nucleic acids (15,16) and relies on statistical and low yield modifications of the potential targets. Reagents are chosen so that modification leads to chain splitting at the modified positions. Using end-labeled molecules it becomes then possible to analyze the positions of modifications by electrophoresis and autoradiography. Three chemical reagents were used : ethylnitrosourea (ENU), dimethylsulfate (DMS) and diethylpyrocarbonate (DEPC). ENU probes the accessibility of phosphates (17,18), while DMS and DEPC probe the N7 positions in purine bases or N3 positions in cytosines (19).

Structural data on anticodon associated tRNA(Asp) molecules (GUC anticodon) in solution were obtained in the presence of Escherichia coli tRNA(Val) (GAC anticodon) and of high concentrations of ammonium sulfate (1.6 M). These conditions, which favor anticodon-anticodon association, were chosen after a systematic study by T-jump relaxation methods (20). Self-splitting patterns of soluble and crystalline tRNA(Asp) were detected after electrophoresis of 3'-end labeled tRNA. End-labeling was done with radioactive pCp and T4 RNA ligase. In the case of crystalline tRNA, the labeling was conducted on washed crystals dissolved in the labeling medium (21).

The structural comparisons of tRNA(Asp) and tRNA(Phe) as well as the interpretation of the chemical modification experiments were done with a graphic interacting color display system (PS300 from Evans and Sutherland) using COMPAR (18), a locally written program, and the graphic modelling program FRODO. Coordinates of tRNA(Phe) are from Quigley et al. (22) and those of tRNA(Asp) from a model of form B currently refined at R = 23 % (8).

SIMILARITIES BETWEEN SOLUTION AND CRYSTAL STRUCTURES

Mapping of tRNA Solution Structures with Chemical Probes

Many experimental data, including chemical accessibility studies as well as spectroscopic measurements, strongly suggest a similarity between the crystal and solution structures of tRNA(Phe) (17, 19, 23 –25). Likewise many solution data obtained on tRNA(Asp) can be explained by its crystal structure (18, 26). Here we report a comparative screening of both tRNAs using ENU, DMS and DEPC as structural probes and show how these probes are sensitive to hydrogen bonding and to base stacking effects. The chemical results are summarized in Figure 2.

Structural Implications : Similarity and Variability in tRNA Conformations

As expected from their functional versatility, tRNA(Asp) and tRNA(Phe) exhibit either similar or differential chemical reactivities towards ENU, DMS and DEPC at common positions in their sequences (see Figure 2). Some of these results were already explained and discussed on the basis of the crystal structures of the two tRNAs (18,27). As a general conclusion it appears that the conformations found in the crystal account for most of the chemical reactivities. So for example the differential behavior of phosphate 22, protected against ENU alkylation in tRNA(Asp) because of the existence of a H-bond with N6 of A46, and accessible in tRNA(Phe) because no H-bond can be formed with m7G46 (18). As to residues A21 and G45, their differential reactivities towards DEPC and DMS are correlated with the different length of the extra-loops in both tRNAs (27). In the present report we will concentrate on two structural aspects : the conformation of the T-loop as probed by ENU and the regularity of helical structures as studied by DMS mapping.

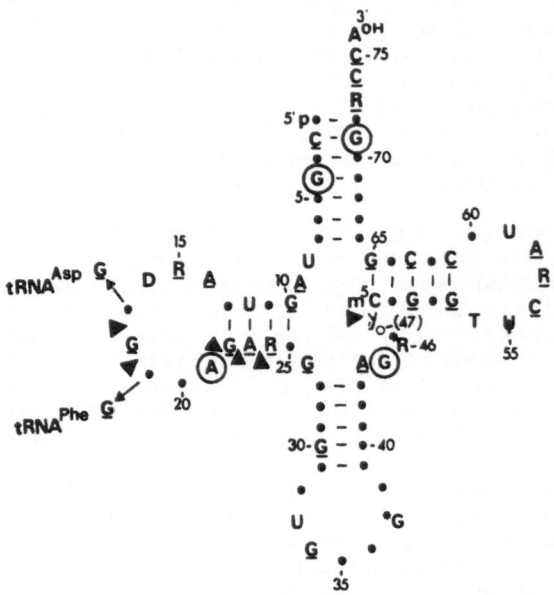

Fig. 2 : Reactivity of bases with DMS and DEPC, and of phosphates with ENU in yeast tRNA(Asp) and tRNA(Phe). The figure represents a composite secondary structure with nucleotides common to both tRNAs; R and Y are for common purines and common pyrimidines, respectively. In tRNA(Asp), residue 47 is missing and the two constant G's in the D-loop are shifted by one position as compared to their location in tRNA(Phe). The bases which react differentially in the two tRNAs are encircled, those which exhibit similar reactivities at common functional groups are underlined; (residues of the U type are not probed by DMS and DEPC); phosphates of differential reactivities are indicated by triangles.

T-loop structure : In tRNAs the T and anticodon-loops are seven nucleotides long and both closed by stems containing 5 base pairs. Their conformations, however, are different (6, 7). These differences are also reflected by ENU mapping (18). While all phosphates are accessible in the anticodon loops of tRNA(Asp) and tRNA(Phe), some of the T-loop residues, particularly phosphate 60, are strongly protected. Interestingly P60 is also protected in other tRNAs, including initiator tRNAs (17,28) as well as the tRNA-like structure of TYMV (29). Since phosphates 58 to 60 are buried inside the tertiary structure of tRNA(Asp) and tRNA(Phe) it could be argued that their protection is the consequence of steric hindrance. Such effect could account for the protection of phosphates 58 and 59, but is not primarily responsible for that of phosphate 60. In that case the two free oxygen atoms of the phosphate group are H-bonded with the N4 atom of C61 and the 2'-hydroxyl group of ribose 58. It is striking to note that only cytosine at position 61 allows hydrogen bonding with P60, and that this C61 residue, beside the CCA 3'-end is the most conserved residue in tRNAs (11). Thus, it is strongly suggested that the similar conformation around phosphate 60, and more generally the T-loop structure found in tRNA(Asp) and tRNA(Phe), represents a common structural feature in tRNAs. This concept is summarized in Figure 3 which represents a consensus T-loop sequence of tRNAs . Note the presence of a reverse Hoogsteen pair between residues 54 and 58 which further contributes to the stabilization of the loop conformation. Such pairing is absent in anticodon loops. Also position 39 in anticodon stems, homologous to C61 in T-stem can be occupied by any of the four nucleotides, and phosphate 38 is not H-bonded like phosphate 60. Considering the structural features which govern the intrinsic T-loop geometry it was interesting to search for similar sequence characteristics in RNAs for which no X-ray data are available. As mentioned before, the tRNA-like structure of TYMV RNA possesses such a loop and exhibits protection against ENU alkylation at the phosphate homologous to P60. Similar sequence features are found in TMV RNA, another plant viral RNA (29,30) and in two loops of 16S ribosomal RNA from Escherichia coli (31). Although no ENU data are yet available for these RNAs it is expected that protection would occur at "position 60".

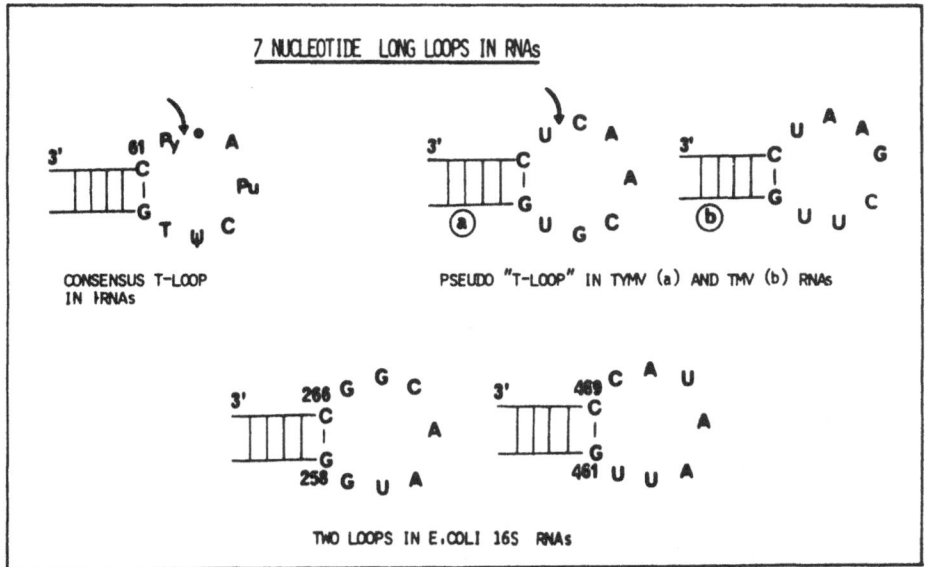

Fig. 3 :   Seven nucleotide long loops in RNAs of the "T-loop" type found in tRNAs. Note the presence of a C residue at the 3'-side of the first base pair. The arrows indicate protection against ENU alkylation.

Base Stacking in Helical Regions : The chemical probe DMS, specific for N7 positions in guanine is sensitive to base stacking (9). As seen in Figure 2, two guanines (G4 and G71) located at similar positions in the amino acid acceptor stem of tRNA(Phe) and tRNA(Asp) exhibit differential reactivities against DMS. The reason for that is linked to the presence of different neighboring bases around G4 in the two tRNAs : in tRNA(Asp), it is surrounded by two pyrimidines (C3 and U5) and in tRNA(Phe) by two purines (G3 and A5). The inverse situation occurs for G71 surrounded by two purines in tRNA(Asp) and by two pyrimidines in tRNA(Phe) (see Figure 1). The structural consequences of such base distributions become apparent in Figure 4 which represents the structural environment of G4 in both tRNAs. In tRNA(Asp) the presence of the two neighboring pyrimidines exposes the N7 position of the guanine, and the base is alkylated by DMS ; in tRNA(Phe), where the pyrimidines are replaced by purines, the guanine is clearly stacked and its N7 position is not alkylated. Similarly it is possible to explain the reactivity of G71 in tRNA(Phe) and its protection in tRNA(Asp). It must however be noted that these structural explanations suppose that the RNA is in a regular helical conformation. Following these lines it should be possible to predict

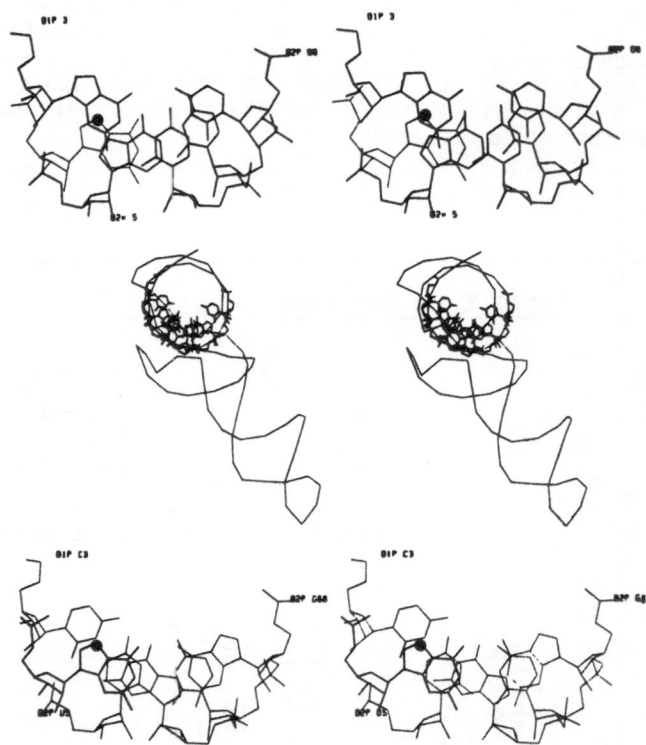

Fig. 4 : Structural environment of G4 in yeast tRNA(Phe) (top) and yeast tRNA(Asp) (bottom). The N7 positions in G4 are indicated by a dot. The stereoscopic view representing the backbone structure of tRNA(Asp) gives the general orientation of the two partial views.

the presence of regular helical regions in tRNAs according to the reactivities of G residues. So for example, in beef liver tRNA(Trp) the reactivities of G42 surrounded by purines and not modified by DMS, and that of G64 surrounded by pyrimidines and slightly modified (28) should reflect the regularity of the RNA stem conformation around these residues. In tRNA(Asp) the crystal structure shows that the anticodon stem is in a regular helical conformation (5-8), and G30 stacked between two pyrimidines is alkylated by DMS. Unexpectedly, in tRNA(Phe) where the pyrimidines 29 and 31 are replaced by purines, G30 is also alkylated. The contradiction, however, is only apparent because in that case the regularity of the RNA helix is broken in the middle part of the anticodon stem (1-4) and consequently G30 is less stacked so that the N7 position of guanine becomes accessible to the probe.

130

DIFFERENCES BETWEEN SOLUTION AND CRYSTAL STRUCTURES : INFLUENCE OF ANTICODON-
ANTICODON INTERACTIONS ON tRNA CONFORMATION

Crystal Packing of tRNA(Asp) and tRNA(Phe)

In the orthorhombic crystal lattice (space group C222(1)) tRNA(Asp)
molecules are associated through a two-fold symmetry axis parallel to the
crystallographic b direction by anticodon-anticodon interactions (5). The two
symmetrically related quasi self-complementary anticodon triplets GUC form a

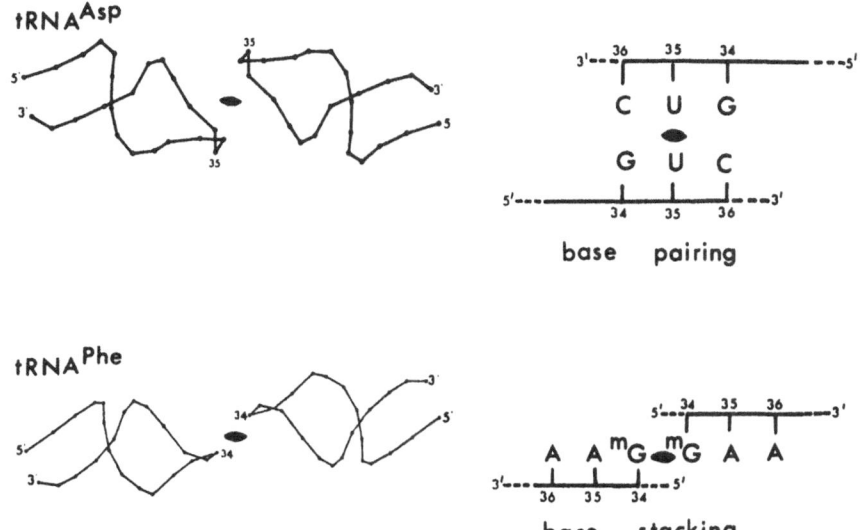

base pairing

base stacking

Fig. 5 : Contacts between anticodon loops in tRNA(Asp) and tRNA(Phe).

mini-helix of the RNA 11 type, with a regular conformation of the backbone.
The two external bases form classical Watson-Crick hydrogen bonded base pairs.
Although at the middle position there is a U/U mismatch, the small helix is
stabilized by stacking of the modified m1G37 base on both sides. This packing
confers a great stability to the dimeric structure and explains the good
quality of the electron density map in the anticodon region (5, 7).

Contacts between the anticodon loops of tRNA(Phe) molecules also exist
in the orthorhombic form of this tRNA. In that case, however, the GmAA
anticodons cannot base pair and they are arranged in a stacked conformation
(Figure 5).

Structural Implications

When comparing the three-dimensional structures of tRNA(Asp) and
tRNA(Phe) three main conformational differences can be noticed : (i) a
widening of the angle formed by the two branches of the L in tRNA(Asp) ; (ii)
a more regular helical conformation in the anticodon stem in tRNA(Asp) and
(iii) differences in the interactions between the D and T-loops (see Figure 1
and ref. 1-8). In what follows we will discuss experimental evidences
indicating that the interactions between the GUC anticodons in tRNA(Asp)

crystals are responsible for some of the structural differences between this
tRNA and tRNA(Phe), in particular that they are correlated with a long range
conformational change in the D and T-loop region of tRNA(Asp).

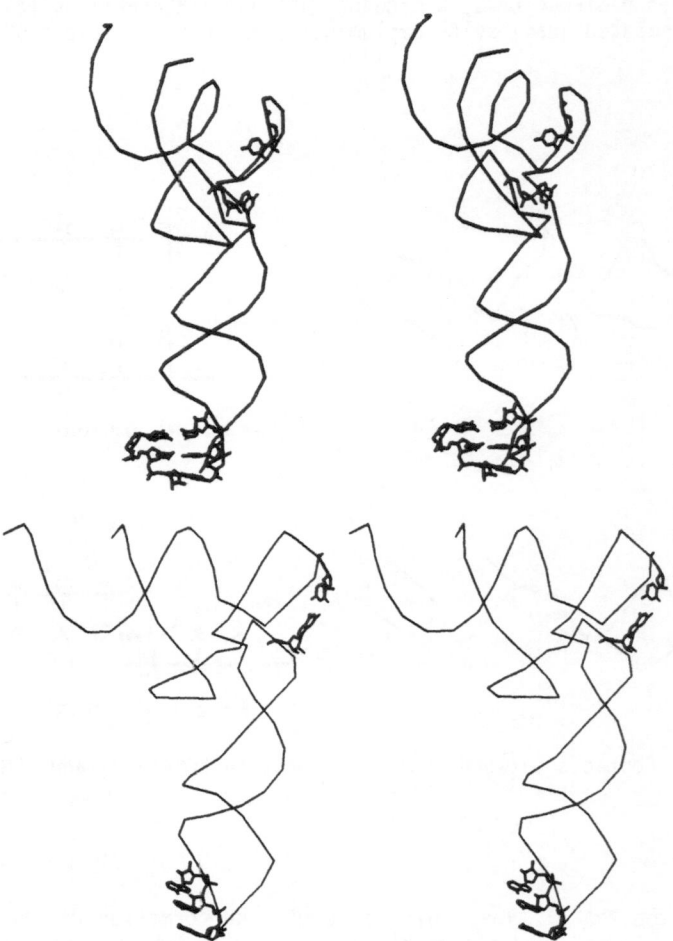

Fig. 6 : Stereoviews of the backbones of yeast tRNA(Asp) (top) and tRNA(Phe)
(bottom) with the bases of the anticodon and the two related
residues G19 in the D-loop and C56 in the T-loop. The drawings
illustrate the long range conformational change induced by the
anticodon-anticodon interactions. Note the Watson-Crick base
pairing between G19 and C56 in tRNA(Phe) and its absence in
tRNA(Asp), mainly due to a movement of residue G19. For tRNA(Asp),
part of the anticodon loop of the symmetrically related molecules
is represented.

A first evidence concerns a conflicting DMS mapping result in solution
of tRNA(Asp). According to the crystal structure of this tRNA, C56 is not
Watson- Crick base-paired with G19 in the D-loop, which implies that the N3

position in C56 should react with DMS. The experiment, however, clearly shows that C56 is non-reactive (32). Control experiments on tRNA(Phe) where C56 is paired with G19 also showed a non-reactivity of the cytosine (19,32). Most likely the data reflect a conformational change between the crystal and the solution structure of tRNA(Asp).

On the other hand T-jump experiments on tRNA(Asp) indicate that under usual solvent conditions most of the tRNA molecules are in their free state (20). When the measurements are done in the presence of high concentrations of ammonium sulfate, under conditions mimicking the crystallization buffer, dimer formation is favored. The stability of the anticodon/anticodon complex is further increased when tRNA(Asp) is mixed with Escherichia coli tRNA(Val) possessing the strickly complementary anticodon GAC. Using these last conditions it was then possible to run DMS mapping on dimerized tRNA(Asp) molecules. Results indicate a partial reactivity of C56 at low temperature ; increasing the temperature up to 50°C, which induces the dissociation of the dimers (20), leads to the non-reactivity of C56. Above 50°C, C56 becomes again alkylated by DMS, most probably as a result of the melting of the tRNA structure.

It is interesting to correlate the structural and biochemical data with the distribution of temperature factors along the ribophosphate backbone in the two tRNAs. These factors, which result from the crystallographic refinement, are an indication for local flexibility and rigidity in macromolecules. For tRNA(Asp) and tRNA(Phe) it soon became apparent that the variation of temperature factors along the ribophosphate backbone was different in the two molecules (7). In particular the anticodon loop and stem region appears rigid in tRNA(Asp) and flexible in tRNA(Phe). Conversely the D and T-loop region presents high temperature factors in tRNA(Asp) and low ones in tRNA(Phe). These differences call for different conformational dynamics in the two tRNAs. The interpretation of the data corresponding to the anticodon region is straightforward : the anticodon-anticodon interactions, by introducing new base pairing and stacking forces, rigidify the anticodon arm. The fact that the core region of the tRNA becomes agitated in dimeric tRNA but more rigid when the tRNA is free, is indicative for a long range conformational change which accounts well for the DMS mapping data of C56. This conclusion is further substantiated by the different self-splitting patterns of solution and crystalline tRNA(Asp). The rationale of the experiments is linked with the intrinsic chemical fragility of RNA due to the presence of the 2'-hydroxyls on riboses. If the ribophosphate backbone exhibits particular flexibility, hydrolytic splitting, involving cyclic intermediates with the neighboring phosphates , is favored (for a more complete discussion see ref. 32). Under normal conditions the extent of nicks is very low, but gel electrophoretic methodologies permit their detection. Results with tRNA(Asp) indicate a great fragility of the anticodon loop in the free molecules ; in the crystalline tRNA this anticodon loop appears more stable but significant nicks appear in the D-loop. Here again the results are best interpreted by different conformations of the tRNA.

Functional Implications

Many biochemical observations from different laboratories have suggested that codon-anticodon interactions induce conformational changes within tRNA (33-38) so that part of the tRNA molecule becomes accessible for interactions with ribosomal components. However, the exact nature of these conformational changes remained controversial (39, 40). The fact, that tRNA(Asp) possesses a self-complementary anticodon GUC and is packed as dimers in the crystal lattice (5) makes its structure a good model for a tRNA interacting with messenger RNA on the ribosomal A site. The biochemical evidence indicating anticodon-induced conformational changes in the D and T-loop region demonstrates that the structural differences between the crystal structures of tRNA(Asp) and tRNA(Phe) are not due to packing effects, but reflects conformational states related with two functions of the molecule (Figure 6). It could be objected that this functional conclusion derives from studies on a model system. But with this simplified system it was possible to approach directly the conformational effects linked to anticodon-anticodon interactions and to demonstrate the dynamic potentialities of tRNA molecules. Further, confident detections of conformational effects would be difficult with more complex experimental systems. It is, however, likely that interactions with ribosomal components would modulate the process. Therefore, more experimental observations are needed to reach a definitive picture for the tRNA conformations in the different sites of the ribosome and during translocation.

ACKNOWLEDGEMENTS

This research was supported by grants from the Centre National de la Recherche Scientifique (CNRS), the Ministère de l'Industrie et de la Recherche (MIR) and Université Louis Pasteur (ULP), Strasbourg. We are most indebted to Drs.H. Grosjean and C. Houssier (Belgium) and Drs. J.C. Thierry, C. Florentz and M. Garret for numerous discussions.

REFERENCES

1. G.J. Quigley, A. Wang, N.C. Seeman, F.L. Suddath, A. Rich, J.L. Sussmann and S.H. Kim, Hydrogen bonding in yeast phenylalanine transfer RNA, Proc. Natl. Acad. Sci. U.S.A. 72 : 4866-4870 (1975).
2. A. Jack, J.E. Ladner and A. Klug, Crystallographic refinement of yeast phenylalanine transfer RNA at 2.5 Å resolution, J. Mol. Biol. 108 : 619-649 (1976).
3. C.D. Stout, H. Mizuno, S.T. Rao, P. Swaminathan, J. Rubin, P. Brennan and M. Sundaralingam, Atomic coordinates and molecular conformation of yeast phenylalanine tRNA. An independant investigation, Acta Crys. B54 : 1529-1544 (1978).
4. J.L. Sussmann, S.R. Holbrook, R.W. Warrant, G.M. Church and S.H. Kim, Crystal structure of yeast phenylalanine transfer RNA : crystallographic refinement, J. Mol. Biol. 123 : 607-630 (1978).
5. D. Moras, M.B. Comarmond, J. Fischer, R. Weiss, J.C. Thierry, J.P. Ebel and R. Giegé, Crystal structure of yeast tRNA^{Asp}, Nature (London) 286 : 669-674 (1980).
6. E. Westhof, P. Dumas and D. Moras, Loop stereochemistry and dynamics in transfer RNA, J. Biomol. Struct. Dyn. 1 : 337-359 (1983).
7. E. Westhof, P. Dumas and D. Moras, Crystallographic refinement of yeast aspartic acid transfer RNA, J. Mol. Biol. 184 : 119-145 (1985).
8. P. Dumas, J.P. Ebel, R. Giegé, D. Moras, J.C. Thierry and E. Westhof, Crystal structure of yeast tRNA(Asp) : Atomic coordinates, Biochimie (Paris) 67 : 597-606 (1985).
9. R.W. Schevitz, A.D. Podjarny, N. Krishnamachar, T.T. Hughes, P. Sigler and J.-L. Sussmann, Crystal structure of an eukaryotic initiator tRNA, Nature (London) 278 : 188-190 (1979).
10. N. Woo, B. Roe and A. Rich, Three-dimensional structure of Escherichia coli initiator tRNA(Met), Nature (London) 286 : 346-351 (1980).
11. M. Sprinzl, J. Moll, F. Meissner and T. Hartmann, Compilation of tRNA sequences, Nucl. Acids Res. 13 : r1-r49 (1985).

12. R. Giegé, D. Kern, J.P. Ebel, H. Grosjean, S. De Henau and H. Chantrenne, Incorrect aminoacylation involving tRNAs or valyl-tRNA synthetase from Bacillus stearothermophilus, Eur. J. Biochem. 45 : 351-362 (1974).

13. H.G. Gassen, Ligand induced conformational changes in ribonucleic acids, Prog. Nucl. Acid Res. Mol. Biol. 24 : 57-86 (1980) and ref. therein.

14. R. Rigler and W. Wintermeyer, Dynamics of tRNA, Ann. Rev. Biophys. Bioeng. 12 : 475-505 (1983).

15. D.A. Peattie, Direct chemical method for sequencing RNA, Proc. Natl. Acad. Sci. U.S.A. 76 : 1760-1764 (1979).

16. A.M. Maxam and W. Gilbert, Sequencing end-labeled DNA with base-specific chemical cleavages, Methods in Enzymology 65 : 499-459 (1980).

17. V.V. Vlassov, R. Giegé and J.P. Ebel, Tertiary structure of tRNAs in solution monitored by phosphodiester modification with ethylnitrosourea, Eur. J. Biochem. 119 : 51-59 (1981).

18. P. Romby, D. Moras, M. Bergdoll, P. Dumas, V.V. Vlassov, E. Westhof, J.P. Ebel and R. Giegé, Yeast tRNA(Asp) tertiary structure in solution and areas of interaction of the tRNA with aspartyl-tRNA synthetase. A comparative study of the yeast phenylalanine system by phosphate alkylation experiments with ethylnitrosourea, J. Mol. Biol. 184 : 455-471 (1985).

19. D.A. Peattie and W. Gilbert, Chemical probes for higher-order structure in RNA, Proc. Natl. Acad. Sci. U.S.A. 77 : 4679-4682 (1980).

20. P. Romby, R. Giegé, C. Houssier and H. Grosjean, Anticodon-anticodon interactions in solution. Studies of the self-association of yeast or Escherichia coli tRNA(Asp) and of their interactions with Escherichia coli tRNA(Val), J. Mol. Biol. 184 : 107-118 (1985).

21. D. Moras, A.C. Dock, P. Dumas, E. Westhof, P. Romby and R. Giegé, The versatile transfer RNA molecule : Crystallography of yeast tRNA(Asp). in "Nucleic Acids : The Vectors of Life", B. Pullman and J. Jortner, eds., Reidel Publishing Company, Dordrecht : 403-414 (1983).

22. G.J. Quigley, N.C. Seeman, A.H.J. Wang, F.L. Suddath and A. Rich, Yeast phenylalanine transfer RNA : Atomic coordinates and torsion angles, Nucl. Acids Res. 2 : 2329-2341 (1975).

23. M. Chen, R. Giegé, R. Lord and A. Rich, Raman spectra and structure of yeast phenylalanine transfer RNA in the crystalline state and in solution, Biochemistry 14 : 4385-4391 (1975).

24. G.T. Robillard, C.E. Tarr, F. Vosman and J.C. Berendsen, Similarity of the crystal and solution structure of yeast tRNA(Phe), Nature (London) 262 : 363-369 (1976).

25. S.R. Holbrook and S.H. Kim, Correlation between chemical modification and surface accessibility in yeast phenylalanine transfer RNA, Biopolymers 22 : 1145-1166 (1983).

26. N. Figueroa, G. Keith, S.L. Leroy, P. Plateau, S. Roy and M. Guéron, NMR study of slowly exchanging imino protons in yeast tRNA(Asp), Proc. Natl. Acad. Sci. U.S.A. 80 : 4330-4333 (1983).

27. R. Giegé, P. Romby, C. Florentz, J.P. Ebel, P. Dumas, E. Westhof and D; Moras, Solution conformation of tRNAs : Correlation with crystal structures, in "Nucleic Acids : The Vectors of Life", B. Pullman and J. Jortner, eds., Reidel Publishing Company, Dordrecht : 415-426 (1983).

28. M. Garret, B. Labouesse, S. Litvak, P. Romby, J.P. Ebel and R. Giegé, Tertiary structure of animal tRNA(Trp) in solution and interaction of tRNA(Trp) with tryphophanyl-tRNA synthetase, Eur. J. Biochem. 138 : 67-75 (1983).

29. C. Florentz, J.P. Briand, P. Romby, L. Hirth, J.P. Ebel and R. Giegé, The tRNA-like structure of turnip yellow mosaic virus RNA : Structural organization of the last 159 nucleotides from the 3'-OH terminus, EMBO J. 1 : 269-276 (1982).

30. H. Guilley, G. Jonard, B. Kukla and K.E. Richards, Sequence of 1000 nucleotides at the 3'end of tobacco mosaic virus RNA, Nucl. Acids Res. 6 : 1287-1308 (1979).

31. P. Carbon, C. Ehresmann, B. Ehresmann and J.P. Ebel, The sequence of Escherichia coli ribosomal 16S RNA determined by new rapid gel methods, Febs Lett. 94 : 152-156 (1978).

32. D. Moras, A.C. Dock, P. Dumas, E. Westhof, P. Romby, J.P. Ebel and R. Giegé, The structure of yeast tRNA(Asp), a model for tRNA interacting with messenger RNA, J. Biomol. Struct. Dyn. submitted.

33. V. Schwartz, H. M. Menzel and H.G. Gassen, Codon-dependent rearrangement of the three-dimensional structure of phenylalanine tRNA, exposing the TUCG sequence for binding to the 50S ribosomal subunit, Biochemistry 15 : 2484-2490 (1976).

34. M. Sprinzl, T. Wagner, S. Lorenz and V.A. Erdmann, Regions of tRNA important for binding to the ribosomal A and P sites, Biochemistry 15 : 3031-3039 (1976).

35. P. Davenloo, M. Sprinzl and F. Cramer, Proton nuclear magnetic resonance of minor nucleosides in yeast phenylalanine transfer ribonucleic acid. Conformational changes as a consequence of aminoacylation, removal of the Y base, and codon-anticodon interaction, Biochemistry 18 : 3189-3199 (1979).

36. R. Wagner and R.A. Garret, Chemical evidence for a codon-induced allosteric change in tRNA(Lys) involving the 7-methylguanosine residue 46, Eur. J. Biochem. 97 : 615-621 (1979).

37. N. Farber and C.R. Cantor, Comparison of the structures of free and ribosome-bound tRNA(Phe) by using slow tritium exchange, Proc. Natl. Acad. Sci. U.S.A. 77 : 5135-5139 (1980).

38. T. Jorgenson, G. Siboskan, F. Wikman and B.F.C. Clark, Studies of the tRNA structure in different ribosomal sites, in:"11th International tRNA Workshop, Banz, Germany", abstract CII-2 (1985).

39. B. Pace, E.A. Matthews, K.D. Johnson, C.R. Cantor, and N.R. Pace, Conserved 5S rRNA complement to tRNA is not required for protein synthesis, Proc. Natl. Acad. Sci. U.S.A. 79 : 36-40 (1982).

40. B. Helk and M. Sprinzl, Interaction of the T-loop of tRNA with the conserved CpGpUmpApApCp sequence of 16S RNA, in "11th. International tRNA Workshop Banz, Germany", abstract CII-6 (1985).

THE SOLUTION STRUCTURES OF RNA FRAGMENTS DETERMINED BY NUCLEAR OVERHAUSER ENHANCEMENT MEASUREMENTS

A. M. Gronenborn and G.M. Clore

Max-Planck-Institut für Biochemie
D-8033 Martinsried bei München
F.R.G.

INTRODUCTION

Over the last few years numerous NMR studies have appeared on the solution structure of small DNA oligonucleotides (see refs. 1-4 for reviews). In contrast, relatively few studies have appeared on RNA duplexes (5-10). This is probably due to two factors, both of which arise from the presence of the 2'-hydroxyl group in RNA. First, RNA is intrinsically more difficult to synthesize than DNA as an additional reactive group has to be protected during the course of the synthesis (11,12). Second, all the sugar resonances, with the exception of the H1' resonances, are superimposed in a very narrow region of the ^{1}H-NMR spectrum only 1 ppm in width, thereby considerably complicating the task of resonance assignment. As part of our continuing studies on the structure and dynamics of oligonucleotides in solution (see for examples refs. 4,8,13-17), we have embarked on a program to investigate the structures of small RNA fragments comprising specific portions of the yeast tRNA[Phe] molecule. This approach allows one to probe the details of the three dimensional solution structures of individual loops and double stranded stems as the spectral complexity exhibited by the whole tRNA molecule is greatly reduced, thereby enabling NOE measurements to be made on a large number of proton types, namely imino, aromatic, ribose and methyl protons.

Essentially there are four NMR approaches which can be used to obtain structural information on an oligoribonucleotide in solution:

(i) The analysis of chemical shifts by calculations of through-space magnetic effects (ring current shift calculations). Although employed in a number of studies, the results are only qualitative in nature and are based on available crystal structure data as input parameters (18,19).

(ii) The use of paramagnetic relaxation effects is well known, but its application is fraught with difficulties since it is based on a considerable number of assumptions, particularly when an external paramagnetic probe is used (20).

(iii) The analysis of three bond spin-spin coupling constants has been widely used to extract information concerning dihedral angles (21-24). It suffers, however, from the fact that the relationship between three bond spin-spin coupling constants and dihedral angles is solely empirical in nature (20,25) and, consequently, the information obtained is essentially of a qualitative rather than a quantitative nature. A further point to consider if one wants to extract structural information from three bond proton-proton coupling constants is the fact that under conditions where small oligonucleotides of 6-12 base pairs are entirely double-stranded, namely at temperatures between 0 and 25 °C, the appropriate coupling constants are difficult to resolve due to fairly large linewidths.

(iv) Potentially the most direct and powerful method of conformational analysis in solution is the use of the proton-proton nuclear Overhauser effect (NOE) which can demonstrate the proximity of two protons in space and can be used to determine their separation (26).

For most cases to date, NOE data have only been interpreted to yield qualitative structural information. With regard to nucleic acids this has proved particularly useful in examining the pattern of secondary and tertiary hydrogen bonding interactions in transfer ribonucleic acids (27-32), in distinguishing A, B and Z DNA (33-35), in examining the effects of base pair mismatching on nucleic acid conformation (36,37) and in monitoring intermolecular contact points in drug-oligonucleotide complexes (38-40). It is, however, also possible to deduce quantitative structural information and, in this manner, the solution structures of a variety of oligonucleotides and oligoribonucleotides have been obtained (8,10,14,17,41,42,43).

SEQUENTIAL RESONANCE ASSIGNMENT

The full potential of NMR spectroscopy for structural studies can only be realized after identification of the individual resonance lines, and a general scheme for obtaining sequential assignments in protein ^{1}H-NMR spectra has been described by Wuthrich and his collaborators (44). In a similar manner, a sequential assignment method, limited to the imino and adenosine H2 proton resonances of adjacent base pairs in tRNA was developed by Redfield et al. (45). Application of essentially the same principles to spectra of oligonucleotides can be used to achieve virtually complete resonance assignments (46). These assignments are a necessary prerequisite for the subsequent structure determination based on the quantitation of NOEs.

The assignment strategy involves basically two different NMR experiments. Firstly, it is helpful to identify the sugar spin system of a particular nucleotide either by decoupling or by two-dimensional J correlated spectroscopy (COSY). Thus the sugar resonances can be grouped into families of signals belonging to the same network of coupled spins via the intranucleotide pathway H1'↔H2'↔H3'↔H4'↔H5/H5" (Fig. 1).

Because of the limited chemical shift dispersion in the
sugar proton region of RNA fragments, it is, however,
necessary to use NOE measurements as the most important step
in the assignment procedure. In this way all protons that are
separated by short distances (< 5 A) within the spatial
structure can be connected either by one-dimensional
pre-steady state NOE measurements or by two-dimensional NOE
spectroscopy (NOESY). Thus, in double-stranded oligoribonucleotides, neighbouring bases can be identified as well as bases

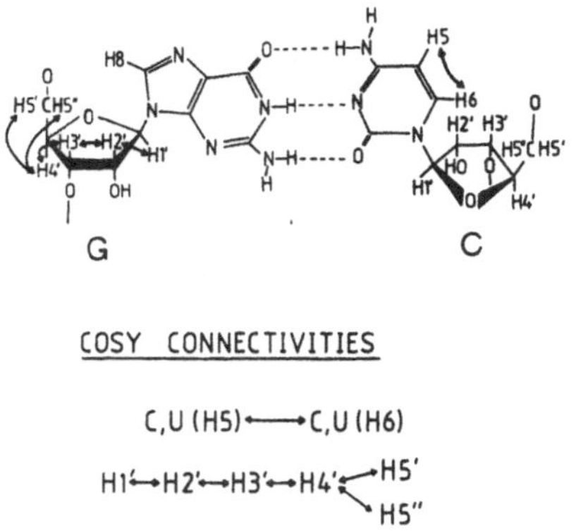

COSY CONNECTIVITIES

C,U (H5) ←→ C,U (H6)

H1'←→H2'←→H3'←→H4'⟨ H5'
                      H5"

Fig. 1 Schematic representation of through bond J-connectivit-
ies in a base pair.

belonging to two different strands which are involved in base
pairing. These NOE measurements provide the main body of
information necessary for assignment, and, in cases where not
all couplings can be resolved, will lead to virtually complete
assignments in their own right. Fig. 2 summarizes a
comprehensive NOE strategy for the assignment of all proton
resonances in right-handed single and double stranded RNA
fragments without making any assumption about an A or B type
geometry.

In addition to providing assignments, NOE measurements can be used to determine interproton distances. For quantitation of the NOE we have used conventional one-dimensional NMR, and pure phase absorption NOESY experiments with small random variations in the mixing time to eliminate zero quantum coherence transfer (47). The one-dimensional NOE experiment involves the saturation of the resonance of proton i and observing the intensity of the other proton resonances. For large molecules (MW > 1000) with long correlation times (τ_c > 5 x 10^{-9} s) such as oligoribonucleotides of 5 base pairs and longer, for which $\omega\tau_c$ > 1, the NOEs observed are negative (48). However, when $\omega\tau_c$ > 1, the NOEs will no longer be selective in the steady state (i.e., following saturation of the resonance of proton i for t → ∞) owing to highly effective cross-relaxation between a large number of protons, a phenomenon known as spin diffusion (49), so that no structural information can be obtained. This problem can be completely circumvented by using only short times (typically < 0.5 s for a molecule of MW ∿ 6000) for either the selective saturation pulse in the one-dimensional experiment or for the mixing time in the two-dimensional experiment (46, 50-53). Under these conditions the pre-steady state NOE between two protons i and j, $N_{ij}(t)$ is given by

$$N_{ij}(t) \sim \sigma_{ij}t \qquad\qquad [1]$$

providing $\sigma_{ij} \gtrsim \sigma_{ik}$ or $\sigma_{ij} \gtrsim \sigma_{jk}$ (where k is any other proton), as the initial build-up rate of the NOE is equal to the cross-relaxation rate σ_{ij} between the two protons i and j (50-53). Distance information can then be obtained since σ_{ij} is inversely proportional to the sixth power of the distance, $\langle r_{ij}^{-6} \rangle$, between the two protons (48). As a result of the $\langle r_{ij}^{-6} \rangle$ dependence, the magnitude of the pre-steady state NOE is very sensitive to interproton distance, decreasing rapidly as r_{ij} increases and becoming virtually undetectable for r_{ij} > 5 A.

The ratio of two interproton distances may thus be obtained from the equation

$$r_{ij}/r_{kl} = (\sigma_{kl}/\sigma_{ij})^{1/6}$$

$$\sim [N_{kl}(t)/N_{ij}(t)]^{1/6} \qquad\qquad [2]$$

providing the correlation times for the two interproton distance vectors i-j and k-l are the same. If one of the distances is known, actual interproton distances can also be calculated. It should be noted that the approximation in Eqn. [2] remains valid up to values of t 3-4-times longer than that in Eqn. [1] (52). In the case of double-stranded oligoribonucleotides, there are two intranucleotide reference distances which are completely independent of the RNA structure: namely, the C(H6)-C(H5) and U(H6)-U(H5) distances which, on the basis of standard bond lengths and angles, have a value of 2.46 A.

In using Eqn. [2] several words of caution should be noted.

First, if proton i is not only close to proton j but to other protons as well, such as say proton k, σ_{ij} can only be accurately determined from the initial slope of the time dependence of the NOE if either $\sigma_{ij} \gtrsim \sigma_{ik}$ or $\sigma_{ij} \gtrsim \sigma_{jk}$ (52). Considerations of stereochemistry indicate that this condition is satisfied for all distances < 3 A. When this condition is not satisfied, a systematic error in the measured value of σ_{ij} and, hence, the calculated value r_{ij} is incurred (52). In particular, the value of σ_{ij} will be overestimated. If the reference distance and cross-relaxation rate σ_{ij} are not part of the cross-relaxation network associated with σ_{ij}, then the value of r_{ij} will always be underestimated. If, on the other hand, the reference distance and cross-relaxation rate σ_{eff}

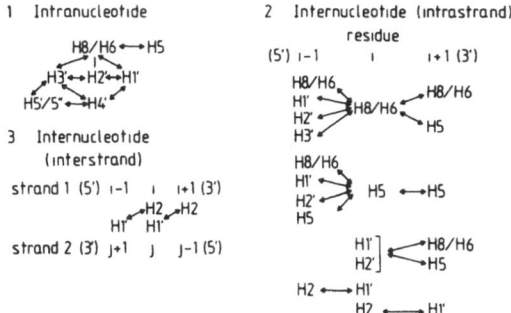

Fig. 2 Schematic representation of through-space connectivities (< 5 A) for right handed RNA.

are part of the cross-relaxation network associated with σ_{ij}, the sign of the systematic error in the estimation of r_{ij} will depend on the relationship between σ_{ij} and σ_{ref}. Namely, r_{ij} will be overestimated if $\sigma_{ij} > \sigma_{ref}$ but underestimated if $\sigma_{ij} < \sigma_{ref}$. In general, however, these effects will be small. The second cautionary note concerns potential variations in the effective correlation time τ_{eff} for different interproton vectors which may also lead to errors in the estimation of r_{ij}. When $\omega\tau_{eff} \gg 1$ (where ω is the spectrometer frequency) as in the cases discussed here, the cross-relaxation rate σ_{ij} is not only proportional to $\langle r_{ij}^{-6} \rangle$ but also to the effective correlation time $\tau_{eff}(ij)$ of the i-j interproton vector. However, because of the $\langle r_{ij}^{-6} \rangle$ dependence of σ_{ij}, quite substantial variations in $\tau_{eff}(ij)$ relative to the effective correlation time τ_{ref} (ref) for the reference interproton vector only lead to relatively small errors in the estimation of r_{ij} using Eqn. [2]. Consider, for example, the case where σ_{ij} has a value of 0.5 s^{-1}, σ_{kl} a value of 1 s^{-1} and r_{kl} a value of 2.5 A. Then for the three cases $\tau_{eff}(ij) = \tau_{eff}(kl)$, $\tau_{eff}(ij) = \tau_{eff}(kl)/2$, $\tau_{eff}(ij) = 2 \tau_{eff}(kl)$, r_{ij} is calculated to be 2.8, 2.5 and 3.1 A, respectively. These three cases of course represent extreme variations in τ_{eff}. Here again certain internal checks can be used to ascertain possible variations in effective correlation times. Thus, for example, within experimental error the intranucleotide H5-H6 base vector for the double stranded RNA pentamer

$$[5'r(CACAG).5'r(CUGUG)]$$

comprising the stem of the TψC loop of yeast tRNA[Phe] exhibits no residue to residue variation in the H5-H6 cross-relaxation rates and hence effective correlation time. Of course, there may still be differences in internal mobility and effective correlation times between different components of each residue, viz. between the base and sugar moieties. Indeed, in the case of DNA oligonucleotides it has been found that the effective correlation time for the intranucleotide H2'-H2'' sugar vector is a factor of approximately 3 times shorter than that of the intranucleotide H5-H6 base vector (54) and that this could lead to errors up to 0.5 A if the inappropriate reference distance and cross-relaxation rate is used to calculate a particular unknown distance (17). In the case of RNA there are no readily available fixed distance interproton vectors in the sugar ring to be able to check this in this manner (note the cross-relaxation rate of the H5'-H5'' sugar vector cannot be measured due to spectral overcrowding). Nevertheless, there still exists a reliable handle with which to probe the mobility of the sugar ring relative to that of the base, namely the intranucleotide H1'-H2' sugar vector which has a minimum value of 2.5 A when the sugar pucker is in the 3'-endo conformation characteristic of A-RNA, and a maximum value of 2.9 A when the sugar pucker is in the 2'-endo conformation. The values we obtain using the H5-H6 base vector as an internal reference lie between 2.5 and 2.7 A. If the effective correlation time of the sugar moieties were significantly shorter than that of the bases, then these values for the H1'-H2' intranucleotide distance would represent overestimates. This, however, cannot be the case on stereochemical grounds. That the internal mobility of the sugar and base moieties is the same in RNA, in contrast to the situation in DNA, is not surprising as the mobility of the ribose ring in RNA would be expected to be considerably reduced relative to that of the deoxyribose ring in DNA, first on account of steric hindrance arising from the presence of the bulky O2'H hydroxyl group on the ribose, and second due to electrostatic interactions between the O2'H group of residue i and the O4' atom of residue i+1 immobilizing one ribose ring with respect to the neighbouring ones. Taking into account both cautionary considerations discussed above as well as experimental errors, we estimate that the errors in the values of the calculated distances is in general < 0.2 A. However, in the case where the estimated distances are greater than about 3.3 A, the errors may be somewhat larger with the values of the calculated interproton distance underestimated by < 0.3 A.

Because of the limited degrees of freedom available for a double-stranded oligoribonucleotide structure, one would expect that a reasonably large number of interproton distances would be sufficient to determine the three-dimensional solution structure with a high degree of confidence. In principle, these structures can be solved by manual model building, and indeed reasonably accurate values of the glycosidic and C4'-C3' (δ) bond torsion angles can be obtained in this manner. However, because of potential cumulative errors inherent in such an approach, only qualitative information can be deduced for the other structural parameters, namely backbone torsion angles, helical twist, helical rise and base tilt. This problem can be overcome using a non-linear restrained least-squares refinement procedure in which all covalent bond lengths, fixed bond angles, van der Waals contacts, and hydrogen bond lengths and geometry are restrained within narrow limits, in order to refine an initial trial model on the basis of the interproton distance data determined from NOE measurements (10,14,43).

The minimum requirement to define both the glycosidic bond torsion angle (χ) and the sugar pucker conformation, defined in terms of the C4'-C3' bond torsion angle (δ), is two out of the three intranucleotide sugar-base distances between the H1', H2' and H3' sugar protons and the H8/H6 base proton. The syn and anti ranges for χ are $60° \pm 90°$ and $240° \pm 90°$, respectively. The distance between the H1' and H8/H6 proton has a minimum value of 2.3-2.5 A at $\chi = 60°$ (syn) and a maximum value of 3.7-3.9 A at $\chi = 240°$ (anti). In addition, each value of the H1'-H8/H6 distance is compatible with two values of χ : $60° < \chi_1 < 240°$ and $\chi_2 = (240° - \chi_1) + 240°$. Given the restricted degrees of freedom imposed by the five-membered sugar ring, the H2'-H8/H6 distance enables one to distinguish between χ_1 and χ_2 and to determine simultaneously the C4'-C3' bond torsion angle (δ). Similar arguments apply to other combinations of these three distances. Naturally, the more distances available, the better the determination of χ and δ. The C4'-C5' bond torsion angle (δ) can also be uniquely defined, providing two out of the three intranucleotide distances between the H3' and H5" protons, the H4' and H5" protons, and the H4' and H5' protons are known.

Once glycosidic bond and sugar pucker conformations are known for each nucleotide, the inter-residue interproton distances enable one to define the position of each individual base pair with respect to its adjacent neighbours in terms of approximate values of the helical rise, helical twist and base tilt. In addition the handedness of the helix can be deduced from the directional specificity of the internucleotide NOEs.

To date we have examined the solution structure of the RNA pentadecamer

$$5'r \; C_{28}A_{29}G_{30}A_{31}C_{m32}U_{33}G_{m34}A_{35}A_{36}Y_{37}A_{38}\psi_{39}m^5C_{40}U_{41}G_{42}$$

comprising the anticodon loop and stem (residues 28-42) of yeast tRNA[Phe] (8) and the double stranded RNA pentamer

```
            5'r C-A-C-A-G 3'

            3'r G-U-G-U-C 5'
```

comprising the stem of the T C loop of tRNA[Phe] (10).

For the pentadecamer extensive and systematic pre-steady
state proton-proton NOE measurements were used to assign all
exchangeable and non-exchangeable base proton resonances, all
H1' ribose resonances and all methyl and methylene resonances
(8). In solution the RNA pentadecamer could in principle adopt
three possible structures: a bulge duplex, a 3' stacked
hairpin loop or a 5'-stacked hairpin loop structure. The
distinction between the bulge-duplex and any hairpin-loop
structure can easily be made on the basis of the correlation
times calculated from the pre-steady state intranucleotide NOE
values involving protons a fixed distance apart, in particular
the H5/H1 and H6 protons of the pyrimidine base residues which
are separated by 2.46 A. In this manner, we obtain correlation
times of 3.2($\pm$0.2) and 1.8($\pm$0.1) ns for these interproton
vectors of the stem and loop pyrimidine base residues
respectively. The difference between the correlation times of
the loop and stem bases is easily explained as one would
expect the amplitude of motion about the glycosidic bond to be
larger for the loop residues than the base ones. These
correlation times compare with values of 3 and 3.4 ns
calculated from the data on a double-stranded DNA hexamer and
octamer respectively (17,54) under similar experimental
conditions. As the correlation time is proportional to
molecular mass, we conclude that the RNA pentadecamer adopts a
hairpin-loop conformation under the experimental conditions
used.

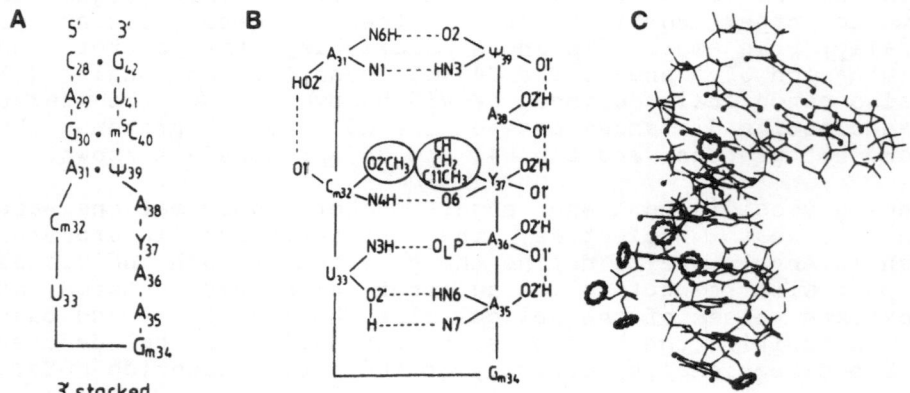

Fig. 3 (A) Sequence of the anticodon loop and stem of tRNA
and schematic representation of the 3' stacked conformation.
(B) Interactions stabilizing the loop structure (polar
electrostatic interactions are represented by dashed lines and
groups involved in hydrophobic interactions are encircled).
(C) Structure of the anticodon loop and stem with those
hydrogen atoms highlighted between which distances were
determined.

144

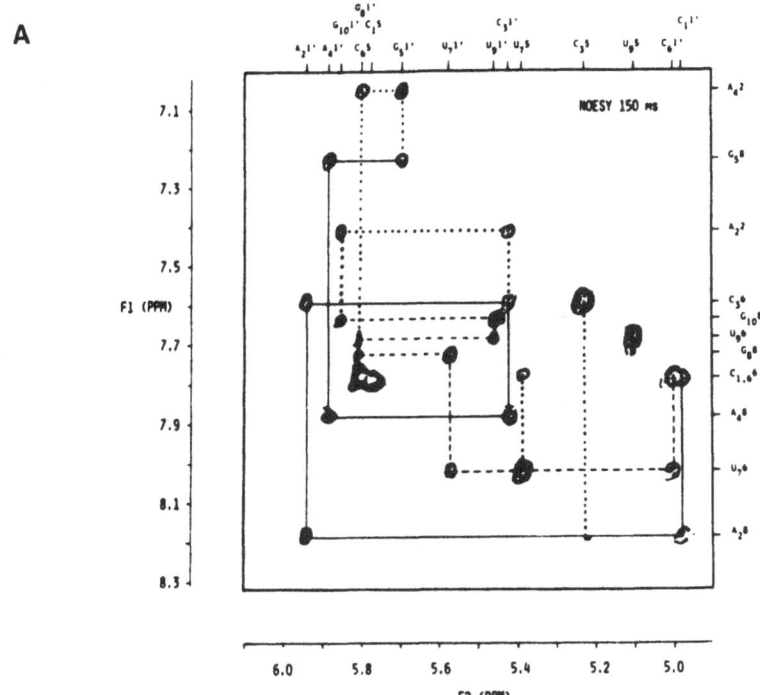

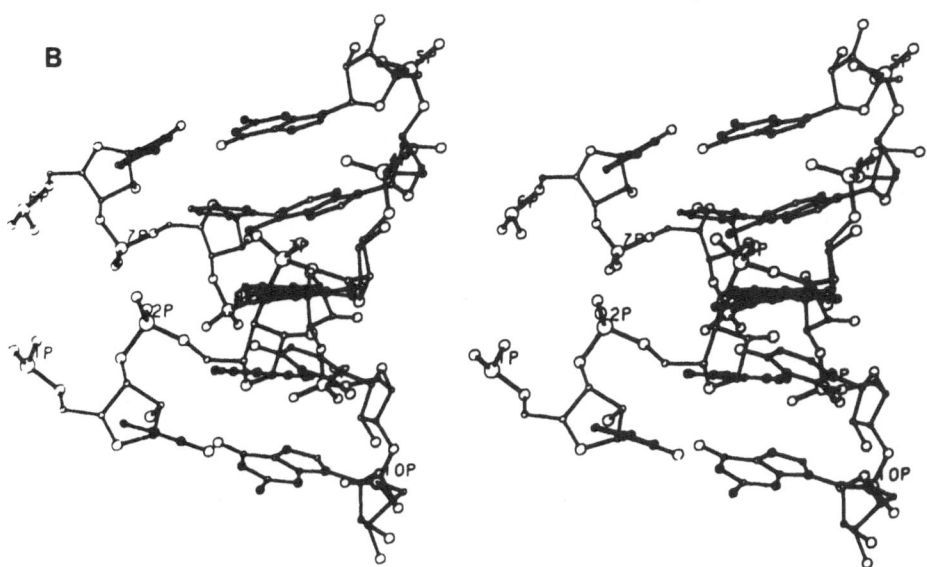

Fig. 4 (A) Pure phase absorption NOESY spectrum indicating H1'(i-1) ↔ H8/H6(i) ↔ H1'(i+1) connectivities for strand CACAG (———) and strand CUGUG (----). H8/H6 ↔ H5 and H2 ↔ H1' connectivities are indicated by dotted lines (....). (B) Stereoview of the refined structure of the pentamer viewed along the helix axis.

In order to arrive at a detailed solution structure of the pentadecamer, 70 intra- and internucleotide interproton distances were determined which were used to solve the structure by model building. It was shown that the loop exists in a 3' stacked conformation since NOE connectivities were observed along the strand from C(28) to U(33) and G_m(34) to G(42) but none between U(33) and G_m(34) (see Fig. 3). In addition, NOEs were observed between the H1' proton of U(33) and the H8 protons of A(35) and A(36). This 3' stacked conformation is stabilized by an array of hydrogen bonds and hydrophobic interactions resulting in a "high" melting temperature of 53 °C. The solution structure is both qualitatively and quantitatively remarkably similar to the orthorhombic (55) and monoclinic (56) crystal structures of yeast tRNA[Phe] with an overall root mean square difference of only 1.2 A between the interproton distances determined by NMR and X-ray crystallography.

For the double stranded pentamer cross relaxation rates were measured from the time dependence of the NOEs using both one- and two-dimensional spectroscopy. A total of 55 interproton distances were determined and the structure of the RNA was refined by restrained least-squares minimization on the basis of distance and planarity restraints (10) using the crystallographic program RESTRAIN (57) modified to include protons. The RMS difference between observed and calculated interproton distances was < 0.2 A, and the RMS difference in the coordinates between two refined structures obtained from two different starting A-RNA model structures was 0.11 A.

The refined structure of the pentamer is clearly of the A-type with the values of the glycosidic bond and backbone torsion angles lying in the range characteristic of A-RNA. However, it is far from regular. For example, propellor twist angles lie in the range 0° to 25°, base roll angles vary between -3° and +35° and glycosidic bond torsion angles vary from -137° to -171° (see Fig. 4). These structural variations are similar in magnitude to those found for the refined solution structures of B-DNA fragments (14,43). However, they cannot as yet be fitted to any generalized set of rules governing their sequence dependence, as too few refined solution structures are available at present.

CONCLUDING REMARKS

With the increasing number of solution structures of nucleic acids now being determined by NOE measurements, it has become clear that NMR data can provide important information complementary to X-ray diffraction data. Thus NMR studies, combined with refinement techniques, will become particularly useful in cases where crystal data are not available, in comparative studies of nucleic acid fragments with related sequences and in the study of transitions between different conformational states.

REFERENCES

(1) Patel, D.J., Pardi, A. & Itakura, K. (1982) Science 216, 581-590

(2) Patel, D.J., Kozlowski, S.A., Ikuta, S., Takura, K., Bhatt, R. & Hare, B.R. (1983) Cold Spring Harbor Symp. Quant. Biol. 47, 197-206

(3) Kearns, D.R. (1984) CRC. Crit. Rev. Biochem. 15, 237-290

(4) Clore, G.M. & Gronenborn, A.M. (1985) FEBS Lett. 179, 187-192

(5) Petersheim, M. & Turner, D.H. (1983) Biochemistry 22, 256-263

(6) Freier, S.M., Burger, B.J., Alkema, D., Neilson, T. & Turner, D.H. (1983) Biochemistry 22, 6198-6206

(7) Sinclair, A., Alkema, D., Bell, R.A., Coddington, J.M., Hughes, D.W., Neilson, T. & Romanink, P.J. (1984) Biochemistry 23, 2656-2662

(8) Clore, G.M., Gronenborn, A.M., Piper, E.A., McLaughlin, L.W., Graeser, E. van Boom, J.H. (1984) Biochem. J. 221, 737-751

(9) Haasnoot, C.A.G., Westerink, H.P., van der Marel, G.A. & van Boom, J.H. (1984) J. Biomolec. Struct. Dynamics 2, 345-360

(10) Clore, G.M., Gronenborn, A.M. & McLaughlin, L.W. (1985) Eur. J. Biochem., in the press

(11) Reese, C.B. (1978) Tetrahedon, 34, 3143-3175

(12) Van Boom, J.H. & Wreesman, C.T.J. (1984) in Oligonucleotide Synthesis: A Practical Approach (Gait, M.M., ed.) pp 153-183, IRL Press, Oxford

(13) Clore, G.M. & Gronenborn, A.M. (1984) Eur. J. Biochem. 141, 119-129

(14) Clore, G.M. & Gronenborn, A.M. (1985) EMBO J. 4, 829-835

(15) Clore, G.M., Gronenborn, A.M. & McLaughlin, L.W. (1984) J. Mol. Biol. 174, 163-173

(16) Gronenborn, A.M., Clore, G.M., McLaughlin, L.W., Graeser, E., Lorber, B. & Giege, R. (1984) Eur. J. Biochem. 145, 359-364

(17) Gronenborn, A.M., Clore, G.M. & Kimber, B.J. (1984) Biochem. J. 221, 723-736

(18) Kan, L.S., Cheng, D.M., Jayaraman, K., Leutzinger, E.E., Miller, P.S. & Ts'o, P.O.P. (1982) Biochemistry 21, 6723-6732

(19) Bell, R.A., Alkema, D., Caddington, J.M., Hadder, P.A., Hughes, D.W. & Neilson, T. (1983) Nucleic Acds Res. 11, 1143-1149

(20) Jardetzky, O. & Roberts, G.C.K. (1981) NMR in Molecular Biology, Academic Press, New York

(21) Mellema, J.R., Haasnoot, C.A.G., Van Boom, J.H. & Altona, C. (1981) Biochim. Biophys. Acta 655, 256-264

(22) Tran-Dinh, S., Neuman, J.M., Huynh-Dinh, T., Gemissel, B., Igolen, J. & Simonot, G. (1982) Eur. J. Biochem. 124, 415-425

(23) Cheng, D.M., Kan, L.S., Leutzinger, E.E., Jayaraman, K., Miller, P.S. & Ts'o, P.O.P. (1982) Biochemistry 21, 621-630

(24) Altona, C. (1982) Recl. Trav. Chim. Pays-Bas 101, 413-433

(25) Karplus, M. (1963) J. Am. Chem. Soc. 85, 2870-2871

(26) Noggle, J.H. & Schirmer, R.E. (1971) The Nuclear Overhauser Effect - Chemical Applications, Academic Press, New York

(27) Johnston, P.D. & Redfield, A.G. (1978) Nucleic Acids. Res. 5, 3913-3927

(28) Roy, S. & Redfield, A.G. (1983) Biochemistry 22, 1386-1390

(29) Hare, D.R. & Reid, B.R. (1982) Biochemistry 21, 1835-1842

(30) Heerschap, A., Haasnoot, C.A.G. & Hilbers, C.W. (1982) Nucleic Acids Res. 10, 6981-7000

(31) Heerschap, A., Haasnoot, C.A.G. & Hilbers, C.W. (1983) Nucleic Acids Res. 11, 4483-4500

(32) Heerschap, A., Haasnoot, C.A.G. & Hilbers, C.W. (1983) Nucleic Acids Res. 11, 4501-4520

(33) Reid, D.G., Salisbury, S.A., Brown, T., Williams, D.H., Vasseur, J.J., Rayner, B. & Imbach, J.L. (1983) Eur. J. Biochem. 135, 307-314

(34) Patel, D.J., Kozlowski, S.A., Nordheim. A. & Rich, A. (1982) Proc. Natl. Acad. Sci. USA 79, 1413-1417

(35) Assa-Munt, N. & Kearns, D.R. (1984) Biochemistry 23, 791-796

(36) Patel, D.J., Kozlowski, S.A., Ikuta, S. & Itakura, K. (1984) Biochemistry 23, 3207-3217

(37) Patel, D.J., Kozlowski,j S.A., Ikuta, S. & Itakura, K. (1984) Biochemistry 23, 3218-3226

(38) Patel, D.J. (1982) Proc. Natl. Acad. Sci. USA 79, 6424-6428

(39) Reid, D.G., Salisbury, S.A. & Williams, D.H. (1983) Biochemistry 22, 1377-1385

(40) Brown, S.C., Mullis, K., Levenson, C. & Shafer, R.H. (1984) Biochemistry 23, 403-408

(41) Gronenborn, A.M., Clore, G.M., Jones, M.B. & Jiricny, J. (1984) FEBS Lett. 165, 216-222

(42) Clore, G.M. & Gronenborn, A.M. (1983) EMBO J. 2, 2109-2115

(43) Clore, G.M., Gronenborn, A.M., Moss, D. & Tickle, I.J. (1985) J. Mol. Biol. 184, in the press

(44) Wüthrich, K., Wider, G., Wagner, G. & Braun, W. (1982) J. Mol. Biol. 155, 311-319

(45) Redfield, A.G., Roy, S., Sanchee, U., Tropp, J. & Figueroa, N. (1981) in: Biomolecular Stereodynamics, vol. 1 (Sarma, R.H. ed.) pp. 195-208, Adenine Press, New York

(46) Gronenborn, A.M. & Clore, G.M. (1985) Progr. in NMR Spec. 17, 1-32

(47) Macura, S. & Ernst, R.R. (1980) Mol. Phys. 41, 95-117

(48) Solomon, I. (1955) Phys. Rev. 90, 559-565

(49) Kalk, A. & Berendsen, H.J.C. (1976) J. Magn. Reson. 24, 343-366

(50) Wagner, G. & Wüthrich, K. (1979) J. Magn. Reson. 33, 675-680

(51) Dobson, C.M., Olejniczak, E.T., Poulsen, F.M. & Ratcliffe, J. (1982) J. Magn. Reson. 48, 87-110

(52) Clore, G.M. & Gronenborn, A.M. (1985) J. Magn. Reson. 61, 158-164

(53) Kumar, A., Wagner, G., Ernst, R.R. & Wüthrich, K. (1981) J. Am. Chem. Soc. 103, 3654-3658

(54) Clore, G.M. & Gronenborn, A.M. (1984) FEBS Lett. 172, 219-225

(55) Holbrook, S.R., Sussman, J.L., Warrant, R.W. & Kim, S.H. (1978) J. Mol. Biol. 123, 631-660

(56) Hingerty, D., Brown, R.S. & Jack, A. (1978) J. Mol. Biol. 124, 523-534

(57) Moss, D.S. & Morffew, A.J. (1982) Comp. Chem. 6, 1-3

MOLECULAR DYNAMICS SIMULATION OF THE ANTICODON ARM OF PHENYLALANINE

TRANSFER RNA

Lennart Nilsson[a] and Martin Karplus[b]

[a]Department of Medical Biophysics, Karolinska Institutet
Box 60400, S-104 01 Stockholm, Sweden
[b]Department of Chemistry, Harvard University
12 Oxford Street, Cambridge, MA 02138

INTRODUCTION

Nucleic acid structures (Saenger, 1984), range from the classical
double helices of A- and B-DNA, the detailed 3 D structures of crystalline
tRNA (Kim et al., 1974; Ladner et al., 1975) to various DNA oligonucleotides
in both left handed and right handed helical forms, for which there
now is a wealth of experimental data (Saenger, 1984). This structural
diversity has prompted numerous theoretical studies of the equilibrium
properties of nucleic acids (see Olson, 1982, and references therein).
By contrast only a very limited number of dynamical simulations of
DNA (Levitt, 1983; Tidor et al., 1983) and tRNA (Harvey et al., 1984)
have been reported.

Molecular dynamics (MD) simulations of biological macromolecules
can provide information concerning the average conformations and the
fluctuations of individual atoms and groups of atoms (Karplus and
McCammon, 1983; Edholm et al., 1984). In spite of the ever increasing
capacity of electronic computers a full scale MD simulation of a molecule
the size of tRNA, which has 1788 atoms, not counting non-polar hydrogens,
poses a formidable computational challenge (a 100 ps trajectory would
require approximately 1500 CPU hours on a VAX 11/780). Methods to reduce
the number of degrees of freedom are therefore of great interest. The
tRNA molecule with its two "active sites", the amino acid binding 3'-end
and the anticodon arm is divided into well defined separate domains.
Thus, an obvious simplification is to simulate just one of these regions.

This paper reports some details of a 40 ps MD simulation of the
anticodon arm of yeast tRNAPhe (nucleotides G26 to A44) in vacuum.
The primary objective of the simulation is to assess the validity
of such an approach, and to obtain an estimate of the internal mobility
of a free anticodon arm in the 5' stacked conformation, as found in
the crystal structure (Kim et al., 1974; Ladner et al., 1975).

The dynamics of the anticodon arm is of interest both as an example
of a structural motif, the hairpin loop, in nucleic acid architecture
and in the light of experimental data concerning anticodon conformations
and dynamics (Rigler and Wintermeyer, 1983; Claesens and Rigler, 1985).
tRNAPhe has a built in mobility reporter in the intrinsically fluorescent

hypermodified base wybutine in position 37, adjacent to the 3' base
of the anticodon triplet, allowing comparisons between simulations
and picosecond fluorescence depolarisation experiments that are planned
(Claesens and Rigler, 1985).

In the 12 ps MD simulation of the whole tRNA molecule by Harvey
et al. (1984) the structural integrity of the molecule during the simulation
was verified, and atomic r.m.s. fluctuations in approximate accord
with crystallographic B-factors were reported. An anticorrelation between
the outermost hydrogen bond lengths in G-C pairs, indicating a rocking
motion of the base pairs, was also found.

The present 40 ps MD simulation of the 485 atoms in the anticodon arm
yields similar results, suggesting that on this timescale there is
very limited interaction between the anticodon loop and the rest of
the tRNA, from which it is separated by the six base pairs in the stem.

METHODS

All calculations were performed using the molecular mechanics program
CHARMM (Brooks et al., 1983).

In the MD simulation the motions of individual atoms of the system
are calculated by numerically integrating (Verlet, 1967) Newton's
equation of motion. The forces on the atoms are obtained as the gradients
of the energy. The energy function used is an empirical function of
the form (Brooks et al., 1983)

$$E(\bar{R}) = E_b + E_a + E_d + E_i + E_{ele} + E_{vdw} + E_{hb} \qquad (1)$$

Here $\bar{R}$ is a vector of the cartesian atomic coordinates, E_b, E_a, and E_d
are the energy terms associated with deviations from ideal bond lengths,
valence and dihedral angles respectively. E_i is a term assigned to so called
"improper dihedrals", which are used to keep chirality and planarity
of certain groups of atoms. The reamaining terms involve non-bonded
atoms (i.e. atoms separated by more than two covalent bonds), E_{ele} is a
term for the electrostatic interactions, E_{vdw} for van der Waals interactions,
and E_{hb} for hydrogen bond interactions.

The parameters and specific functional form of the energy function
were taken from Nilsson and Karplus (1985). with some extensions to
accomodate the modified bases in the tRNA. Only polar hydrogens were
treated explicitly (CH, CH2 and CH3 groups are treated as extended
carbon atoms).

The system simulated consists of the 485 atoms in the 19 nucleotides
from G26 to A44 in tRNA[Phe] (see Fig. 1), without water or counterions.
The absence of solvent was partially accounted for by reducing the
net charge on the phosphates to -0.32e (Tidor et al., 1983), and by
using a 1/r shielding function multiplying the Coulomb interactions
in E_{ele}. Non-bonded interactions were computed only for atoms closer than
10.5 Å, taken from a list that was updated every 20 steps, as was the
list of possible hydrogen bonding atoms. The hydrogen bond cutoff distance
was set to 7.5 Å.

With this choice of cutoff distances the number of non-bonded
interactions to calculate in each energy evaluation is reduced from
a total of 115 000 non-bonded pairs in the anticodon arm to 38 000.

The simulation was started from the atomic coordinates (G.J. Quigley, personal communication) of tRNA[Phe] in the orthorombic form (Kim et al., 1974). The anticodon arm from this structure was first subjected to 1000 steps of energy minimization using the standard CHARMM minimizer ABNR (Brooks et al., 1983) to relieve any strains. Velocities were then assigned to the atoms according to a Maxwellian distribution with T=48K, and the Verlet integration (Verlet, 1967) was started with a timestep of 0.001 ps. Bonds involving hydrogens were kept at fixed length using the SHAKE algorithm (Ryckaert et al., 1977). Velocities were reassigned every 0.1 ps increasing T by 25K until T=298K was reached. This heating was followed by an equilibration period during which the temperature was monitored every 0.5 ps. If the temperature deviated more than 5K from the nominal T=298K the atomic velocities were scaled accordingly. The last such scaling occurred at t=17.5 ps, and the temperature had stabilized at t=25 ps (Fig. 2). Coordinate sets were saved every 0.005 ps during the 40 ps from t=26 to t=66 ps that were used in the analysis.

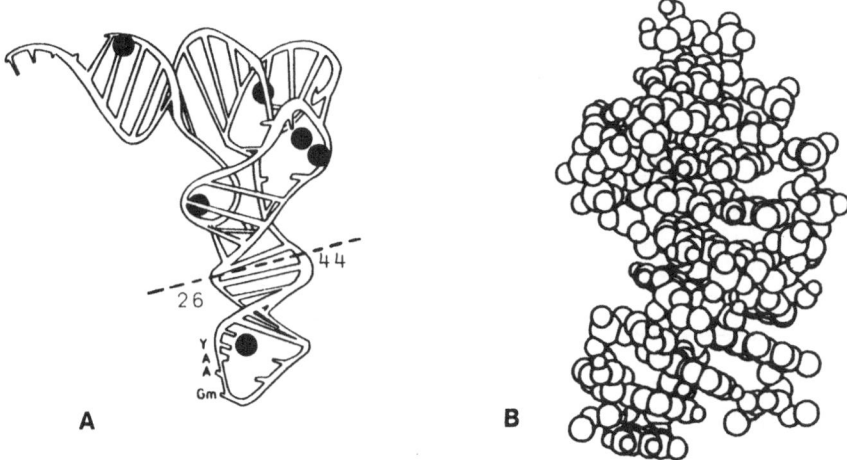

Fig. 1. Outline of the 3D structure of tRNA[Phe], from X-ray crystallography (Quigley et al., 1978). A) The whole molecule with the anticodon (G-A-A) and adjacent to it the wybutine (Y) indicated. The black circles are Mg binding sites. The dashed line shows where the part used in the simulation was cut off. B) Space-filling model of the anticodon arm, from G26 to A44, used in the simulation. The anticodon itself is in the lower right hand part of the figure.

RESULTS AND DISCUSSION

The simulation required 3.5 CPU hours of VAX 11/780 time per ps, a reduction by a factor of 4 compared with a simulation of the whole molecule. During the 40 ps analysis period the average temperature was 300 K, and the r.m.s. fluctuations in the total and kinetic energies were 0.1 and 11 kcal/mole, respectively. With a fluctuation in total energy less than 1% of the fluctuation in kinetic energy the numerical integration is considered stable and well-behaved. For analysis purposes average coordinate sets were calculated for i) eight consecutive 5 ps periods, ii) four consecutive 10 ps periods, and iii) for the whole 40 ps trajectory.

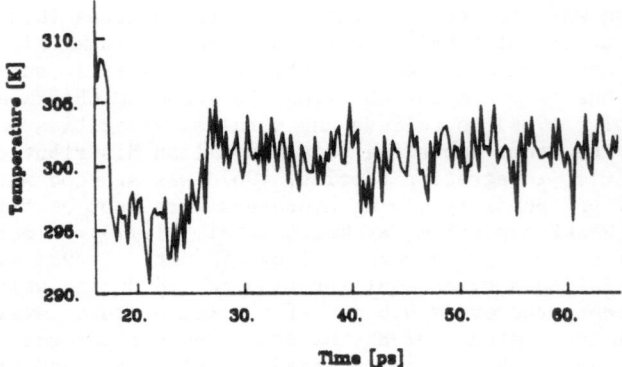

Fig. 2. The temperature during the simulation, from the last velocity
adjustment at 17.5 ps to the end of the run.

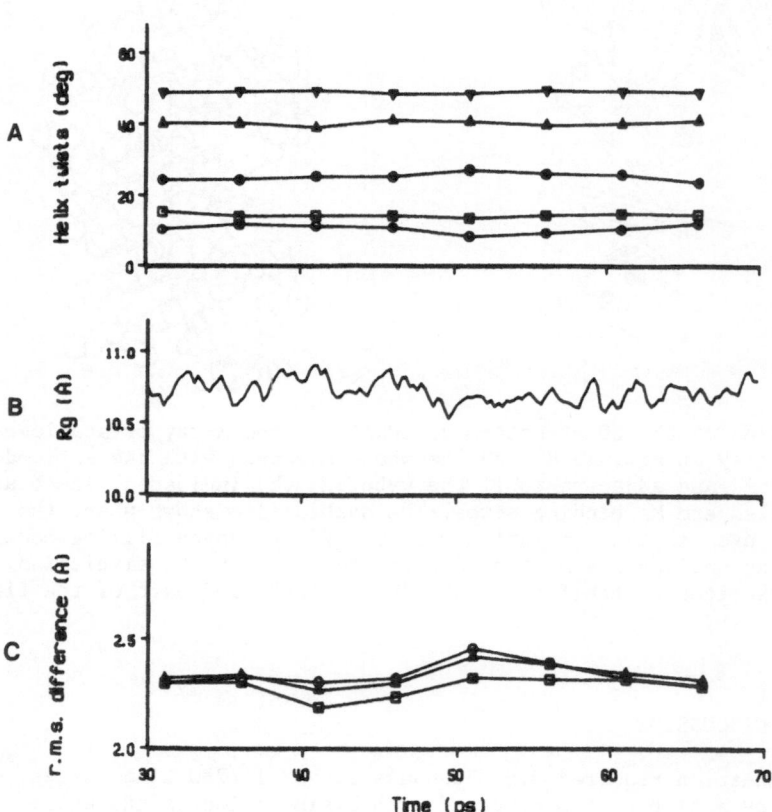

Fig. 3. Time dependence of global parameters of the molecule:
A) Global helix twist for base pair steps in the stem (●)
G26 to C27, (△) C27 to C28, (▽) C28 to A29, (□) A29 to G30,
(○) G30 to A31. From consecutive 5 ps averages.
B) Radius of gyration. C) r.m.s. difference from the initial ·
X-ray structure for (△) all atoms, (□) bases, and (○) backbone.
From consecutive 5 ps averages.

The stem keeps its overall helical character, with an average global helix twist of 28°, compared to 33° for the initial structure. From the twists at each step for the 5 ps average structures in Fig. 3A it is clear that the stem helix is stable during the simulation. The radius of gyration, Fig. 3B, is also stable around 10.8 Å, slightly lower than the 11.4 Å of the anticodon arm in the X-ray structure. The overall r.m.s. atomic displacements between the initial structure and the 5 ps average structures in Fig. 3C show that the backbone and base atoms contribute equally to the total, which is close to 2.3 Å; the deviation in the simulation by Harvey et al. (1985) was somewhat larger (3.1 Å). The r.m.s. displacements between the 40 ps average structure and the crystal structure averaged per nucleotide (Fig. 4) show that U41 in the middle of the stem is the nucleotide with the largest deviation.

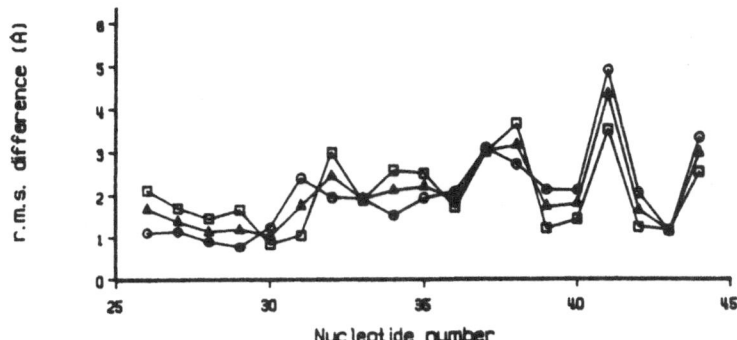

Fig. 4. r.m.s. difference between the initial X-ray structure and the 40 ps average structure for (△) all atoms, (□) bases, and (○) backbone averaged for each nucleotide.

The r.m.s. atomic fluctuations calculated in the 5, 10 and 40 ps averaging windows are given in Fig. 5. The results indicate that the fluctuations in the stem region converge to a maximum in 5 ps or less. In the loop, especially for the base atoms of G34 at the turn, the fluctuations increase with increasing averaging time. The large fluctuations for G26 and A44 are clearly artifacts due to the removal of the rest of the tRNA. Apart from these two nucleotides, no unexpected fluctuations are seen. While the trends in the fluctuations are the same as in the MD simulation of the whole tRNA by Harvey et al. (1984), the amplitude in this simulation is approximately half the amplitude in the full simulation. This could be due partly to some large domain motions, possibly of a hinge bending type (Harvey and McCammon, 1982; Nilsson et al., 1982), that are absent in a simulation of a reduced system.

Linear correlation coefficients for the hydrogen bond lengths in the stem were computed and are listed in Table 1 together with the corresponding coefficients from Harvey et al. (1984). The agreement between the two sets of correlation coefficients is generally good, and close to the loop it is almost perfect. This is true in spite of the fact that the two simulations differed not only in the size of the system used but also in the details of the potential function.

Correlation functions due to internal mobilities of individual nucleotides can be computed and compared with fluorescence depolarization experiments (Ichiye and Karplus, 1983). If the overall rotational tumbling of the system is approximately isotropic and much slower than the internal mobility, the time dependent fluorescence anisotropy can be written

$$r(t) = 0.4 \exp(-t/\tau) <P_2[\hat{\mu}_A(t') \cdot \hat{\mu}_E(t'+t)]>_{t'}, \qquad (2)$$

where τ is the rotational correlation time for the isotropic overall motion, and $\hat{\mu}_A$ and $\hat{\mu}_E$ are unit vectors parallel to the absorption and emission dipole moments of the chromophore. The angular brackets denote an average over the trajectory and P_2 is the second order Legendre polynomial

$$P_2(x) = \frac{(3x^2-1)}{2} \qquad (3)$$

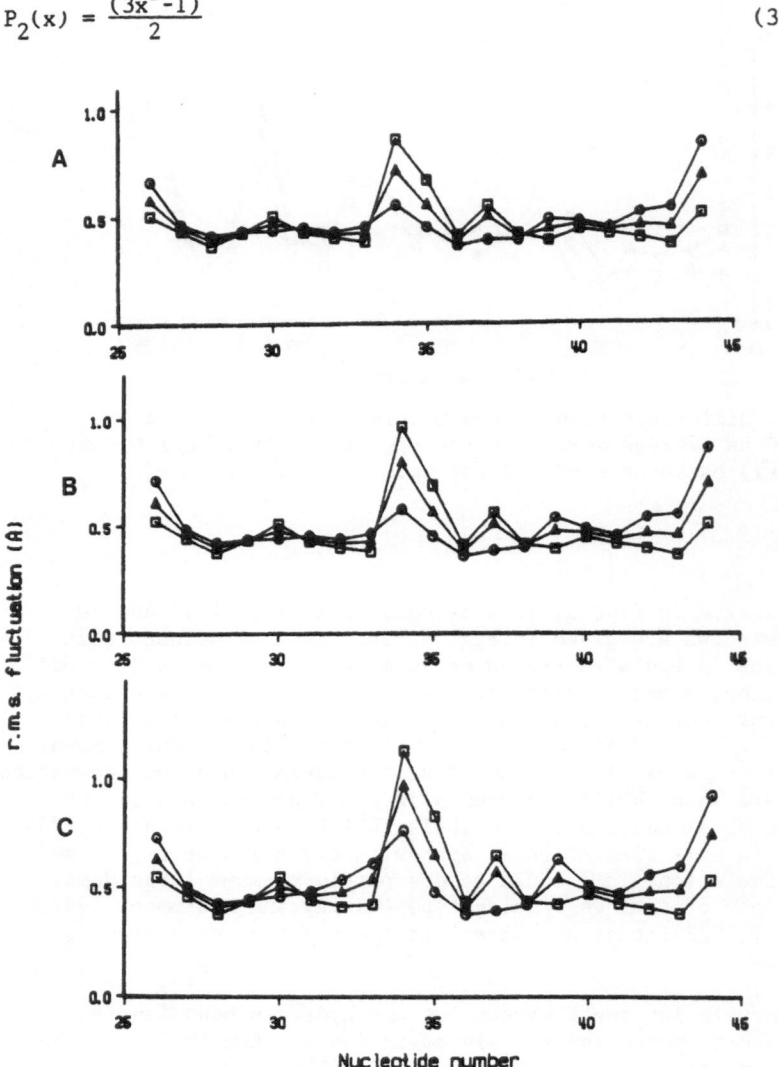

Fig. 5. r.m.s. fluctuations for ($\triangle$) all atoms, ($\square$) bases, and ($\bigcirc$) backbone in A) 5 ps, B) 10 ps, and C) 40 ps segments of the simulation averaged for each nucleotide.

Table 1. Linear Correlation Coefficients for Hydrogen Bond Lengths
in Anticodon Stem

Base pair	r_{12}^a	r_{23}^a	r_{13}^a
C27 - G43	0.71 (0.37)[b]	0.26 (0.65)	0.00 (-0.01)
C28 - G42	0.45 (0.33)	0.48 (0.22)	0.09 (-0.10)
A29 - U41	0.02 (0.19)		
G30 - C40	0.47 (0.34)	0.29 (0.29)	-0.12 (-0.13)
A31 - Ψ39	0.25 (0.26)		

In G-C pairs bond 1 is between N4(C) and O6(G), bond 2 is in the
middle and bond 3 is between O2(C) and N2(G).
[b] Values in parentheses are from the simulation by Harvey et al. (1984).

Correlation functions were calculated for the wybutine base at
position 37 assuming that the dipole moments are either parallel or
perpendicular to the long axis of the three ring system of the wybutine,
since the actual orientations are not known. In both cases the correlation
functions have a rapid initial decay (Fig. 6) and then levels out at
a plateau value of 0.97 - 0.98. The decay time is approximately 0.3
ps, which is much faster than the fastest observed fluorescence anisotropy
decay times of 0.37 ns (Claesens and Rigler, 1985). This means that
in the course of the simulation the wybutine oscillates rapidly in
a spatially very restricted region. A possible explanation of the
experimentally observed anisotropy decay is that the wybutine jumps
between different local conformations on a longer timescale than that
of the present simulation. With the wybutine stacked on top of the
anticodon triplet as seen in the crystal there is hydrogen bonding
between the anticodon bases and the wybutine (Fig. 1B), which also
restricts its flexibility in the simulation. The hydrogen bonding may
be different in solution and other anticodon loop conformations may
be present, leading to a less confined motion of the wybutine.

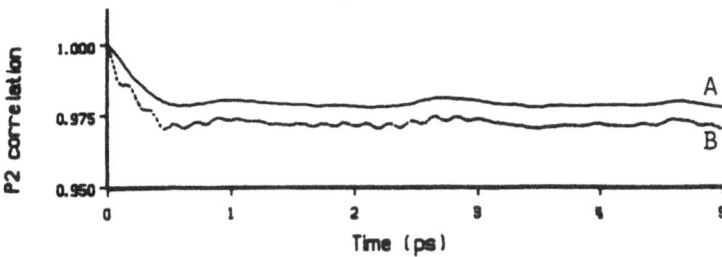

Fig. 6. The second order Legendre polynomial correlation function
computed from the motions of the wybutine assuming absorption
and emission dipoles A) parallel and B) perpendicular to the
long axis of the ringsystem of the wybutine.

We have shown that no serious artifacts are introduced for a molecular
dynamics simulation of the anticodon loop of tRNA[Phe] when only the
nineteen nucleotides from G26 to A44 are included in the simulation.
It is planned to extend the investigation of the anticodon dynamics
to longer times in the presence of the surrounding solvent. This will
be achieved by using a stochastic boundary simulation method recently
applied to the active site of RNAse (Brunger et al., 1985). More detailed
comparisons will be made with experimental anisotropy data, where processes

on a time scale within reach of MD simulations are seen.

ACKNOWLEDGEMENTS

We wish to thank Drs. B.R. Brooks, C.L. Brooks, M. Pettitt, C.B. Post, and R. Rigler for many helpful discussions. L.N. was the recipient of a Swedish Natural Science Research Council Post-Doctoral Fellowship for his stay at Harvard University where this study was initiated.

REFERENCES

Brooks, B. R., Bruccoleri, R. E., Olafson,B.D., States, D. J., Swaminathan, S., and Karplus, M., 1983, CHARMM: A Program for Macromolecular Energy, Minimization, and Dynamics Calculations, J. Comp. Chem., 4:187.

Brunger, A. T., Brooks, C. B., Karplus, M., 1985, Proc. Natl. Acad. Sci. USA, in press

Claesens, F., and Rigler, R., 1985, Conformational Dynamics of the Anticodon Loop in Yeast tRNA^{Phe} as Sensed by the Fluorescence of Wybutine, Eur. Biophys. J., in press

Edholm, O., Nilsson, L., Berg, O., Ehrenberg, M., Claesens, F., Gräslund, A., Jönsson, B., and Teleman, O., 1984, Biomolecular dynamics. A report from a workshop in Gysinge, Sweden, October 4-7, 1982., Quart Rev. Biophys., 17:125.

Harvey, S. C., and McCammon, J. A., 1982, Macromolecular Conformational Energy Minimization: An Algorithm Varying Pseudodihedral Angles, Computers & Chemistry, 6:173.

Harvey, S. C., Prabhakaran, M., and McCammon, J. A., 1985, Molecular-Dynamics Simulation of Phenylalanine Transfer RNA. I. Methods and General Results, Biopolymers, 24:1169.

Harvey, S. C., Prabhakaran, M., Mao, B., and McCammon, J. A., 1984, Phenylalanine Transfer RNA: Molecular Dynamics Simulation, Science, 223:1189.

Ichiye, T., and Karplus, M., 1983, Fluorescence Depolarization of Tryptophan Residues in Proteins: A Molecular Dynamics Study, Biochemistry, 222:2284.

Karplus, M., and McCammon, J. A., 1983, Dynamics of Proteins: Elements and Function, Ann. Rev. Biochem., 53:263.

Kim, S. H., Suddath, F. L., Quigley, G. J., McPherson, A., Sussman, J. L., Wang, A. H. J., Seeman, N. C., and Rich, A., 1974, Three-Dimensional Tertiary Structure of Yeast Phenylalanine Transfer RNA, Science, 185:435.

Ladner, J. E., Jack, A., Robertus, J. D., Brown, R. S., Rhodes, D., Clark, B. F. C., and Klug, A., 1975, Structure of Yeast Phenyl-alanine Transfer RNA at 2.5 Å Resolution, Proc. Natl. Acad. Sci. USA, 72:4414.

Levitt, M., 1983, Computer Simulation of DNA Double-helix Dynamics, Cold Spring Harbor Symp. Quant. Biol., 47:251.

Nilsson, L., and Karplus, M., 1985, Empirical Energy Functions for Energy Minimizations and Dynamics of Nucleic Acids, Submitted for publication.

Nilsson, L., Rigler, R., and Laggner, P., 1982, Structural Variability of tRNA: Small-Angle X-ray Scattering of the Yeast tRNA^{Phe} - Escherichia coli tRNA$^{Glu}_2$ Complex, Proc. Natl. Acad. Sci. USA; 79:5891.

Olson, W. K., Theoretical Studies of Nucleic Acid Conformation: Potential Energies, Chain Statistics, and Model Building, 1982, in: "Topics in Nucleic Acid Structure, Part 2", S. Neidle, ed., MacMillan Press, London.

Quigley, G.J., Teeter, M. M., and Rich, A., 1978, Structural Analysis of Spermine and Magnesium Ion Bidning to Yeast Phenylalanine Transfer RNA, Proc. Natl. Acad. Sci. USA; 75:64.

Rigler, R., and Wintermeyer, W., 1983, Dynamics of tRNA, Ann. Rev. Biophys. Bioeng., 12:475.

Ryckaert, J. P., Cicotti, G., and Berendsen, H. J. C., 1977, J. Comp. Phys., 23:327.

Saenger, W., 1984, "Principles of Nucleic Acid Structure", Springer-Verlag, New York.

Tidor, B., Irikura, K. K., Brooks, B. R., and Karplus, M., 1983, Dynamics of DNA Oligomers, J. Biomol. Struct. Dyn., 1:231.

Verlet, L., 1967, Phys. Rev., 159:98.

ANTICODON-ANTICODON INTERACTIONS AND tRNA SEQUENCE COMPARISON:

APPROACHES TO CODON RECOGNITION

H. Grosjean[*], C. Houssier[**] and R. Cedergren[***]

*Laboratoire de Chimie Biologique, Université Libre
de Bruxelles, B-1640 Rhode St-Genèse, Belgique
**Laboratoire de Chimie Physique, Université de Liège
Sart-Tilman, B-4000 Liège, Belgique
***Département de Biochimie, Université de Montréal
Montréal, Québec H3C 3J7, Canada

INTRODUCTION

The intricate series of events which produce protein from nucleic
acid structural information is called translation. The simplicity of this
term largely misrepresents the complexity of the underlying processes
which involve precise interactions between many proteins and nucleic
acids. Although the importance of the interplay of each component of the
system is not well known, it is generally recognized that two molecules
play a key role : transfer RNA and its trinucleotide anticodon which
recognizes a trinucleotide codon in messenger RNA. There is now ample
evidence that the efficiency and accuracy of this system depend on the
RNA-RNA and RNA-protein interactions (for a review see ref. 1 and 2).
Given the complexity of the ribosomal context in which RNA is decoded
we are forced to implement simpler experimental models from which information, hopefully related to the biological system, can be obtained.
We present here the tack which has been taken and some critical results
which lead to important structural correlations dealing with translation
and its control.

STRATEGY

In order to develop an experimental strategy, it is useful to
consider the origin and historical developments of the translation apparatus. Evolutionists have generally agreed that the origin of translation was much simpler than the present-day system (3,4). Differences in
translation factors required in eukaryotes and eubacteria as well as the
apparent simplicity of mitochondrial protein synthesis are consistent with
the notion of the evolution of translation complexity. Taken to the
extreme one could suggest that the origin of translation involved only
RNA. Recent discoveries on the enzymatic activity of RNA (5,6) has gone
a long way to fan speculation on the primacy of RNA in genetic expression.
The hypothesis of an RNA origin for translation implies however that at
some very early point in evolution, probably before the divergence of
the existing kingdoms, a primitive ribosome was introduced. The subsequent historical development of translation was concomitant with higher
efficiency and overall accuracy to attain the present level (7). If this

scenario is correct, information on the basic recognition or interaction between mRNA and tRNA resides in the three dimensional structure and dynamics of these nucleic acids.

We have attacked this complex problem by developing a model system based on the interaction of RNA molecules. In particular, when it was shown that tRNAs having complementary anticodons could form dimers (8,9), a very useful representation of codon-anticodon interactions was devised using anticodon-anticodon interactions (10, reviewed in 11). The data on anticodon-anticodon interactions determined by fast relaxation techniques (T-jump) showed an unexpectedly high stabilization energy in anticodon associations, a result which would not have been seen, if simple oligo-nucleotide models had been used. The following section summarizes our data with those anticodon tRNA-tRNA complexes having strictly complementary anticodons. Figure 1 shows an example of the relaxation signal observed for the dissociation of one such complex after a temperature jump of $4.2^{\circ}C$. Kinetic and thermodynamic parameters as well as spectral characteristics of the anticodon-anticodon complex are easily obtained from such experiments performed at the different temperatures and/or concentrations of interacting tRNAs molecules.

FEATURES OF THE ANTICODON-ANTICODON COMPLEX

The association constant of two tRNAs due to interaction of three complementary anticodon base pairs is of the order of 10^{6} M^{-1} at $0^{\circ}C$. This value is two to three orders of magnitude higher than that of the tRNA anticodon-codon triplet interaction (12,13 and references therein). On the other hand, two complementary triplets do not associate to any significant extent even at $0^{\circ}C$ (14). It is evident that the tRNA molecule has some built-in features which considerably stabilize base pairing between anticodons. A first tentative explanation proposed that the accrued energy was due to the fact that the anticodon was frozen in a particularly favourable geometry, thereby reducing the unfavourable entropy term (8). This explanation however is not the complete story. First, physical measurements of tRNA by various techniques clearly point to a non-rigid anticodon in solution (15 and review in ref. 16, 17a and 17b). Secondly, an extensive study on the stability of different tRNA pairs showed that there is no correlation with the G-C content. In fact the relative stability of different pairs of complementary tRNAs, as measured by the lifetime of the complexes fall within a range of a factor of only 10 (ref. 18). From data on double helical ribonucleotides, one would expect an association involving three G-C to be at least 1000 times longer lived than one with three A-U pairs (19). Apparently, the architecture of anticodon loops in natural tRNA species is such that not only is the binding energy of complementary nucleotides increased by several orders of magnitude compared with binding between unstructured oligonucleotides but also the intrinsic difference between G-C and A-U pairs has been dramatically reduced ("optimal binding energy hypothesis"; ref. 18).

Detailed analysis of thermodynamic and kinetic parameters of different tRNA-tRNA complexes (10,18,20-25) reveals three major effects which enhance the stability of the interaction : a) the decreased flexibility of the loop due to constraints afforded by the base-paired stem. b) the presence of non-complementary nucleotides outside the triplet interaction (the dangling end effect). c) the presence, in the loop, of modified nucleotides which have stronger stacking interactions. Our results argue against a major contribution from the remaining part of the tRNA molecule (10,23), but we can not rule out a long range effect in ribosomal decoding. The relative importance of these three categories of effects depends on the nucleotide sequence and environment or

"context" (particularly the 3' side) of the anticodon. For the complex
between yeast tRNA (Phe) (anticodon GmAA) and E. coli tRNA (Glu) (anticodon mnm^{5}S^{2}UUC), loop constraints have been evaluated to a factor of 44,
a value obtained by comparing the kinetic and thermodynamic values of tRNA
pair interaction with RNA fragments (10 and 12 nucleotides long) containing
the anticodon sequences, but which are unable to adopt a loop structure
(10). Thus, in tRNA pairs having at least two A-U pairs in their anticodons the major effect comes from stacking of adjacent nucleotides with
the paired triplets. On the contrary, for tRNA pairs having G-C rich
anticodons such as complex between E. coli tRNA (Gly) (anticodon GCC) and
E. coli tRNA (Ala) (anticodon GGC) it is more the total standard entropy
term which favours the formation of the complex (25,and unpublished
results).

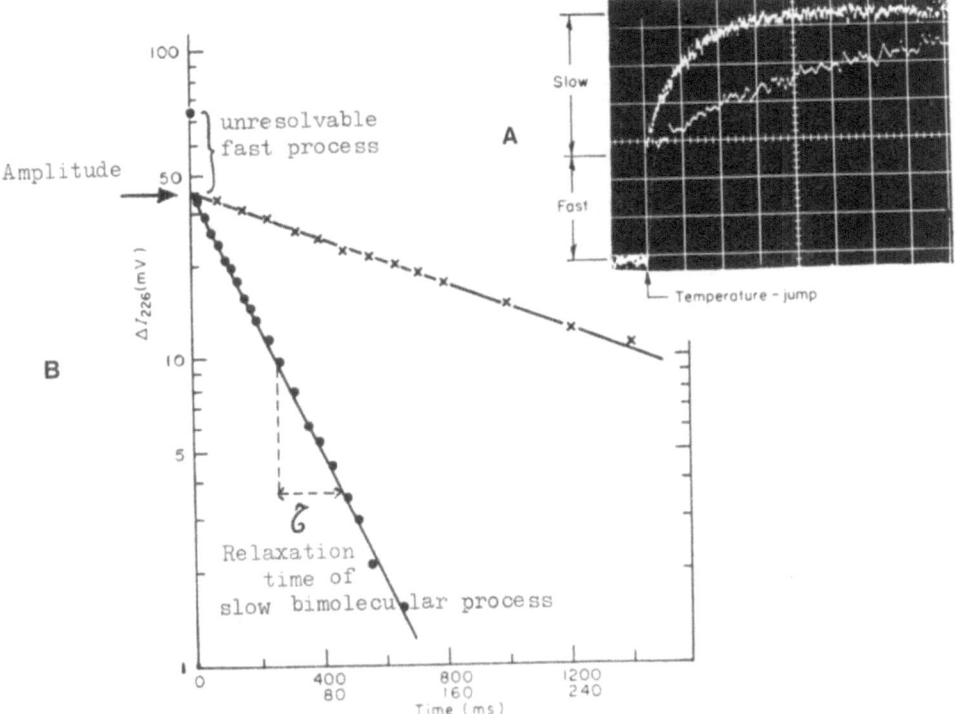

Fig.1   Relaxation signal for the dissociation of an anticodon-anticodon
complex, observed in the U.V. range.
Data correspond to the dissociation of the complex [yeast tRNA(Phe).
E.coli tRNA(Glu)] after a T-jump of 4.2°C (initial temp. is 13.5°C).
The oscilloscope trace (upper part A) is displayed at 2 different
time scale (200 ms/div. and 40 ms/div.); the vertical scale corresponds to a variation in amplitude of only 10^{-3}A$_{266}$/div. The initial
very rapid change of absorbance after the T-jump (in the microsecond
range) corresponds to base unstacking in single stranded regions of
tRNA; the characteristic slower variation of absorbance (in the milli-
second range) corresponds to the dissociation of a fraction of the
tRNA-tRNA complex. From the semilogarithmic plots of these relaxation
signals (lower part B) one can obtain the relaxation time from the
slope (here ζ = 210 ms) and the corresponding amplitude from the
intercept at t = 0 (here 4.10^{-3}at 266nm). Other experimental conditions are : 1.5 10^{-6} M in each tRNA,   10 mM cacodylate buffer,pH 7,
10 mM MgSO$_4$, 100 mM Na$_2$SO$_4$.(for more details see ref. 10).

Concerning the kinetic parameters, the recombination rate constants observed for the tRNA anticodon-anticodon association are very high ca. 10^6 $M^{-1}sec^{-1}$ at 0°C, the expected value in reference to oligonucleotide duplex formation (19). Their variation with temperature is small. It has been suggested that the rate determining step in the process of recognition depends on the zipping up from one base pair (26). The activation energy for complex formation may become highly negative when a mismatch is present in the middle of the anticodon or when only A-U base pairs are implicated. Although it was shown that the recombination rate was relatively salt independent, some compounds such as ammonium sulfate do increase the rate constant for complex formation (23). It is possible that an alternate conformation of the loop is favoured under these conditions, since Wye base fluorescence in yeast tRNA (Phe) is also affected (27, 28).

In contrast with the recombination rate constants, the dissociation rate constants are low and show large sequence dependent variations. Oligonucleotide duplexes (29) and tRNA anticodon-anticodon associations display very similar behaviour in that the changes in dissociation rate constants are responsible for most of the variations in stability caused by mismatching or nucleotide modifications (18,20,21,23-25).

It is well known that the dominant stabilizing effect in double stranded nucleic acids is the energy from base stacking. Base pairing provides only the proper orientation of base rings to favour this stacking. Hydrogen bonds between bases hardly compensates for the energy loss by removing solvent molecules. An early structural model of anticodon-anticodon interaction suggested the stacking of the triplets in an almost continuous A-form helix with the stem region (10,11). Confirmation of these suggestions has recently been obtained from NMR on a yeast tRNA (Phe) complex with the UUC codon (30a,30b) and from X-Ray crystallography of the yeast tRNA (Asp) dimer (16,45). From the latter studies it is clear that the so-called wobble position (position 34) is involved not only in intra-molecular stacking with its neighbouring anticodon base but also in intermolecular stacking with the opposing base in the anticodon-anticodon complex. Thus, 3' adjacent nucleotides of both anticodons are equally involved in stacking stabilization whereas the 5' adjacent pyrimidines are outside of the interaction and may be considered to be dangling. In other words a short double helix formed between the two (out-of-three) complementary anticodons is greatly stabilized if it is "sandwiched" between stacked, but not necessarily complementary nucleotides (see below). This would account for the large enthalpy change, the slow dissociation rate, the effect of loop constraint, modified nucleotides and dangling ends.

Interestingly enough the strong binding that characterizes the complex between anticodons is restricted to the three central bases of the seven membered loop. An additional pairing with U on the 5' side was found to be less stable by a factor of 100 (ref. 10) even though one would expect increased stability as compared to oligonucleotide model compounds. These and other observations taken together suggest that the triplet nature of the genetic code is fundamentally related to the seven membered anticodon loop. They also suggest that one function of the ribosome would be to provide the environment so that the mRNA might adopt a conformation similar to that of an anticodon loop.

FEATURES OF THE ANTICODON LOOP AND STEM ARCHITECTURE

As seen above interactions involving the anticodon loop of tRNA lead to unusually stable complexes. However examination of anticodon loop sequences shows additional pecularities. A comparison of 260 tRNA sequences from various origins (165 cytoplasmic, chloroplastic and phage tRNAs, 53 fungal mitochondrial tRNAs and 20 archaebacterial tRNAs) reveals certain characteristic environmental features of the anticodon nucleotides (Fig. 2) (ref. 31,32). With few exceptions, the three anticodon nucleotides are localized in the middle of a seven nucleotide loop which is closed by a five base-pair stem. On the 5' side of the anticodon (positions 32 and 33) only pyrimidines are permitted : base 33 is uracil in all elongator tRNAs and base 32 is occasionally modified. The base 3' adjacent to the anticodon is always a purine and is frequently modified in a characteristic way. Next to it, position 38 usually contains an unmodified adenosine, although a pyrimidine, sometimes modified, and a guanosine are occasionally tolerated. Like in the anticodon loop, the 3' half of the anticodon stem is characterized by a non-random distribution of nucleotides. No adenosine is found in position 40 of any tRNA except from mitochondria. The base pair 29-41, in the middle of the anticodon stem, is a strict Watson-Crick complement containing no modified nucleotides. The non-random distribution of nucleosides in the loop and stem is even more prominent when isoacceptor tRNAs from a wide variety of organisms are considered (33; see also Fig. 3 in ref. 1). Although a common origin for these tRNAs is likely (4,31,33), the highly conserved nature of distinctive bases must reflect a strong structural requirement related to the efficient and accurate reading of mRNA codons by the present-day translation ribosomal system (1,7,31,34).

Striking correlations are also evident between positions within the anticodon loop and stem. For example, many correlations have been observed between the presence of a particular modified purine in position 37 and the type of base in the third position of the anticodon, also between the presence of a particular modified base in position 34 and the type of base in the middle position of the anticodon (see Fig. 2c,2d and ref. 35a,35b). The precise modification depends on the type of organism and on its level of differentiation (31,36); thus nucleotide modification can be related to evolutionary principles of selection. We believe that one function of these modified nucleotides in the anticodon loop is to modulate the binding energy of the anticodon for its conformationally complementary codon (since Watson-Crick complementarity is not strictly required in decoding, we prefer the term conformational complementarity). In consideration of these correlations and of the non-random distribution of nucleotides in this region, we would like to define the "anticodon conformational region" shown in Fig. 2b involving the anticodon loop and the stem up to and including the 29-41 base pair. This region includes the region defined as the "extended anticodon" of Yarus (34) but it emphasizes the importance of conformational contributions of each nucleotide rather than their identity.

As stated above, the conformation of the anticodon loop is known to be quite flexible (15-17,27,28). This flexibility depends on experimental conditions such as the presence of complementary oligonucleotides. With yeast tRNA (Phe), the interaction with the anticodon complementary oligonucleotide was shown to produce an important structural reorganization around the modified nucleoside in position 37, corresponding to a stronger stacking with the neighbouring base (30a,30b); an accompanying reorganization takes place in the rest of the molecule and also leads to the dimerization of the tRNA molecules (13,37). Clearly, the importance of a given nucleotide is not necessarily related to the chemical nature of functional groups, but rather to the interplay of these groups to produce and define a standard conformation and a flexibility or dynamics existing between a limited number of conformers.

FINE TUNING IN DECODING

Modulation role for modified bases 3'-adjacent to the anticodon

One of the most distinctive structural features of tRNAs is the presence of a significant proportion of post-transcriptional modifications of nucleosides. As shown in Fig. 2a, most of these modifications of bases are found either at the first position of the anticodon (position 34), or adjacent to its 3' side (position 37). In this latter case, it was observed (Fig. 2c and ref. 35) that anticodons ending with A or U were almost always flanked on the 3' side by a hypermodified purine base (base Y in eukaryotic tRNA (Phe), i^6 A or ms^2i^6A, t^6A, mt^6A or ms^2t^6A in other systems). On the contrary, in the case of anticodons ending with C or G or containing two of these bases, the 3' adjacent base is an unmodified A or an A or G modified by a simple methylation (m^2A, m^6A, m^1G, m^2G or m^1I). These modifications are specific not for the tRNA per se but for a class of anticodons and are therefore relevant to the codon tRNA recognition process (31).

We have started to evaluate the effect of these hypermodified bases on the intrinsic decoding properties of the anticodon loop in tRNAs. To this end, we have examined the influence of structural variations on the thermodynamic, kinetic and spectral properties of various pairs of tRNAs having complementary (or quasi complementary) anticodons (20,23,25).

Results of Figure 3 (ref. 25) indicate that methylthioisopentenyl (ms^2i^6) modification of the adenine residue in position 37 has a strong stabilizing effect arising mainly from a decrease of the dissociation rate constant by a factor of eight (23). Most, if not all, of the effect may be attributed to the ms^2 part of this hypermodification of A37 (ref. 25). Since it is detected in the differential relaxation spectrum in the 300 nm region, we can conclude that the stabilizing effect occurs by stacking of the hypermodified purine residue in position 37 on the anticodon triplets in the tRNA-tRNA complex (see also ref. 22).

Significant stabilization effects of the N^6-threonincarbamoyl group (t^6) on A37 (ref. 20) as well as of the Wye base (10) has also been measured. As discussed above, these modifications may reinforce the intra-strand stacking of an anticodon loop but also the inter-strand stacking with the nucleoside 34 of the complementary anticodon. They may also impede base pairing either with other bases of the anticodon loop (as found in the T-loop) or with the first base of the next codon of mRNA on the ribosome (16).

Another remarkable result is the very high stability of the complex between E. coli tRNA (Ser) (anticodon GGA) and E. coli tRNA (Gly) (anti-codon UCC) (Fig. 3 and ref. 25), in which no modified adenosine is present at position 37 (ref. 38). The most plausible rationalization of these data is to hypothesize that the affinity between two anticodons is very dependent on their anticodon sequence. The presence of three consecutive purines in the anticodon triplet, together with the two purines on its 3' side may yield sufficient stability for the A37 hypermodification to be useless. This kind of intra-strand stacking interaction should stabilize the preassociation conformation of the anticodon loop.

The stabilization effect of the methylthio group on anticodon-anti-codon interactions in certain tRNAs with A-U rich anticodons is of particular interest in relation to the recent demonstration that, in E- coli and in Salmonella, a deficiency in methylthiolation of i^6A37, due to iron starvation during growth, results in the derepression of the aromatic biosynthetic enzymes as well as in an increased synthesis of enterobactin, a high affinity iron chelator (39,40). This phenotype is identical to that of the E. coli miaA mutants which lack the isopentenyla-ting enzyme and therefore do not synthesize the entire ms^2i^6 group on A37.

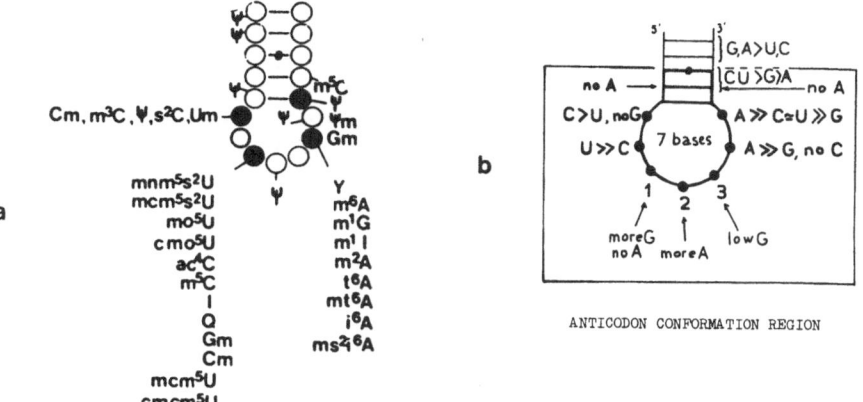

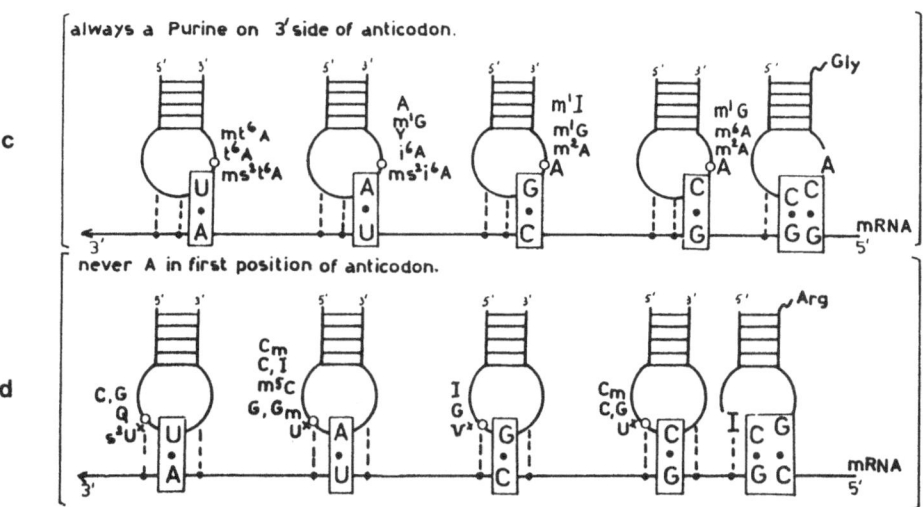

Fig.2 Built-in features of anticodon loop and stem as deduced from a sequence compilation of 260 tRNAs. This includes elongation and initiator
tRNAs from eubacteria and eukaryotes but not from phage T$_4$, mitochondria
or chloroplasts. Abbreviations for the modified bases are those used
in ref. 53. The dot in the middle of the anticodon stem is a strict
Watson-Crick base pair in the tRNA sequences compiled. "Anticodon conformation region" is as defined in the text. Part a gives positions
and types of posttranscriptional base modifications; part b gives an
overall distribution of the four canonical bases A, G, C and U.
part c shows the relationship between the third anticodon base and the
nucleotide in position 37; part d shows the relationship between the
base in position 34 and the middle base of the anticodon (position 35).
For more details see refs 31-36, 54.

In this latter case, undermodified tRNAs (Trp) were clearly shown to induce derepression of aromatic aminoacid operons by attenuation of transcription because of impaired efficiency of translation of tandemly repeated tryptophan codons in the mRNA corresponding to the leader peptide (41). Clearly, the control of the degree of 2-methylthiolation in a given tRNA could be one major determinant for the regulation of certain operons by attenuation of transcription. This does not exclude the possibility that the bulky hydrophobic isopentenyl group, or other parts of the tRNA molecule (1,2) may also contribute to the modulation of the codon-anticodon interaction during the translation process, upon interaction with the "decoding system" on the ribosome (peptidyl-tRna in P site, ribosomal RNA or proteins) (51, 52).

Modulation role for the wobble base

The kind of base in the first position of the anticodon (wobble base) is generally related to ambiguity of the decoding process : for example, tRNAs that read the four family codons ending with A, G, C or U have unmodified uridine in position 34 while tRNAs that read pairs of family codons ending with A or G, as well as those reading only one codon ending with A have some modified form of uridine in the same anticodon position (5-substituted uridines or their 2-thiouridine derivatives) (see ref. 35a and 35b). It is believed that these modified bases influence the intrinsic pairing properties by alteration of the stacking tendency of the whole anticodon making it more or less rigid (42 - 44). However, as we will argue below, it may also be that the kind of base (modified or not) is mostly related to the modulation of the binding energy between codon and anticodon.

Information about the specificity of pairing in anticodon-anticodon complexes was derived from T-jump relaxation experiments on more than sixty pairs of tRNAs with partly complementary anticodons, i.e. which can make only two correct base pairs out of three (18). Results indicate that a "correct" base triplet interaction (as determined by the genetic coding rules) has a long lifetime. Among those stable anticodon-anticodon complexes are those involving the G.U or Q.U base pair between the first and the third base of the anticodon (see Fig. 6 in ref. 11). Yet, these complexes are less stable than the same ones involving the G.C base pair instead of G.U by about one order of magnitude depending on the kind of base pair (G.C or A.U) in the middle position of the anticodon (compare the results from ref. 20 and 21). The complexes involving a G.U base opposition in the middle of the anticodon as well as most "genetically incorrect" complexes with mismatching bases, are marginally stable. One remarkable exception is the yeast tRNA (Asp) duplex involving a U.U mismatch in the middle position of the pseudo-complementary GUC anticodons (24,45). The special situation of the U.U pair between two G.C is probably one main determinant of this self-association.

These results demonstrate two important things : first, standardization of binding energy between two anticodons do not necessarily produce high selectivity in complex formation; second, the intrinsic potentiality of an anticodon to mispair is dependent on its immediate "context". If such RNA loop-loop interaction is related to the mechanism of genetic translation, then, to avoid errors, the ribosome must amplify the small differences in lifetime between various codons and anticodons to produce a greater distinction between "genetically correct" and "genetically incorrect" pairings in protein synthesis. Specific mechanisms for such an amplification have been discussed under the heading of kinetic proofreading (reviewed in ref. 2). Also, because differences in codon binding energy by a single tRNA recognizing more than one codon in its codon family, one may expect, as first suggested by Weiss (46) that mispairing at the wobble base is a possible target for regulation of the

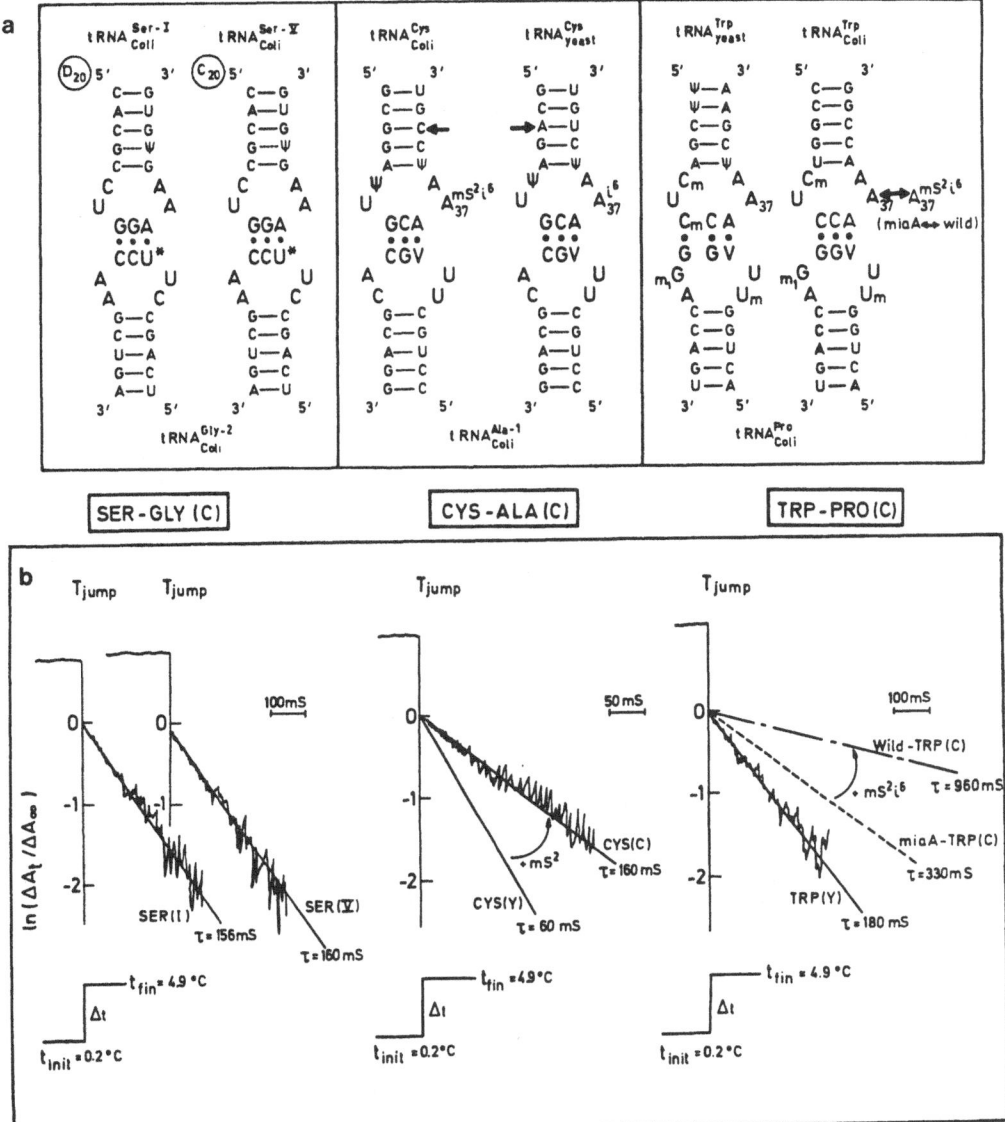

Fig.3 <u>Part a</u> : Nucleotide sequences of the anticodon stems and loops for different pairs of tRNAs having complementary anticodons.
Most symbols are the same as in Fig. 2, see also ref. 53. D_{20} and C_{20} point out the only difference in two species of E.coli tRNA(Ser) with a dihydrouridine in species I and a cytosine in species V(ref.38); mia A corresponds to a tRNA(Trp) from a mutant of E.coli lacking the isopentenylating enzyme; tRNA (cys) from yeast and E.coli differ by the type of modification in position 37, by one base pair in the anti-codon stem (see arrow) and by several nucleosides in the rest of the tRNA molecule.
<u>Part b</u> : Semilogarithmic plots of the slow components of relaxation signals observed at 260 nm for the various tRNA pairs shows in Fig 3a. Temperature jump size is 4.2°C; initial temperature : 0.2°C; tRNA concentration 1.6 10^{-6}M in the same buffer as indicated in legend of Fig.1. Results show that ms^2 and ms^2i^6 modification of A_{37} in tRNA stabilize the tRNA-tRNA complex which dissociates more slowly after a T-jump (as indicated by the arrows). Data taken from ref. 25, with permission.

rate of translation. Indeed, we observed that the choice between a U or C
at the third (degenerate) position of a codon in highly expressed mRNA
from E. coli is consistently determined by the nature of the other two
bases of the codon. If the first two interactions involve G.C pairs, the
third interaction tends to be the wobble pair G.U rather than G.C. Conversely, the interaction in the wobble position tends to be G.C if it is preceeded by two A.U pairs (the optimal codon-anticodon hypothesis; ref. 10,
47,48). This optimal binding energy is probably of the same order of magnitude for all codon-anticodon pairs in a given organism but not necessarily
the same for all organisms. In agreement with this idea, yeast tRNAs often
form less stable anticodon-anticodon complexes than their corresponding E.
coli pairs(25). This is probably related to the particular architectural
organization of eukaryotic tRNAs as well as to the types of hypermodified
bases found in their anticodons (31,35a,36). One can note, for example, the
systematic absence of a stabilizing ms^2 group on i^6A37 and the presence of
a characteristic inosine at position 34 in most major species of these yeast
tRNAs. From T-jump relaxation work , we know that complexes involving a
I.C pair are notably less stable than those involving a G.C pair, and
are of the same relative stability as the complexes containing a wobble
G.U or I.U pair (18).

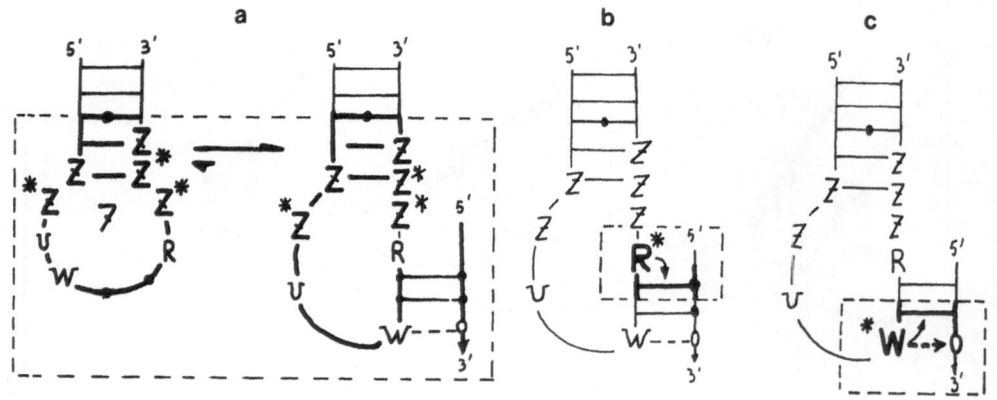

Fig.4 Aspects of codon - anticodon interactions discussed in this paper
        a) modulation of the strenght of binding by loop constraint:
here the kind of bases (Z) in the proximal anticodon stem and loop
play a role in the flexibility and hence the preferential conformation
adopted by the anticodon when it binds to a complementary codon; the
3'stacked conformation (refs 16,30,42) is schematically represented.
A major contribution of the ribosomal milieu is probably also important.
        b) modulation by the immediate "context":here the purine (R)
3' adjacent to the anticodon (and possibly the base 3' adjacent to the
codon) plays a role in the stabilization of the base pair between the
third anticodon base and the first codon base; certain modified bases
such as t^6A, ms^2i^6A and "Y" base are particularly efficient because of
their stacking potential.
        c) modulation by the so-called wobble base (W) : here the
strength of binding depends on the possibility to form a stable Watson-
Crick type base pair rather than a less stable "mismatched" pair(modulation by third codon choice); it also depends on the stabilizing
effect (stacking) of the wobble base (modified or not)on the strenght
of pairing in the middle position of the codon-anticodon complex.
Asterisks show positions where the bases are most frequently modified.

These arguments clearly support the idea of a cellular strategy in selecting particular codons and anticodons to optimize translation efficiency. An important target of the modulation of codon-anticodon interactions is careffully built-in features of tRNA molecules such as the base modifications in the anticodon loop and the proximal stem (as shown by asterisks in Fig. 4). These may be part of a coordinated structure-modification strategy that allows cells and organelles to regulate the translation of their messenger RNA in response to physiological stress (36,54).

EVOLUTION OF DECODING

The results presented can also be related to the evolution of decoding. The constant nemesis of the primordial translation system has concerned the lack of sufficient binding energies in triplets to effect protein synthesis (3,4). However the stability of triplet binding within seven member loops is probably more than enough to promote protein synthesis in primordial "soups" (7,10). In this way codons of no more than the current three nucleotide codons are necessary. This intrinsic property of RNA structures to form loops which can strongly interact with other molecules may be the basis of the enzymatic activity of RNA as well (49).
The argument presented above is related to the most stable arrangement of the anticodon loop, i.e. a 3' stack as originally described by Fuller and Hodgson (42). In this view, the stacking energies of participant bases are crucial to the definition of the conformation of the anticodon loop and of the codon-anticodon complex. This might be an important feature of the present genetic code, because it must be at the origin of the non-complementarity permitted in the so-called wobble position or "two out of three" decoding (50). However, if one takes this point one step further, that is to say if at some earlier date the 5' stack was permitted, the loop could have a choice of stacking depending on the sequence of the anticodon. This "best two out of three" choice of conformations would simplify the aminoacid code and could have preceeded the present code. Subsequent refinement by the use of modified nucleotides in the 3' adjacent position for example could eventually freeze the preferred arrangement of the loop (for more details see ref. 31 and 50).

EPILOGUE

The purpose of this essay is to present a simplified model of the interactions which are involved between mRNA and tRNA during message decoding. This model is based on studies of anticodon-anticodon interactions between two tRNAs by a fast relaxation technique. The T-jump technique permits the collection of data related to kinetic, thermodynamic and spectral parameters of a variety of triplet interactions between tRNA pairs. Although a model is by definition an approximation of the true phenomenon, we have given reasons showing that much of the data can be applied directly to in vivo decoding. As a complement to this experimental approach we have used the theoretical approach of sequence comparison and evolutionary interference to define the framework in which the above data is to be evaluated. The combination of the two approaches have permitted us to formulate the following concepts : i) the notion of an "anticodon conformation region" in tRNA which defines the dynamics of the tertiary structure of the anticodon triplet and thereby the structural constraints to codon-anticodon binding; ii) an optimal energy criterion which explains the relative sequence independence of the stability of the codon-anticodon interaction by loop constraints, stacking and modified nucleosides;

iii) a mechanism of translational control (Coordinated Modulation) relying on the interdependence of metabolic pathways, modified nucleotide synthesis and codon usage; iv) the evolutionary significance of these findings.

Although this work has been centered on the codon-anticodon interaction, it must not be forgotten that in vivo translation is much more complicated. Decoding is related to the more global phenomenon of translation which include contributions from the rest of the tRNA molecule as well as the messenger RNA, the ribosome, the peptidyl-tRNA and various protein factors (reviewed in 1,2). All these factors can contribute to the stability of the codon-anticodon interaction and could even contribute differently in various organisms. Thus evolutionary refinements of the translation process has led down many paths.

ACKNOWLEDGEMENTS

The T-jump work was initiated at the laboratory of Prof. D. M. Crothers (Yale University, USA), pursued at the laboratory of Prof. M. Eigen and Dr. D. Pörschke (Max Planck Institute for Biophysical Chemistry, Germany) before a T-jump instrumentation was completely set up in one of the authors laboratory in Belgium with the help of grants from the "Fonds National de la Recherche Scientifique" (FRFC contract nb. 2.4520.81). Collaboration with our colleagues in Canada (University of Montréal) was greatly facilitated by travel grants from the "Fonds de la Coopération pour les échanges culturels entre la Belgique et le Québec" (project nb. 3.1.10). H.G. benefited of short-term fellowships from NIH, NATO and EMBO. We thank many collaborators who worked with us and/or provided some of the pure tRNA samples; their names appear in the references list.

REFERENCES

1. R.H. Buckingham and H. Grosjean, in : "Accuracy in Molecular Processes" D.J. Galas, T.B. Kirkwood and R. Rosenberg, ed., Chapman and Hall, London, in press (1985).
2. C.G. Kurland and M. Ehrenberg, Progr. Nucl. Ac. Res. Mol. Biol. 31 : 192-217 (1984).
3. F.H.C. Crick, S. Brenner, A. Klug and G. Piecznik, Origins of Life, 7 : 389-397 (1976).
4. M. Eigen and P. Schuster, in : "The Hypercycle", Springer-Verlag, Berlin (1979).
5. K. Kruger, P.J. Grabowski, A.J. Zaug, J. Sands, D.E. Gottschling and T.R. Cech, Cell 31 : 147-157 (1982).
6. C. Guerrier-Takada, K. Gardiner, T. Marsh, N. Pace and S. Altman, Cell 35 : 849-857 (1983).
7. C.R. Woese, in : "Ribosomes : Structure, Function and Genetics", G. Chambliss, G.R. Craven, J. Davies, K. Davis, L. Kahan and M. Nomura, eds, University Park Press, Baltimore, p. 357-373 (1979).
8. J. Eisinger, Biochem. Biophys. Res. Commun. 43 : 854-861 (1971).
9. J. Eisinger and N. Gross, J. Mol. Biol. 80 : 165-174 (1974).
10. H. Grosjean, D.G. Söll and D.M. Crothers, J. Mol. Biol. 103 : 499-519 (1976).
11. H. Grosjean and H. Chantrenne, in : "Molecular Biology, Biochemistry and Biophysics", F. Chapeville and A. Haenni, eds, Springer-Verlag, Berlin (1980).
12. J. Eisinger, B. Feuer and T. Yamane, Nature 231 : 126-128 (1971).
13. D. Labuda, G. Striker and D. Pörschke, J. Mol. Biol. 174 : 587-604 (1984).

14. S.R.Jaskunas, C.R. Cantor and I. Tinoco, Jr., Biochemistry 7 : 3164-3178 (1968).
15. D.G.Gorenstein and E.M. Goldfield, Biochemistry 21 : 5839-5849 (1982).
16. E. Westhof, P. Dumas and D. Moras, J. Biomol. Struct. Dyn. 1 : 337-344 (1983).
17a. R. Rigler and W. Wintermeyer, Ann. Rev. Biophys. Bioeng. 12 : 475-505 (1983).
17b. C.W. Hilbers, A. Heerschap, C.A.G. Haasnoot and J.A.L.I. Walters, J. Biomol. Struct. Dyn. 1 : 183-207 (1983).
18. H. Grosjean, S. de Henau and D.M. Crothers, Proc. Natl. Acad. Sci. USA 75 : 610-614 (1978).
19. D. Pörschke, O.C. Uhlenbeck and F.A. Martin, Biopolymers 12 : 1313-1335 (1973).
20. J. Weissenbach and H. Grosjean, Eur. J. Biochem. 116 : 207-213 (1981).
21. D. Labuda, H. Grosjean, G. Striker and D. Pörschke, Biochim. Biophys. Acta 698 : 230-236 (1982).
22. D. Labuda and H. Grosjean, Biochimie 63 : 77-81 (1981).
23. J. Vacher, H. Grosjean, C. Houssier and R.H. Buckingham, J. Mol. Biol. 177 : 329-342.(1984).
24. P. Romby, R. Giégé, C. Houssier and H. Grosjean, J. Mol. Biol. 183 : in press (1985).
25. C. Houssier and H. Grosjean, J. Biomol. Struct. Dyn. 4 : in press (1985).
26. S.M. Freier, D.D. Albergo and D.H. Turner, Biopolymers 22 : 1107-1131 (1983).
27. D. Labuda and D. Pörschke, Biochemistry 21 : 49-53 (1982).
28. B.D. Wells, Nucleic Acids Res. 12 : 2157-2170 (1984).
29. N. Tibanyenda, S.H. De Bruin, C.A.G. Haasnoot, G.A. Van der Marel, J.H. Van Boom and C.W. Hilbers, Eur. J. Biochem. 139 : 19-27 (1984).
30a. G.M. Clore, A.M. Gronenborn and L.W. McLaughlin, J. Mol. Biol. 174 : 163-173 (1984).
30b. G.M. Clore, A.M. Gronenborn, E.A. Piper, L.W. McLaughlin, E. Graeser and J.H. Van Boom, Biochem. J. 221 : 737-751 (1984).
31. R. Cedergren, D. Sankoff, B. Larue and H. Grosjean, C.R.C. Critical Reviews in Biochemistry 11 : 35-104 (1981).
32. H. Grosjean, R. Cedergren and W. McKay, Biochimie 64 : 387-397 (1982).
33. B. Larue, R. Cedergren, D. Sankoff and H. Grosjean, J. Mol. Evol. 14 : 287-300 (1979).
34. M. Yarus, Science 218 : 646-652 (1982).
35a. S. Nishimura, in : "tRNA : Structure, Properties and Recognition" P.R. Schimmel, D. Söll and J.A. Abelson, eds, Cold Spring Harbor Laboratory USA, Monograph 9A : 59-79 (1979).
35b. T.H. Tsang, M. Buck and B.N. Ames, Biochim. Biophys. Acta 741 : 180-196 (1983).
36. G. Dirheimer, in : "Recent Results in Cancer Research", Springer-Verlag, Berlin 84 : 15-46 (1983).
37. H.A.M. Geerdes, J.H. Van Boom and C.W. Hilbers, J. Mol. Biol. 142 : 219-230 (1980).
38. H. Grosjean, K. Nicoghosian, E. Haumont, D. Söll and R. Cedergren, Nucleic Acids Res. 13 : in press (1985).
39. M. Buck and E. Griffiths , Nucleic Acids Res. 10 : 2609-2624 (1982).
40. M. Buck and B.N. Ames, Cell 36 : 523-531 (1984).
41. S.P. Eisenberg, L. Soll and M. Yarus, J. Mol. Biol. 135 : 111-126 (1979).
42. W. Fuller and A. Hodgson, Nature 215 : 817-821 (1967).
43. E. Egert, H.J. Lindner, W. Hillen and M.C. Bohm, J. Am. Chem. Soc. 102 : 3707-3713 (1980).
44. N. Yokoyama, Y. Yamamoto, T. Miyazawa, K. Watanabe, S. Higuchi, Z. Yamaizumi and S. Nishimura, Nucleic Acids Res., Symp. Ser. 10 : 155-156 (1981).

45. D. Moras, M.B. Comarmond, J. Fisher, R. Weiss, J.C. Thierry, J.P. Ebel and R. Giégé, Nature 288 : 669-674 (1980).
46. G.B. Weiss, J. Mol. Evol. 2 : 199-204 (1973).
47. H. Grosjean, D. Sankoff, M.J. Jou, W. Fiers and R. Cedergren, J.Mol. Evol. 12 : 113-119 (1978).
48. H. Grosjean and W. Fiers, Gene 18 : 199-209 (1982).
49. A.J. Zaug, P.J. Grabowski and T.R. Cech, Nature 301 : 578-583 (1983).
50. U. Lagerkvist, American Scientist 68 : 192-198 (1980).
51. G.R. Björk, J. Ericson, Th. Hagervall, Y. Jönsson; reported at the 11th Internat. tRNA workshop-Banz-Germany, abstract DII-6,May 1985
52. M. Faxén, L.A. Kirsebom, E. Palmcrantz and L. A. Isaksson, 11th Internat. tRNA workshop-Banz-Germany, abstract DII-12, May 1985.
53. M. Sprinzl, J. Moll, F. Meissner and T. Hartmann, Nucleic Acids Res., r1 - r49 and r51 - r104 (1985).
54. G.R. Björk, in : "Processing of RNA", D. Apirion Ed., CRC Press,Inc., Boca, Raton, Florida. pp 291-330 (1984).

STUDIES ON [1]5N LABELLED 5S RNA: ASSIGNMENTS IN THE HELIX V REGION OF 5S RNA, AND IN THE 5S/L25 COMPLEX

M. Jarema[1] and P.B. Moore[2]*

[1]Biochemistry Department, Brandeis University
Waltham, MA, 02254
[2]Department of Chemistry, Yale University, New
Haven, CT., 06511

INTRODUCTION

The discovery that RNAs possess catalytic capabilities
(Kruger et al, 1982; Guerrier-Takada et al, 1983), and the
growing realization of the importance of ribosomal RNAs in
ribosome function (for review see Noller, 1984) have stimulated
renewed interest in the structural and chemical properties of
RNA. Of the ribosomal RNAs only one is simple enough and small
enough to lend itself to detailed physical characterization in
solution, 5S RNA.

5S RNA has a chain length of 120 bases or so, a number which
is species-dependent (Erdmann et al, 1984). The version from
E.coli binds three ribosomal proteins in vitro, L5, L18, and L25
(see Monier, 1974), and there is evidence that this complex
constitutes a subassembly of the 50S ribosomal subunit (Rohl and
Nierhaus, 1982; Stoeffler and Stoeffler-Meilicke, 1984). The
molecular weights of the 5S proteins are 20,171, 12,770, and
10,694 respectively (Wittmann, 1982). Thus the complex of L25
and 5S RNA, for example, constitutes a reasonably low molecular
weight ribonucleoprotein, suitable for physical study.

We have been investigating the 5S RNA from E. coli and its
complexes with ribosomal proteins by high field NMR for several
years. Most of our effort has concentrated on the interpretation
of the downfield portion of the proton spectrum of these materials (10-15 ppm) where hydrogen bonded nucleotide base imino
protons contribute resonances. In this paper we present some new
data on assignments in 5S RNA and its complex with L25 produced
by investigations done using [1]5N labelled samples.

---

*Address all correspondence to P.B.M.

MATERIALS AND METHODS

5S RNA

Methods for purifying 5S RNA from ribosomes and from the 5S overproducing strain HB101/pKK5-1 (Brosius et al, 1981) have been described elsewhere (Kime and Moore, 1983a). To obtain adequate quantities of ^{15}N labelled 5S RNA a strain was constructed carrying the pKK5-1 plasmid competent to grow on minimal medium, which HB101 will not. The plasmid was isolated from whole cell DNA extracted from HB101/pKK5-1 and transferred in two stages through E. coli strain LE392, to alter the plasmid's restriction properties, to NG135, a recA⁻ strain capable of growing on mininmal media. Both strains were provided to us by Drs. Nigel Grindley and Catherine Joyce whose assistance we gratefully acknowledge. The methods used were those described by Maniatis et al (1982). NG135/pKK5-1 was grown on M9 medium (Anderson, 1946) with the NH_4Cl content reduced to 0.3g/l, the minimum level found necessary to sustain growth to normal saturation levels. ^{15}N NH_4Cl (99%) was purchased from Merck,- Sharp and Dohme (Canada). Cells were grown in the presence of ampicillin (to maintain the plasmid in the strain) until an apparent optical density at 550_{nm} of 2.0 was reached, chloramphenicol was then added, and the growth continued for 3 more hours to maximize overproduction of 5S RNA. Both the 5S RNA in the cytosol, the over produced material, and that incorporated into ribosomes were recovered by standard techniques.

5S Fragment and protein L25

Methods for purifying a fragment of 5S RNA comprising bases (1-11,69-120) have been detailed elsewhere (Kime and Moore, 1983a). Protein L25 was purified from E. coli ribosomes and complexes formed between it and 5S RNA as described previously (Kime et al, 1981; Kime and Moore, 1983b)

NMR

Conventional ^1H NMR experiments were done using the Bruker WH500 spectrometer and the home-built 490MHz spectrometer at the Northeast Regional NMR Facility (Yale). $(45°-t-45°)$ pulse sequences were used for H_2O supression (Kime and Moore, 1983c). $^{15}N/^1H$ spectra were obtained using the 500 MHz spectrometer at Brandeis University. Three kinds of $^{15}N/^1H$ spectra will be discussed below: (1) difference decoupling spectra, which enable one to identify protons bonded to ^{15}N which decouple at a single ^{15}N frequency (Kime,1984a), (2) decoupled echo difference spectra which display the resonances of protons bonded to ^{15}N atoms decoupling over a range of ^{15}N frequencies as singlets in a one dimensional spectrum, and (3) forbidden echo spectra which display $^{15}N-^1H$ chemical shift correlations in two dimensions. Techniques (2) and (3) have been described by Griffey et al (1985), and are discussed in this volume in the article by Redfield and his colleagues.

RESULTS

Prior Assignments

Figure 1 shows the sequence of the 5S RNA of E. coli

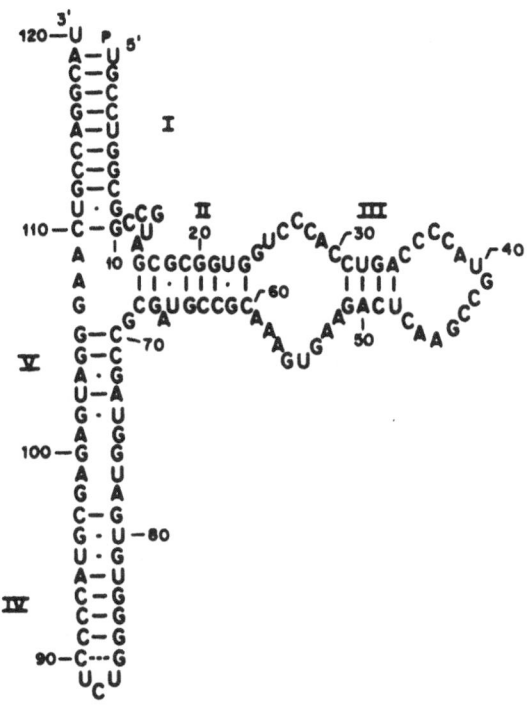

Figure 1. The sequence of E. coli 5S RNA. The sequence of E. coli 5S RNA is displayed in the standard three-stem form (see Delihas et al., 1984, for review). Helical regions are designated by Roman numerals for reference purposes.

displayed in the secondary structure which has been deduced for 5S RNAs primarily on the basis of comparative sequence studies (for review Delihas et al, 1984). Figure 2 shows the downfield spectrum of the 5S fragment, bases 1-11,69-120 of the intact molecule. The spectrum of fragment is a subset of the spectrum of its intact parent molecule (Kime and Moore, 1983c).

The resonances in the spectrum of 5S RNA were assigned initially using ^{1}H-^{1}H nuclear Overhauser effect (NOE) methods, concentrating on fragment (Kime and Moore, 1983a). These experiments provided strong evidence that resonances I,N,A,and (P,Q) connect by NOE in that order and represent helix IV. A is the AU base pair at (82,95), and P and Q represent the imino protons of the adjacent GU. Similarly resonances J,C,F,B,M,E,H, and S were found to come from helix I, B being A115U5, and J the penultimate GC base pair in the standard model (see Figure 1). At all temperatures and pHs, resonances O and Q were found to represent a GU base pair related by NOE to resonance D, probably a GC base pair. At temperatures below 20°C or pH's below 6.0, a second such juxtaposition could be demonstrated involving an

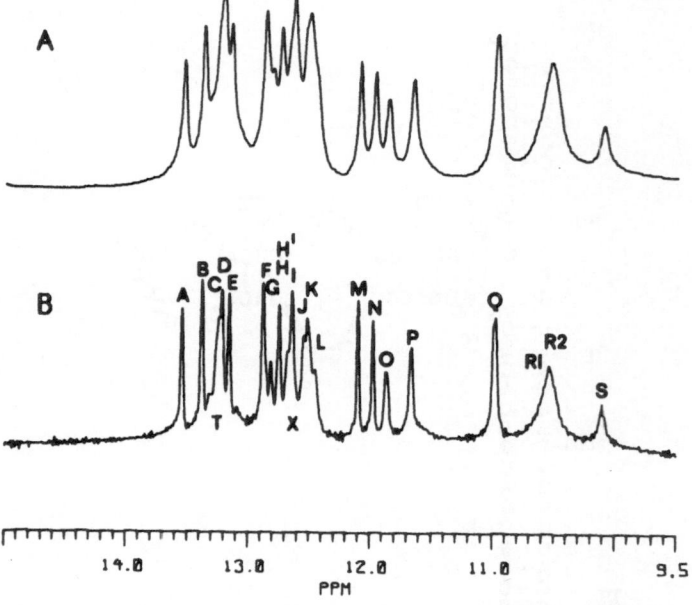

Figure 2. The Downfield Proton Spectrum of The
RNase A-resistant Fragment of 5S RNA. The spectrum
shown in (A) is that of a 1mM sample of fragment
in 4mM $MgCl_2$, 0.1MKCl, 5mM cacodylate, pH 7.2, 5%
D_2O at 30°C. The lower spectrum (B) is the same data
as in (A) resolution enhanced. The letters over the
resonances give the naming convention for this spectrum.

upfield component of resonance R called R2, and resonance P
(which doubles in intensity under these conditions), constituting
a GU, and another resonance in the C,D,E part of the spectrum.

In the standard 5S model, helix I ends with a (GU)(GC)
sequence (bases 9,10,110,111) and helix IV does the same (bases
79,80,96,97). The only other GU is the one at (74,102) which is
surrounded by non-Watson-Crick base pairs and so should be less
stable. The O,Q,D (GU)(GC) cannot be placed directly by NOE
methods; repeated attempts to elicit connectivities between this
set of resonances and others have failed.

Helix I is palindromic around its central AU for several
base pairs on either side. In order to allign resonances in the
sequence in essence one has to decide whether resonance F or
resonance M belongs on the terminal side of the AU resonance,
resonance B. This question has been answered by two means: (1)
the application of ring current shift rules, and (2) quantitative
NOE measurements (Kime et al, 1984). Ring current shift rules
show that the imino proton resonance of G6C114 should have a
chemical shift near that of resonance M, not resonance F.
Consistent with this finding, quantitative NOE determinations
demonstrate that the imino proton of resonance M is further away
from the proton of resonance B than is the proton of resonance F
as is required by the right-handed geometry of the RNA helix and
the base sequence. Alligned in this way, the NOE connectivities

178

seen indicate that S must be the imino proton of either G10 or
U111. Furthermore, we have studied the spectrum of molecules
consisting of bases 69-120, i.e. fragment missing bases 1-11
entirely. These molecules display the O,Q,D connectivity as
usual, as well as the R2-P connectivity which proves that
neither of these GU´s can have anything to do with G10 or U111.
On the grounds of thermal stability it is reasonable to assign
O,Q, and D to G96U80 and G79C97. These assignments for D,O, and
Q imply that the R2-P GU is U74G102.

Four comments can be made about the O,Q,D assignment.
First, when L25 binds to fragment (or 5S RNA), resonances
I,N,A,P, and Q all change in chemical shift indicating that the
geometry of helix IV alters slightly in response (Kime and
Moore, 1983b). It is interesting that the chemical shifts of
resonances O,Q and D are also affected, suggesting again an
association of these resonances with helix IV. (Resonances R2,
and P are unaffected by L25 binding.) Second, all the NOE
connectivities discussed so far in the context of fragment are
demonstrable in intact 5S RNA; fragment is a domain of 5S RNA
(Kime and Moore, 1983b,c). Third, the failure of helix I
to continue to give a normal GU-GC sequence involving bases G9,
G10, C110 and U111 is quite unexpected. All bacterial 5S RNAs
have sequences which accomodate base pairing at these positions
(Erdmann et al, 1984), and recent studies on the 5S RNA from B.
stearothermophilus, which also has a (UG)(GC) sequence at the
homologous location, show that the base pairs in question exist
(D. Gewirth, unpublished observations). Finally, from its
chemical shift and the absence of an imino resonance giving a
strong NOE to it, S is most unlikely to represent a standard GU
base pair. Under many conditions, when NOEs from S are sought a
strikingly strong effect is seen to a resonance at 8 ppm,
suggesting relaxation of the S proton by an aromatic proton which
must be that of another base.

$^{15}N/^{1}H$ Results. UN3 Resonances

When an RNA is uniformly labelled with ^{15}N, all the resonances due to protons directly bonded to nitrogen become doublets
with a splitting of approximately 90 Hz. Decoupling of a given
proton resonance can be achieved by irradiating the sample at
^{15}N frequencies while observing the proton spectrum. What is
found experimentally is that the ^{15}N frequency at which protons
decouple in RNA is dominated by their chemical identity and only
secondarily affected by interactions with non-bonded groups in
the neighborhood (Gonnella et al, 1982). All GN1 imino protons
decouple over a narrow frequency range, which does not overlap
with the range for UN3 protons. Thus the ^{15}N chemical shift at
which a downfield proton decouples identifies it chemically.

The initial experiments done using ^{15}N labelled samples
were of the difference decoupling design (Kime, 1984a,b). ^{1}H
spectra are collected as the decoupler frequency is stepped
through the ^{15}N spectrum, irradiating a narrow band of frequencies at each step. ^{1}H difference spectra are then calculated to
identify the resonances whose decoupling is affected at each
step. This is a tedious way to obtain $^{15}N/^{1}H$ chemical shift
correlations, but is easy to carry out and has excellent sensitivity, irrespective of the linewidths of the ^{1}H resonances being
observed. An important disadvantage of the technique is that a
decoupled resonance in the difference spectrum is a central peak

with two flanking peaks of opposite sign. If there are many resonances of similar chemical shift in the ^{1}H domain which decouple at the same frequency in the ^{15}N domain, interpretation of the resulting difference spectrum can be difficult.

These problems notwithstanding, the initial difference decoupling spectra were quite useful in identifying UN3 resonances. Resonances A,B, O and P all decouple at nearly the same ^{15}N frequency, a frequency consistent with their being UN3 imino protons, as expected. Thus A is U82, B is U5 and O and P are the U side of GU base pairs, U80 and U95 respectively. Quite surprising was the discovery of two other UN3 resonances, one identical in chemical shift with H, called "H´" and a second slightly upfield of I, designated "X" (see Figure 2) (Kime, 1984a).

The sample used to obtain the ^{15}N/^{1}H correlations just mentioned was made by digesting overproduced pKK5-1 product with RNase A at a weight ratio of 100 parts RNA to 1 part RNase under standard conditions (see Materials & Methods). To get further information about the assignments of H´ and X a second sample of ^{15}N 5S fragment was prepared using ribosomal material digested at a weight ratio of RNA to RNase of 900:1 under otherwise similar conditions. At 100:1 only 10% of the fragment molecules survive digestion with the loop at 87,88,89 intact whereas at 900:1, the intact fraction exceeds 50%. The difference is easily verified on acrylamide gels run under denaturing conditions (7M urea), and was in this case.

Both forms of fragment were sequenced. In the ribosomal sample the 5´ terminus was G2, not U1 as is the case normally. (Perhaps during chloramphenicol treatment 5S incorporated in ribosomes suffers some breakdown.) The rest of the ribosomal fragment sequence was normal. The cleavages separating it from the rest of 5S were after C11 and C68 as usual, and the molecule terminated at U120, the normal 3´ end. The overproduced material had U1 and G2 as its 5´ termini in almost equal amounts. Its 3´ termini included A119 and U120. (We have seen such fraying at the ends of overproduced 5S RNA before, but our previous experience was with material produced in E. coli HB101 not NG135. In HB101 the 5´ termini are U(-2), U(-1) and U1 and the 3´ termini are U120, C121 and A122 (Kime et al, 1983). Clearly there is a strain-dependent difference in the fate of the overproduced material). The separation points were after C11 and C68, as usual.

Figure 3 compares the decoupled echo difference UN3 spectra of the two fragment samples. The decoupled echo difference method is a multiple pulse technique which enables one to examine the resonances of all protons bonded to ^{15}N which decouple over a range of ^{15}N frequencies. The ^{15}N range was chosen in this case to cover UN3 frequencies. As is clear in Figure 3, each resonance is represented as a single, decoupled peak, not as a central peak flanked by two peaks of opposite sign as is the case when difference decoupled spectra are taken. This feature of decoupled echo difference spectra gives them a distinct advantage in resolving crowded spectra. What is also clear in Figure 3 is that the spectra of the two samples are essentially identical.

It will be noted that within the sequence of fragment, there are in fact only a few unassigned Us. Two are at the termini, U1 and U120. Given the sequences of the molecules examined, neither

of these can possibly give rise to resonances H′ and X. Another
pair of potential candidates are U87 and U89, unlikely a priori
because of their involvement in a loop, but were either responsible for H′ or X, the difference in cleavage at that loop between
the two specimens would surely have resulted in spectral changes;
none are seen. Thus H′ and X must come from the helix V portion
of the molecule and are likely to represent U77 and U103 since
U74 is involved in the R2-P base pair. U11 could be, but is
unlikely to be H′ or X (see below).

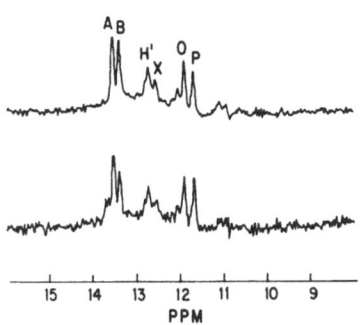

Figure 3. UN3 Decoupled Echo Difference Spectra of
100:1 and 900:1 Fragment Compared. The upper spectrum is that of a 1.2mM sample of 100:1 5S fragment
in 100mM KCl, 5mM cacodylate, 4mM $MgCl_2$, pH 7.1, 5%
D_2O at 28°C. The lower spectrum is that of a 0.7mM
sample of 900:1 fragment under the same conditions.
Both spectra were accumulated identically, and both
display UN3 resonances.

It is also clear from the spectra in Figure 3 that H′ and X
are both weaker and broader than the other resonances, the
difference being particularly obvious for X. Both must have
shorter T2′s than the other UN3 resonances in the spectrum, which
may account for our failure to recognize them in the course of
the ^{1}H-^{1}H NOE studies described earlier.

GN1 Imino Resonances

A spectrum similar to that shown in Figure 3 was obtained
of the GN1 resonances in the 900:1 fragment sample. This sample

was chosen because the intensity of resonance G in the fragment spectrum increases as the fraction of molecules in the preparation with unbroken helix IV loops goes up. G was not assigned in the original difference decoupling experiments due to its weakness in the sample used (Kime, 1984a). The difference decoupling assignments of resonances C,D,E,F,H,I,J,K,L,M,N, and Q were confirmed. All are GN1 imino protons. G was shown to be a GN1 proton also. For obscure reasons resonances upfield of Q were poorly visualized in these echo difference spectra. Fortunately both the difference decoupling data and the forbidden echo data (see Figure 4) clearly show that the two resonances upfield of Q, R and S, are both GN1´s. S must therefore represent the GN1 proton of G9.

$^{15}N/^1H$ Chemical Shift Correlations in the L25-Fragment Complex

Figures 4 and 5 are two-dimensional, decoupled forbidden echo difference spectra of 5S fragment and 5S-L25 complex. These spectra were obtained using a multiple pulse technique which yields detailed information on $^{15}N/^1H$ chemical shift correlations. Each strip displays across its width ^{15}N chemial shift from 187.46 to 88.74 ppm left to right and is lowfield in ^1H at its bottom and highfield at its top. The bottom of the leftmost strip is about 13.8 ppm and the top of the rightmost strip is about 10 ppm. The presentation is identical in both figures. In effect these spectra display the information one would have if one took a large number of one dimensional difference decoupled spectra collected at regular intervals in ^{15}N frequency and arranged them in a stacked plot. Note, however, that the direction of the specta in the stack would be perpendicular to the direction shown here. Note also that each resonance in Figures 4 and 5 is a singlet demonstrating that forbidden echo difference technique has the same advantage in resolution the decoupled echo difference technique enjoys. The data in these figures in fact do replicate in two-dimensional form the difference decoupling data of Kime (1984a,b). The fragment preparations used were 100:1 products in both cases. All the normal resonances in fragment can be distinguished in Figure 4, with the exception of G which is too weak to detect. To the left of each strip, the UN3 resonances are identified, and to the right the GN1´s are shown.

In Figure 5, the identification of resonances from A to L is made difficult by the fact that the L25 occupancy of this sample is only 50% or so. The result is that resonances which change their ^1H chemical shifts upon L25 binding are represented by twice. Thus A and B which are clearly resolved in Figure 4 become a broad single feature due to the fact that the chemical shift of A in the L25 complex is halfway between that of A and B in the free RNA. In Figure 5 the only resonances identified are those unique to the complex, all of them designated by arabic numerals, following the convention established earlier (Kime and Moore, 1983b), and those whose response to L25 binding is clearly visible. Note the obvious shift of H´ intensity downfield and the shifting and splitting of Q into two new resonances, Q1 and

Q2. It appears that L25 binding enhances the detectability of H´ and X, presumably by stabilizing them against exchange.

DISCUSSION

The finding of two UN3 imino protons in the midst of the GN1 group around 12.5 ppm brings to 3 the number of UN1 resonances assignable to helix V, the third being the UN1 of the R2-P GU base pair. There are only three U´s in this sequence; the problem is how to assign them.

The standard model for 5S RNA suggests that helix IV and V form an uninterrupted loop and helix, without bulged bases (see Delihas et al, 1984). This being the case, the R2-P GU has to be G102-U74. R2-P clearly relate by NOE to another base whose imino proton is called "T", and which underlies the C,D,E region. Conditions which enhance the R2-P connectivity increase the visibility of R2-T and P-T connectivities. Since there are only GN1 resonances in the C,D,E region, T must represent G75 which should be part of an AG base pair. We have examined our data for evidence of an NOE from T to an AH2 or AH8 proton resonance in the aromatic region, connectivities anticipated in AG base-pairs. No such NOE is found. It should be noted, however, that it is impossible to devise any simple way of pairing the strands of helix V to generate a GU-GC juxtaposition, the only alternative if T is not an AG resonance.

In the past, some weak NOEs connecting R to broad resonances underlying C,D,E and I have been observed (Kime and Moore, 1984a). The C,D,E resonance is accounted for by T. The broad resonance under I has about the right chemical shift to be X, but the available data fail to support the hypothesis of an R2 to X connectivity. First when pH and/or temperature favor the R2-P-T connectivity there is no enhancement of the NOE between R and the resonance under I. Furthermore, at acid pH, R breaks up into three components. It is the upfield member which gives the R2-P-T connectivity, but the R to "I" NOE, still weak, seems more strongly associated with the middle component.

A working hypothesis for the assignments of H´ and X can be formulated nonetheless. The chemical shift of H´ responds when L25 binds to 5S RNA and that of X does not. It is tempting, therefore, to assign H´ to U77, the only unassigned U between the end of helix IV, all of whose resonances respond to L25, and A101G75 and G102U74 which do not. X which is likewise insensitive to the presence of L25 then assigns to U103.

Again older data have been examined for evidence of strong NOEs from H´ and X to sharp aromatic resonances consistent with their involvement in AU pairs. No such NOEs are evident.

The assignments within fragment we can propose using the available information to its utmost, are shown in Figure 6. (The R1 resonance assigned to G85 should be regarded with care. The NOE on which the assignment depends is seen only in preparations of bases 69-120 whose structure is not identical to that of ordinary 5S RNA (Kime et al, 1984). Further, the chemical shifts of the several resonances represented in the R region are sensitive to environmental conditions. R1 in the 69-120 preparations may not be the same as the resonance called R1 in ordinary fragment.) Means are being sought to test these assignments further, especially the assignments for H´, T, and X. Whatever

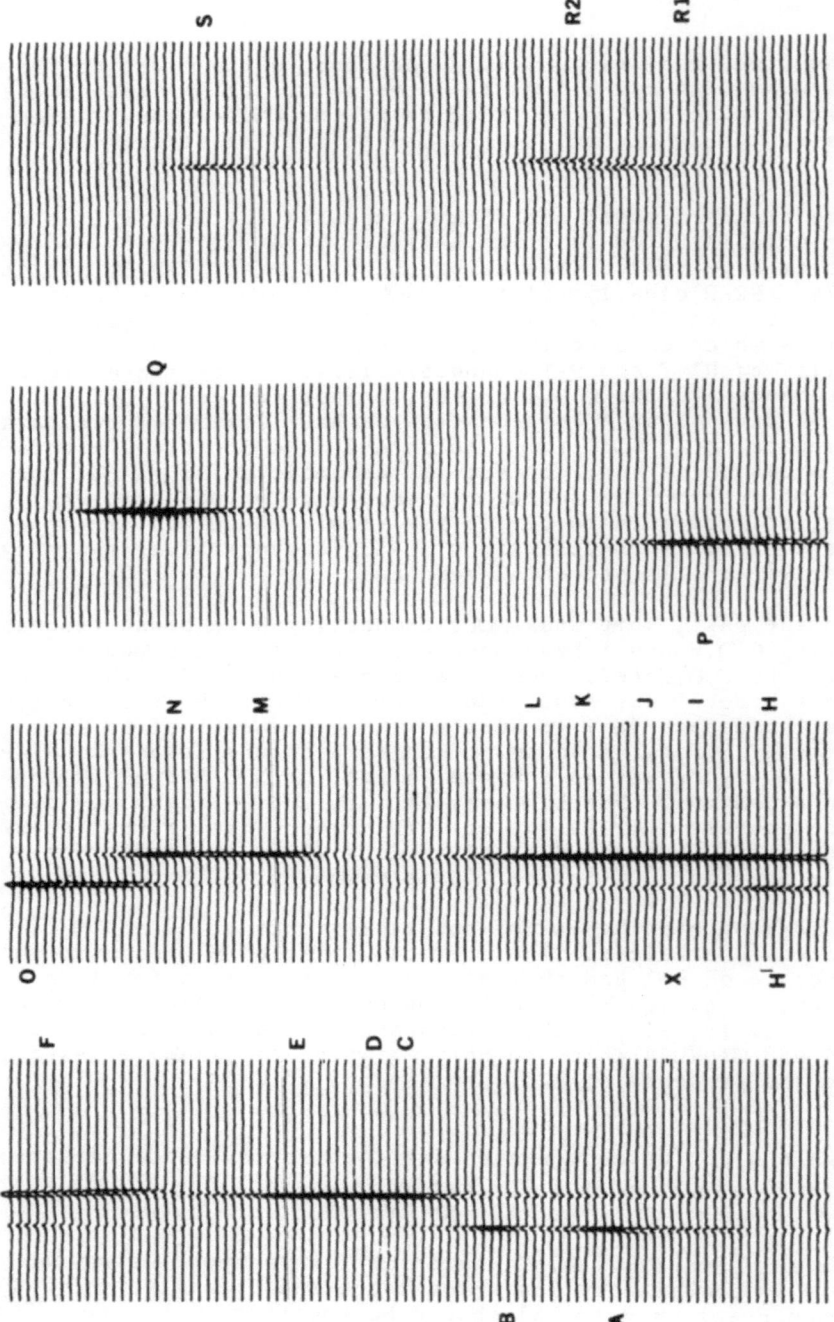

Figure 4. A Two Dimensional Forbidden Echo Spectrum of 5S
Fragment. A decoupled forbidden echo experiment was carried
out on the 100:1 fragment sample described in the legend to
Figure 3 at 28° C. Conditions were chosen so that both UN3
and GN1 resonances appear as upright singlets (see Materials
and Methods). The horizontal dimension of each strip
displays ^{15}N chemical shift running from 187.46 to 88.74
ppm, left to right. The vertical dimension represents ^{1}H
chemical shift, the botttom of the leftmost strip being
13.8ppm and the top of the rightmost strip being 10ppm.
UN3 resonances are identified to the left of each strip,
and GN1's to the right.

the outcome, the existence of hydrogen bonded structure in helix
V is clear, but it would be hard to make a case from the NMR data
that a regular $(GC)_2(AG)(AU)(GU)$ stem exists as drawn in Figure
1. It should be noted that resonances G, K, and L, all GN1
protons are still unassigned, and very likely to fall in helix V.

When L25 complexes with 5S or 5S fragment a number of new
resonances appear between 9.5 and 12.5 ppm (see Figure 5)
primary assignments of which have been by ^{15}N/^{1}H correlation
methods. That new information coupled with earlier NOE data
involving these protein-specific resonances (the ones designated
with numerals in Figure 5) leads to some new assignments in the
L25 complex.

In our first analysis of the spectrum of fragment/L25
complex we noted that complex formation leads to a loss of
intensity in resonance H and an increase in intensity of resonance G (Kime and Moore, 1983b). This change is the result of the
downfield shift of resonance H′ in response to L25 binding,
clearly apparent in Figure 5. Contrary to our original view,
neither H nor G is affected by the protein; H′ is. Recognizing
this fact, the NOE we reported between "G" and resonance 7, a GN1
resonance at 10.1 ppm, can equally well be interpreted as
an H′ to 7 NOE. H′ (U77) has three G residues on either side of
it, anyone of which could be responsible for 7.

An NOE connectivity exists in the complex running from "I"
to 6, a GN1, to 3, another GN1, to 1, a weak UN3 resonance (see
Figure 6). There is a second connectivity involving "I" which
runs "I" to 8, a UN3, to 2, a second UN3. Since I has N as a
neighbor on one side, the existence of two other connectivities
appears awkward. It must be remembered, however, that I,J, and X
all have the same chemical shift in the L25 complex, and that 8
has a component which is not split by the labelling of the RNA
moiety of the complex with ^{15}N (Kime, 1984b).
J, the penultimate GC in helix I is unlikely to connect
onwards to anything, and L25-induced connectivities involving J
are improbable on the basis of evidence that L25 does not interact with the terminal end of helix I(see below). Therefore, the
"I"to 6 to 3 to 1 connectivity must originate with I or with X.
Assigning X to any of the 3 U's in helix V one discovers no way
of continuing onward in a GN1,GN1,UN3 sequence which does not
require the UN3 to be a known resonance with a chemical shift
different from that of 1, either P or H′. (Note that the R2-P
connectivity exists in the complex identifying P unambiguously.)

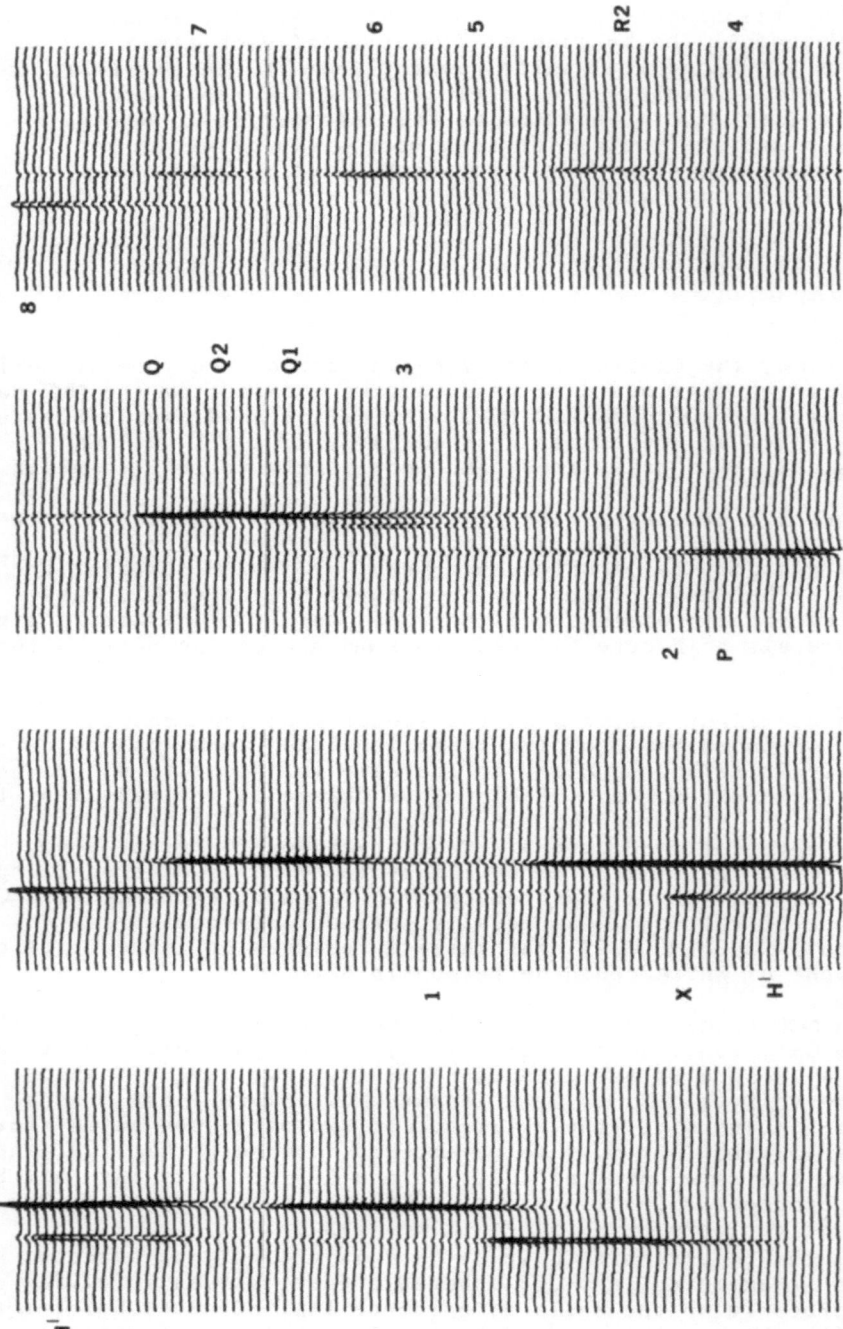

Figure 5. A Two Dimensional Forbidden Echo Spectrum of 5S Fragment-L25 Complex. A decoupled forbidden echo spectrum was taken of a sample of ^{15}N labelled fragment at a concentration close to 1mM, complexed with unlabelled L25. The molar ratio of RNA to protein in the complex was about 2:1. The buffer and data collection conditions are as given for Figure 4. Complex-related resonances are identified following the convention of Figure 4.

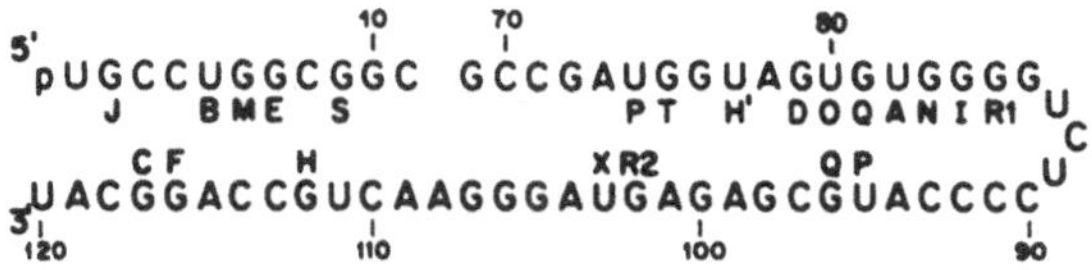

```
              10      70              80
          pUGCCUGGCGGC GCCGAUGGUAGUGUGGGG
   5'        J   BME S          PT H' DOQANIR1 U
                                                 C
            CF      H          X R2      Q P    U
   3' UACGGACCGUCAAGGGAUGAGAGCGUACCC
      120          110            100          90
```

Figure 6. Assignments of Fragment Resonances to the Fragment Sequence. (See text.)

The conclusion is that the "I" to 6 to 3 to 1 connectivity does not fit in helix V. It can, however, be accomodated as a continuation from I (G84) to 6 (G85) to 3 (G86) to 1 (U87).

The NOE connectivity "I" to 8 to 2 has to represent transfer of magnetization from protein amides to RNA. First, there are only UU juxtapositions in the RNA: one at the extreme termini, another at U74,U103 and the last at U80,U95. The first is unlikely to be called into play in the L25 complex. The other two are already accounted for by assigned resonances. Certainly I has no U near it from which it could receive magnetization. J is out of consideration because of the position of the base it represents, and X, whatever its assignment either has no neighboring U or adjoins a U already assigned. 8 to 2 must be an NOE from protein to an unassigned UN3 in fragment and 8 to I must be a similar transfer from protein to resonance I or resonance X. Consistent with this view is the fact that when 8 is saturated a spin diffusion-like transfer of magnetization to L25 is observed (Kime and Moore, 1983b).

With the addition of 3 new UN3 resonances to the spectrum upon L25 binding, there is evidence for downfield contributions from 10 UN3's: resonances A, B, O, P1, P2, H', X, 1, 2, and 8. Excluding the two terminal U's, U1 and U120, both unlikely

187

to contribute to the spectrum, there are exactly 10 U's in the sequence, only two of which are unassigned, U89 and U111. Resonances 2 and 8 must correspond to these two residues. It is interesting to note that there is no H to S connectivity in fragment/L25 complex nor any indication of where resonances S has gone (Kime and Moore, 1983b). It is hard to escape the conclusion that the geometry of the molecule around G9,U111 is altered by L25 binding, and that a U111 hydrogen bond replaces the G9 bond characteristic of the free 5S molecule.

The only resonances whose assignments are totally unclear are the two GN1 resonances, 4 and 5. If one takes the view that these resonances too are likely to arise in the helix IV -helix V region, again one finds oneself compelled to recognize that in the complex virtually every imino proton in that region contributes to the spectrum.

Two points should be made in closing. First, a variety of chemical and enzymatic protection experiments have been done to identify the binding site of L25 on 5S RNA. The site runs from G72 around to A108, but does not include helices I,II or III (see Huber and Wool, 1984, and references therein). Second, L25 binding to 5S RNA does not protect the 87,88,89 loop from RNase attack; it enhances cleavage at that point , a surprising result since the NOE data alluded to above might be taken to suggest a protection of that region instead (Garrett et al, 1984; P.B.M. unpublished results). The point of cleavage in the complex is heavily biased towards U89; in the free RNA bases 87,88, and 89 are cleaved more or less at random (Garrett et al, 1984). L25 does indeed protect some of the loop residues as the NMR data suggest it should.

ACKNOWLEDGEMENTS

We wish to thank Prof. A.G. Redfield for encouraging the pursuit of these studies and his assistance in taking spectra. We thank Mrs. Betty Freeborn for her help in preparing ^{15}N labelled RNA. This work was supported by grants from the NIH to P.B.M. (GM-32206) and to Prof. A. G. Redfield (GM-20168). Some of the spectra shown were obtained at the Northeast Regional NMR Facility which is supported by grant from NSF (CHE-7916210).

REFERENCES

Anderson, E.H., 1946, Growth requirements of virus-resistant mutants of Escherichia coli strain B, Proc. Nat. Acad. Sci. USA, 32:120.

Brosius, J., Dull, T.J., Sleeter, D.D., and Noller, H.F., 1981, Gene organization and primary structure of a ribosomal RNA operon from Escherichia coli, J. Mol. Biol., 148:107.

Delihas, N., Anderson, J., and Singhal, R.P., 1984, Structure, function and evolution of 5S ribosomal RNAs, Prog. Nucl. Acid Res. Mol. Biol., 31:161.

Erdmann, V.A., Wolter, J.H., Haysmans, E., Vandenberghe, A., and DeWachter, R., 1984, Collection of published 5S and 5.8S RNA sequences, Vol. 12 supplement Nuc. Acids Res. r133.

Garrett, R.A., Vester, B., Leffers, H., Sorensen, P.M., Kjems, J., Olesen, S.O., Christensen, A., Christensen, J., and

Douthwaite, S., 1984, Mechanisms of protein-RNA recognition and assembly in ribosomes, in: "Gene Expression," Alfred Benzon Symp. 19, Clark, B.F.C. and Petersen, H.V., eds., Muriksgard, Copenhagen, p. 331.

Gonnella, N.C., Birdseye, T.R., Nee, M., and Roberts, J.D., 1982, ^{15}N NMR study of a mixture of uniformly labelled tRNAs, Proc. Nat. Acad. Sci. USA, 79:4834.

Griffey, R.H., Redfield, A.G., Loomis, R.E., and Dahlquist, F.W., 1985, Nuclear magnetic resonance observation and dynamics of specific amide protons in T4 lysozyme, Biochemistry, 24:817.

Guerrier-Takada, C., Gardiner, K., Marsh, T., Pace, N., and Altman, S., 1983, The RNA moiety of ribonuclease P in the catalytic subunit of the enzyme, Cell, 35:849.

Huber, P.W. and Wool, I.G., 1984, Nuclease protection analysis of ribonucleoprotein complexes: Use of the cytotoxic ribonuclease alpha-sarcin to determine the binding sites for Escherichia coli ribosomal proteins L5, L18 and L25 on 5S rRNA, Proc. Nat. Acad. Sci. USA, 81:322.

Kime, M.J., 1984a, Assignment of resonances in the Escherichia coli 5S RNA fragment proton NMR spectrum using uniform nitrogen-15 enrichment, FEBS Letters, 173:342.

Kime, M.J., 1984b, Assignment of resonances of exchangeable protons in the NMR spectrum of the complex formed by Escherichia coli ribosomal protein L25 and uniformly nitrogen-15 enriched 5S RNA fragment, FEBS Letters, 175:259.

Kime, M.J., Gewirth, D.T., and Moore, P,.B., 1984, Assignment of resonances in the downfield proton spectrum of Escherichia coli 5S RNA and its nucleoprotein complexes using components of a ribonuclease-resistant fragment, Biochemistry, 23:3559.

Kime, M.J. and Moore, P.B., 1983a, Nuclear Overhauser experiments at 500 MHz on the downfield proton spectrum of a RNase-resistant fragment of 5S RNA, Biochemistry, 22:2615.

Kime, M.J. and Moore, P.B., 1983b, Nuclear Overhauser experiments at 500 MHz on the downfield proton spectra of 5S ribonucleic acid and its complex with ribosomal protein L25, Biochemistry, 22:2622.

Kime, M.J. and Moore, P.B., 1983c, Physical evidence for a domain structure in E. coli 5S RNA, FEBS Letters, 153:199.

Kime, M.J., Ratcliffe, R.G., Moore, P.B., and Williams, R.J.P., 1981, A proton NMR study of ribosomal protein L25 from Escherichia coli, Eur. J. Biochem., 116:269.

Kruger, K., Grabowski, P.J., Zaug, A.J., Sands, J., Gottschling, E. and Cech, T.R., 1982, Self-splicing RNA: autoexcision and autocyclization of the ribosomal RNA intervening sequence of Tetrahymena, Cell, 31:147.

Maniatis, T., Fritsch, E.F., and Sambrook, J., 1982, "Molecular Cloning, A Laboratory Manual," Cold Spring Harbor Laboratory.

Monier, R., 1974, 5S RNA in ribosomes, in: "Ribosomes," Nomura, M., Tissieres, A., and Lengyel, P., eds., Cold Spring Harbor Laboratory, pp. 141-168.

Noller, H.F., 1984, Structure of ribosomal RNA, Ann. Rev. Biochem., 53:119.

Rohl, R. and Nierhaus, K., 1982, Assembly map of the large subunit (50S) of Escherichia coli ribosomes, Proc. Nat. Acad. Sci. USA, 79:729.

Stoffler, G. and Stoffler-Meilicke, M., 1984, Immunoelectron microscopy of ribosomes, Ann. Rev. Biophys. and Bioengin., 13:303.

Wittmann, H.G., 1982, Components of bacterial ribosomes, Ann. Rev. Biochem., 51:155.

EQUILIBRIA IN RIBOSOMAL RNA SECONDARY STRUCTURE

Rupert De Wachter

Departement Biochemie
Universiteit Antwerpen (UIA)
Antwerpen, Belgium

INTRODUCTION

Secondary structure models have been elaborated for each of the RNA constituents of the ribosome. Documentation on models for the larger ribosomal RNAs, i.e. small subunit RNA (including eukaryotic 18 S rRNA, bacterial 16 S rRNA and their mitochondrial equivalents) and large subunit rRNA (eukaryotic 28 S rRNA, bacterial 23 S rRNA and mitochondrial equivalents) can be found in a review by Noller (1984). Some references to models for the small rRNAs i.e. 5 S rRNA, eukaryotic 5.8 S rRNA and chloroplast 4.5 S rRNA, can be found in a sequence compilation by Erdmann et al. (1985). These secondary structure models are constructed primarily by a comparative approach, i.e. by alignment of the available RNA primary structures in different species, and selection of a base pairing scheme applicable to all these sequences. Further evidence for a postulated model is collected by a search for compensating substitutions (Woese et al., 1983) i.e. evolutionary changes in one strand of a postulated helix compensated by a concomitant change in the opposite strand which preserves base pairing. Experimental approaches are usually employed in a subsequent stage, in attempts to verify or falsify proposed models or to make a choice among alternative possibilities.

The facts exposed below tend to demonstrate the existence of dynamic equilibria between alternative secondary structures, which may allow rRNA molecules to change their shape continuously by switching from one structure to the other. Just like the derivation of static secondary structure models, the search for dynamic models is based on comparison of the largest possible number of different primary structures for a given RNA type. For this reason, the discussion below is concerned mainly with 5 S rRNA, since it is the rRNA molecule for which the largest number of sequences - over 300, August 1985 - has been published, and hence for which the secondary structure has been derived in greatest detail. Sites of potential equilibria similarly exist in the larger ribosomal RNAs, but their occurrence has been hitherto examined only superficially on the existing models for small subunit rRNA, hence it will be discussed only briefly.

5 S rRNA SECONDARY STRUCTURE MODELS

Probably the first attempt to reconstruct 5 S RNA secondary structure by comparison of a sizable number of nucleotide sequences was made by Fox and Woese (1975a,b). They proposed a model containing 4 helices, applicable

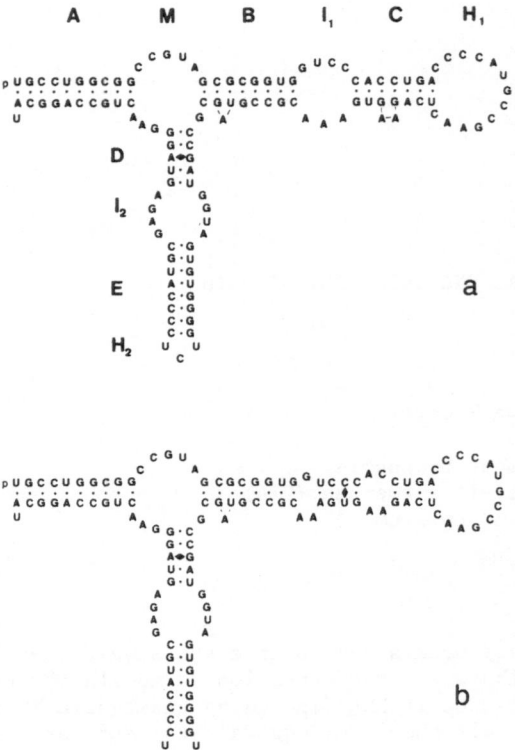

Fig. 1. Five-helix model for 5 S RNA secondary structure.
The model in (a), now adopted by most 5 S RNA students, was first
proposed by Schwartz and Dayhoff (1978) as an adaptation of the
Nishikawa-Takemura (1978) model shown in (b). Both models are applicable to all 5 S RNAs except that bulges may be absent, or
additional bulges present, on certain helices. The application of
the model in (a) to 238 published 5 S RNA sequences can be found
in Erdmann et al. (1985). The lettering of helices and loops in
(a) is the one adopted in this paper, but some authors use different nomenclatures. Base pairs G·C, A·U and G·U are symbolized by
dots, odd base pairs (see text) by losenges.

to prokaryotic 5 S RNAs. It was not clear at that time if the model was
transposable entirely or only partly to eukaryotic 5 S RNAs. After a numbe
of eukaryote-specific models had been proposed (reviewed by De Wachter et
al., 1982), several research groups independently came to the conclusion
(Böhm et al. 1982, De Wachter et al. 1982, Delihas & Andersen, 1982) that
both prokaryotic and eukaryotic 5 S RNAs can be folded in essentially the
same 5-helix model, illustrated in Fig. 1a with the Escherichia coli 5 S
RNA sequence. There is no established convention for the nomenclature of
helices and loops, but it seems logical to letter or number the helices and
loops in their order of occurence when the sequence is scanned from 5'- to
3'-end. In this paper, helices are labeled A to E, and loops are designate
M (multibranched), I (internal) and H (hairpin). Bulges are not labeled
because their presence is somewhat less universal. As an example, the 5 S
RNAs from plant mitochondria and from thermo-acidophilic archaebacteria mis
the bulge on helix B.
    The model shown in Fig.1a is now accepted by most students of 5 S RNA
structure, although there remains some argument about the size of helix D

Table 1. Previously proposed dynamic models for 5 S RNA structure.
In order to facilitate comparison of the models the nomenclature of Fig.1a
for helices and loops is used, although it may differ from that used by the
authors.

Structural change assumed	Authors
1. Coiling-uncoiling of a helix	
Helix B coils and uncoils	Fox & Woese (1975a,b)
Segment of helix E neighbouring loop H_2 coils and uncoils	MacKay et al. (1982)
2. Relative orientation of helices	
Helix A is coaxial alternatively with helix B or D	Luehrsen & Fox (1981)
Angle between helices A and D changes	Rabin et al. (1983)
3. Alternative base-pairing	
Helix E is dismantled and its 5'-proximal strand pairs with a sequence from loop H_1	{ Weidner et al. (1977) Trifonov & Bolshoi (1983)

in eubacterial 5 S RNAs. The existence of at least two base pairs adjoining
loop M is proven by compensating substitutions. The helix can be extended
in most species via an A♦G odd base pair (*) by an A·U and G·U pair. However, this extension is so strongly conserved and hence supported by few if
any compensating substitutions, that its existence is put into question by
some investigators (e.g. Studnicka et al., 1981).

Among the numerous models proposed before the more systematic comparative study of Fox and Woese (1975a,b), the model of Nishikawa and Takemura
(1974,1978) deserves mention. The latter authors advocated the uniformity
of secondary structure in prokaryotic and eukaryotic 5 S RNA at a time when
this idea was not yet accepted, and the 1978 version of their model, represented in Fig.1b, differs from the presently accepted model (Fig.1a) only in
area I_1-C. An amended version, coming even closer to the scheme shown in
Fig.1a was adopted by Schwartz and Dayhoff (1978) for an alignment of 24 sequences known at that time.

The idea that 5 S RNA may switch between different conformations, and
that such a switch may be correlated with some function in protein synthesis,
has been put forward several times. Table 1 gives a list of previously proposed "dynamic" models for 5 S RNA secondary structure, based on different
types of conformational switches. On the other hand, different conformers
of 5 S RNA, distinguishable on the basis of chromatographic or electrophoretic mobility, have been isolated from certain bacteria (Aubert et al., 1968)
and eukaryotes (Toots et al., 1982). Some of the aforementioned dynamic
models (e.g. Weidner et al., 1977) have been associated with the existence of
experimentally observed conformers, but other studies suggest that the latter
conformers rather represent native and denatured states of the 5 S RNA (reviewed by Christensen et al., 1985).

The dynamic models proposed below are different from those listed in

---

*The term odd base pair or non-standard base pair is used to designate any
of the 7 possible base pairs other than G·C, A·U and G·U. The possible
existence of odd base pairs in RNA secondary structure has been discussed,
among others, by Ninio (1979) for tRNA and by Traub and Sussmann (1982) and
Noller (1984) for rRNA. The potential occurrence of such pairs is indicated in the secondary structure models in this paper by a losenge, e.g.
A♦G or A♦C. For the possible structure of such pairs, see e.g. Topal and
Fresco (1976) and Traub and Sussmann (1982).

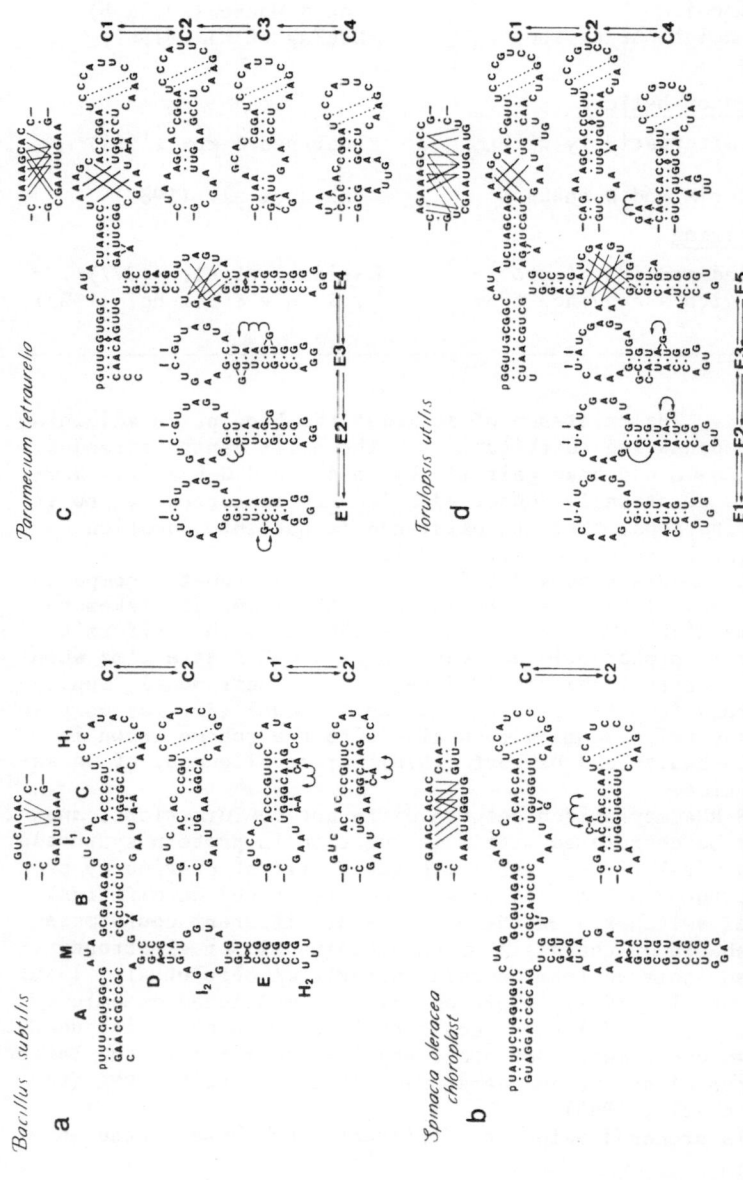

Fig. 2. Equilibria in 5 S RNA secondary structure.
The equilibrium between structures C1 and C2 for area I₁-C is possible in all 5 S RNAs. It arises from the ambiguous complementarity indicated on the scheme above area I₁-C in eubacterial 5 S RNAs (a and b) have a more complex pattern of complementarities in area I₁-C, and can therefore adopt structures C3 and C4 in addition to structures C1 and C2. Helix E can adopt a number of structures, E1 to E5, distinguishable by the relative disposition of bulges. The internal loops show criss cross complementarity indicated by lines connecting complementary bases (see Fig.3b for structures resulting from base pairing in loop I₂). Potential base pairs in loop H₁ of all four models are indicated by dotted lines, structures that could result (C1' and C2') are shown in case (a) only. Curled arrows indicate the movement of bulges migrating along helices.

194

Table 1 in the following respects. The structural changes proposed do not involve complete dismantling or uncoiling of helices, rather a partial rearrangement of base pairs within existing helices. No attempt is made to predict tertiary structure features, such as the angle between helices or the interaction between different loops. The dynamic models proposed here are thought to represent alternative states of the molecule occurring in rapid succession within the ribosome, and to be unrelated with the existence of different conformers that can be experimentally isolated.

EQUILIBRIA COMMON TO ALL 5 S RNAs

The two models of the type represented in Fig.1a and b for E. coli 5 S RNA can in fact be constructed for all of the 300-odd known 5 S RNA sequences. This has been illustrated repeatedly for sequences of eukaryotic, eubacterial and archaebacterial origin (De Wachter et al., 1982; Fang et al., 1982; Dams et al., 1982,1983a,1983b; Huysmans et al., 1983; Chen et al., 1984; Vandenberghe et al., 1984,1985). This suggests that two conformations for area I_1-C may actually be in dynamic equilibrium in the 5 S RNA secondary structure, as illustrated in Fig.2a and b for a bacterial and a chloroplast 5 S RNA. The two postulated equilibrium forms are labeled C1 and C2. The possibility of a switch between structures of nearly the same free energy (De Wachter et al., 1984) arises from the fact that a sequence in one strand of a hairpin structure faces a repeated complementary sequence in the opposite strand. As an example, the sequence G U faces A C A C in area I_1-C of Bacillus subtilis 5 S RNA, as indicated on top of Fig.2a. A similar though not identical equilibrium can be considered in chloroplast 5 S RNAs, where a more complicated set of complementarities exists in the same area (Fig.2b). Shape C2 in the chloroplast 5 S RNA shows the phenomenon of a "migrating bulge" (De Wachter et al., 1984), i.e. a bulge that can occupy several alternative positions along a helix.

The existence of forms C1 and C2 in eukaryotic 5 S RNAs is illustrated in Fig.2c and d with a protist and a yeast structure. In eukaryotes, however, the existence of additional base pairing opportunities in loop I_1 accounts for the extra potential equilibrium forms labeled C3 and C4. These structures are discussed below.

In many, but not all, 5 S RNAs, the possibility exists for two additional base pairs to form at the expense of loop H_1 (Thompson et al., 1981). This would convert structures C1 and C2 into C1' and C2' (shown only in the case of B. subtilis 5 S RNA, Fig.2a). In the latter structures helix C is lengthened by two base pairs and carries an additional, migrating bulge on its 3'-strand.

EQUILIBRIA SPECIFIC FOR EUKARYOTIC 5 S RNAs

Helix E

Helix E of eukaryotic 5 S RNAs is usually represented in one of two ways : either carrying a bulge on the 5'-proximal strand, or carrying no bulge but comprising an odd base pair. These two possible shapes are represented in Fig.2c by forms E2 and E4 of helix E in Paramecium tetraurelia 5 S RNA. However, helix E may actually take 5 different shapes in eukaryotic 5 S RNAs, as demonstrated in Fig.2c and d for a protist and a yeast. In any single 5 S RNA only a subset of this set of 5 shapes is possible, e.g. E1, E2, E3 and E4 in protists, E1, E2, E3, E5 in yeasts. An inventory of the subsets of structures possible in different eukaryotic taxa has been made (De Wachter et al., 1984). If the different forms actually occur, they would be interconnected by reversible "reactions", which can be described as follows in the case of Paramecium 5 S RNA (Fig.2c). The transition from form E1 to E2 would consist in a bulge switching from the 3'-strand to the 5'-strand of the hairpin. A bulge on the 3'-strand of form E2 could then

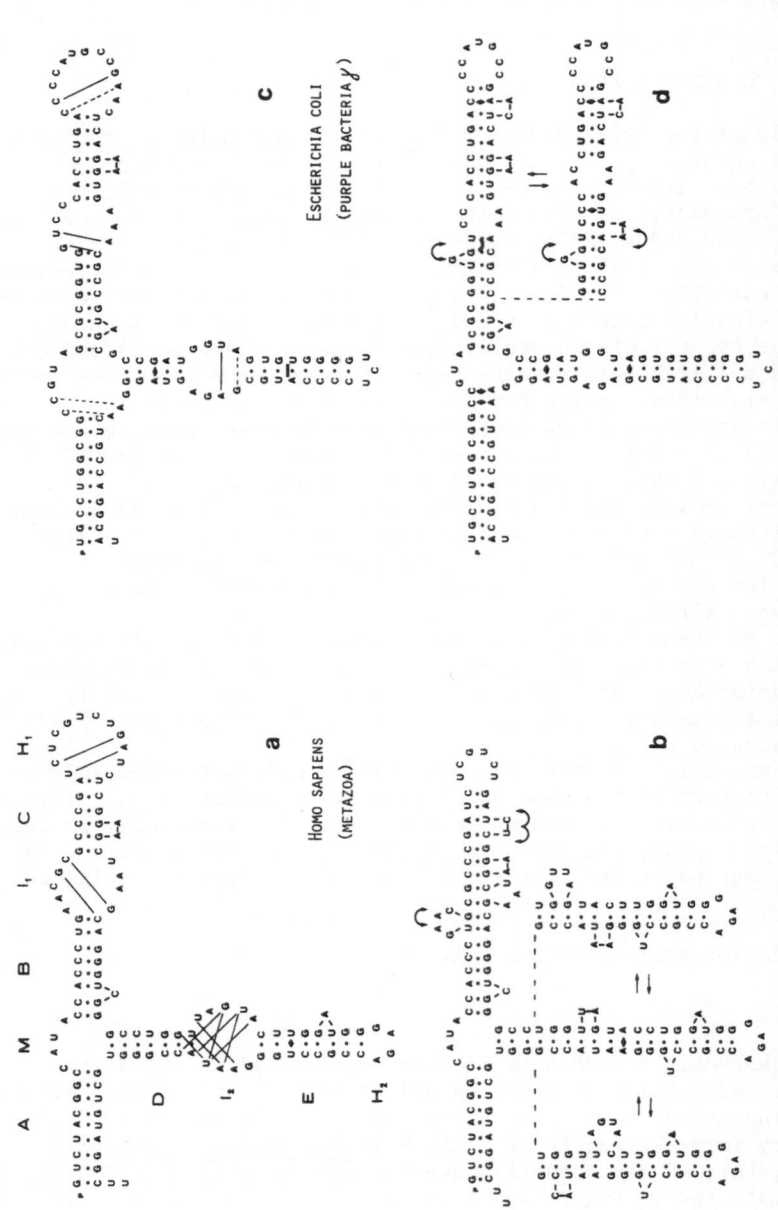

Fig. 3. Conventional and compact secondary structure models for 5 S RNA.
The conventional models are drawn for human 5 S RNA in (a) and for E. coli 5 S RNA in (c). Pairing between bases connected by lines would convert model (a) into the more compact model (b). The three equilibrium structures in area D-I₂-E of
model (b) would result from alternative base pairing within loop I₂ as indicated in model (a). Additional equilibria are
possible in area I₁-C and in helix E (cf. Fig.2) but are not shown here. Helix E is drawn in form E2 in model (a) and in
form E3 in model (b). In the case of E. coli 5 S RNA, the asymmetric loop I₁ in model (c) becomes symmetrical in model
(d). Formation of odd base pairs (indicated by broken lines in model (c) and by losenges in model (d) could reduce loop M
to the same size it has in B. subtilis 5 S RNA (cf. Fig.2a). The symmetrical internal loops remaining in model (d) would
in fact disappear if formation of stacks of two odd base pairs is assumed.

196

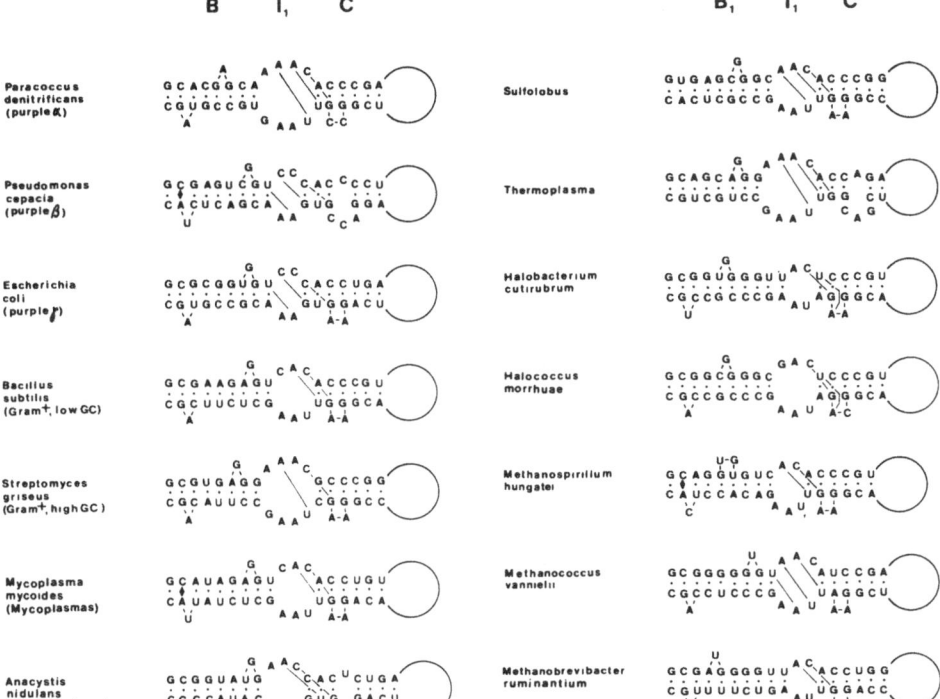

Fig. 4. Making internal loop I₁ symmetrical in bacterial 5 S RNAs.
Elongation of helix B at the expense of loop I₁ as demonstrated in
Fig.3c and d for E. coli 5 S RNA is shown here to be possible in
eubacteria (left column) and archaebacteria (right column) belonging to the most diverse taxa. Internal loop I₁ becomes symmetrical
and a bulge, which occupies a variable position and can usually
migrate, appears on helix B. Base pairing according to lines drawn
at the I₁-C boundary would result in the Cl ⇌ C2 transition illustrated in Fig.2 and 3d.

arise by migration starting at the boundary with loop I₂ and moving in the
3'- to 5'-direction. This would lead to form E3. The bulge on the opposite,
5'-strand could then move in the 3'- to 5'-direction until it "meets" the
bulge on the 3'-strand and neutralizes it forming and odd G◆A base pair
in form E4. The reverse sequence of events would lead from form E4 to E1.
In the case of a yeast 5 S RNA (Fig.2d), the set of reactions would be similar, except for the interconversion E3-E5, which would consist of a "double
bulge switch", discussed below and illustrated in Fig.5.

Internal loops I₁ and I₂

Another feature which seems to be peculiar to eukaryotic 5 S RNAs as
opposed to eubacterial 5 S RNA, is the presence of "criss cross complementarity" in the internal loops. This phenomenon is illustrated in Fig.2c
and d, by lines connecting sets of complementary bases within loop I₁ and
I₂. As a result of the complementarities in loop I₁, eukaryotic 5 S RNAs
not only show the possible equilibrium forms Cl and C2 which also exist in
eubacterial 5 S RNAs (Fig.2a and b). Additional shapes become possible, labeled C3 and C4 in Fig.2c and d. Similarly, a number of different structures
could result from alternative formation of the complementary sets of bases
in loop I₂. The number of potential structures would be multiplied because

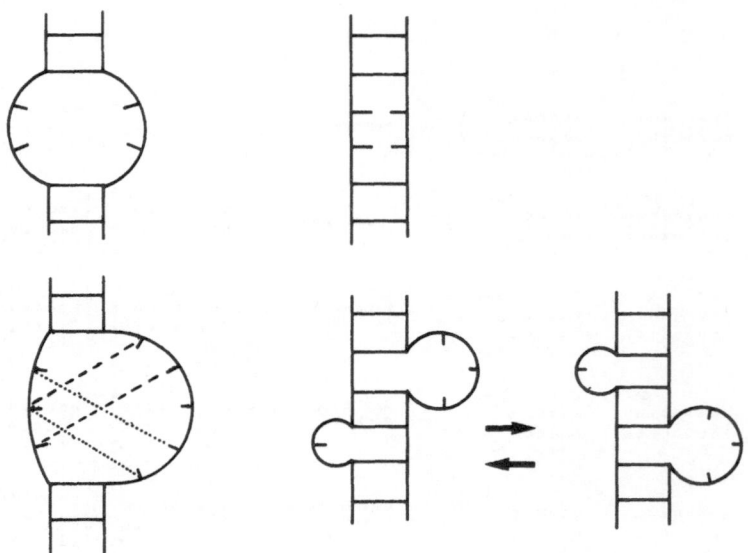

Fig. 5. Two types of internal loops.
The symmetrical internal loop (top left) can be considered as a
stack of odd base pairs (top right). It occurs in eubacterial 5 S
RNAs. The asymmetrical internal loop (bottom left), if it shows
criss-cross complementarity as it does in eukaryotic 5 S RNAs,
could adopt two structures in equilibrium via a double bulge switch
(bottom right).

each could be combined with alternative shapes of helix E discussed above.
The secondary structure that would result if all these base pairing possibilities alternated would not only be extremely mobile, but also much more compact (i.e. comprising more base pairs) than in the conventional model of
Fig.1a. A few of the possible structures are shown for human 5 S RNA in
Fig.3b. The variants drawn for area D-I_2-E result from three alternative
pairings possible in loop I_2, combined with the shape E3 for helix E.

THE STRUCTURE OF INTERNAL LOOPS IN BACTERIAL 5 S RNAs

In eukaryotic 5 S RNAs, both internal loops I_1 and I_2 are always asymmetrical, i.e. the opposing strands contain different numbers of nucleotides
(see e.g. Fig.2c,d and 3a). In contrast, loop I_2 is symmetrical in 5 S RNAs
of eubacteria, including the chloroplasts (see e.g. Fig.2a,b and 3c). It
has been suggested (Stahl et al., 1981) that the latter internal loop may
consist of a stack of a few odd base pairs intercalated between helices D
and E. Area D-I_2-E-H_2 would then in fact form an uninterrupted hairpin.
Internal loop I_1 is asymmetrical in eubacteria as well as in eukaryotes.
However, a closer look at eubacterial 5 S RNA models shows that helix B can
be extended by a variable number of base pairs at the expense of loop I_1,
which then becomes symmetrical, while an extra bulge appears on the 5'-
strand of helix B. This possibility is illustrated in Fig.4 for eubacteria
of various taxa. The structure is conceivable in all presently sequenced
eubacterial 5 S RNAs except that of Rhodospirillum rubrum. The same type of
structure can be contemplated in archaebacterial 5 S RNAs, as also shown in
Fig.4. The more compact type of eubacterial 5 S RNA secondary structure that
would result from formation of the additional base pairs hitherto discussed
is shown in Fig.3d for Escherichia coli 5 S RNA.

ARE THERE TWO STRUCTURAL TYPES OF INTERNAL LOOP ?

A comparison of the "compact" models for 5 S RNA secondary structure (Fig.3b and d) with the "conventional" models (Fig.3a and c) shows that the internal loops of the conventional models can be converted either into helix segments carrying bulges (eukaryotic 5 S RNAs, Fig.3b) or into small symmetrical internal loops (eubacterial 5 S RNAs, Fig.3d). This suggests that what is conventionally called an internal loop may in fact be a more elaborate structure, and that two types may be distinguished, a concept which is illustrated in Fig.5. The symmetric internal loop may in fact be a helix segment consisting of odd base pairs. Hence it would have a weaker structure than the surrounding helix segments, would more easily be disrupted, and could serve as a temporary bending point in the helix during movement of the RNA. The asymmetric internal loop on the contrary cannot form a smooth helix segment. Even if base pairing between the strands is assumed, at least one bulge must be present to accomodate the bases present in excess in the longest strand. However, if the asymmetric internal loop shows crisscross complementarity as seems to be the rule in eukaryotic 5 S RNAs, it can alternate between two structures, each having two bulges, by a "double bulge switch". A third structure, showing a bulge on the longest strand only, is conceivable in some cases, such as human 5 S RNA (Fig.3b). The assymmetric internal loop would also form a weak spot in the helix due to the presence of bulges but it would form a bending point of a type different from the symmetrical internal loop, since the relative movements of the surrounding helix segments during a switch would probably be different.

PRIMARY STRUCTURE FEATURES FAVORING SECONDARY STRUCTURE EQUILIBRIA

Each of the potential equilibria described above arises from the existence of multiple, but mutually exclusive, opportunities for base pairing in some area of the RNA molecule. These "ambiguous complementarities" are of two types : (1) a sequence faces a repeat of its complement in the opposite strand, and (2) there exists "criss cross complementarity" between opposing strands. The ambiguous complementarity can exist within a structural irregularity connecting two stable helix segments (an internal loop) or it can occur in or near a hairpin loop. An inventory of ambiguous complementarities occuring in various structural surroundings, and of the secondary structure switches that may result, is given in Fig.6 taken from De Wachter et al. (1984). Most of the topologies are seen in 5 S RNA, e.g. case (b) can account for additional structure in hairpin H_1, case (d) for migrating bulges in eukaryotic helix E, case (g) for switches in eukaryotic loop I_2.

EQUILIBRIA IN SMALL SUBUNIT rRNA

There is no reason to assume that secondary structure switches would occur more frequently in 5 S rRNA than in other ribosomal RNAs. However, the present models for large ribosomal RNAs are probably less well defined and less extensively explored than those of 5 S RNA since they rest on comparison of a smaller number of sequences. One possibility for a secondary structure switch, mentioned by Woese et al. (1983), exists in the small subunit RNA area extending from nucleotide 1068 to 1107 in E. coli 16 S rRNA, and is illustrated comparatively in Fig.7. The switch would occur at a multibranched loop and would consist in the "siphoning" of two base pairs from one hairpin to the other.
A possibility for a structural switch also exists in the area comprising nucleotides 557 to 575 and 811 to 887 in E. coli 16 S RNA, where different secondary structures have been proposed for bacterial (Woese et al., 1983) and eukaryotic (Ohlsen et al., 1983; Atmadja et al., 1984) small subunit RNA. This possibility has been discussed elsewhere (Nelles et al., 1984).

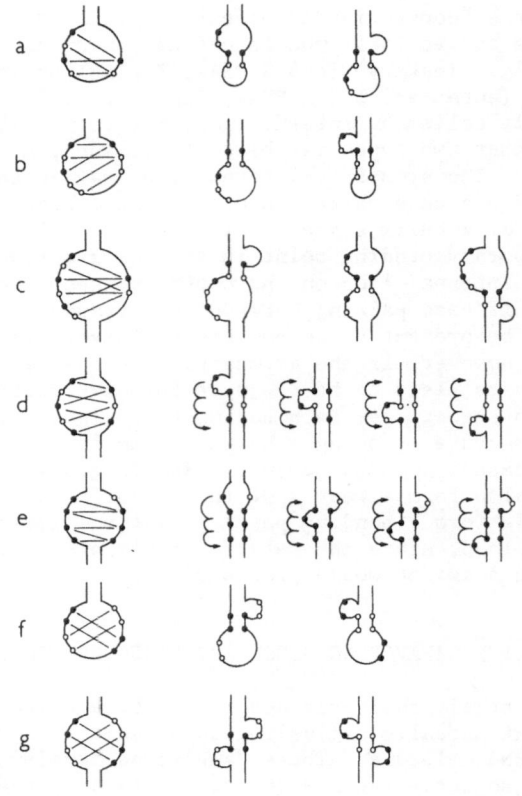

Fig. 6. Primary structure topologies resulting in secondary structure switches.

Each leftmost drawing in a row defines a topology with ambiguous base-pairing potential. Complementary bases on opposite strands are symbolized by dots of the same kind, connected by lines. The following drawings in the row depict resulting secondary structure equilibria. Arrows indicate potential movements of migrating bulges. Four types of topologies causing secondary structure switches can be distinguished :

Case	Base-pairing ambiguity	Location
a,b	repetitive complementary sequences	near hairpin
c,d,e	repetitive complementary sequences	between two helix segments
f	criss-cross complementarity	near hairpin
g	criss-cross complementarity	between two helix segments

Seven types of secondary structure switches can result :

Case	Type of switch
a	bulge/internal loop switch
b	bulge leaves hairpin loop or fuses with hairpin loop
c	double bulge/internal loop switch
d	bulge migrates along helix
e	internal loop (or odd base-pair) dissociates into two bulges, or two bulges meet by migration to form an odd base pair
f	single bulge switch : a bulge switches to the opposite strand
g	double bulge switch

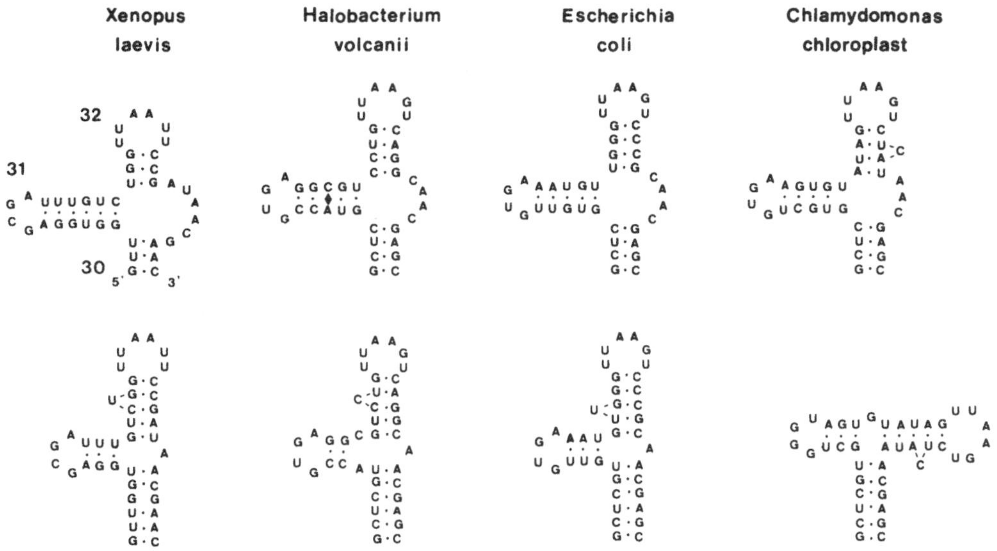

Fig. 7. Possible switch structure in small subunit rRNA.
Alternative secondary structures possible for helices 30, 31 and 32 in the
numbering system of Nelles et al. (1984) (nucleotides 1068 to 1107 in E.
coli 16 S rRNA) are compared in (from left to right) a eukaryote, an archaebacterium, a eubacterium and an algal chloroplast. The sequence in this
area is very conserved. The structure drawn for E. coli is applicable to
all examined eubacteria and most chloroplasts, but not that of Chlamydomo-
nas. The structure for eukaryotes is that found in rat, Xenopus and Artemia, with compensating substitutions occurring in S. cerevisiae and D. discoideum.

DISCUSSION

As stated in the introduction, there now exists a certain consensus
among different research groups that the topology of 5 S RNA secondary structure is essentially the same in eukaryotes, eubacteria and archaebacteria.
To put it more precisely, the same number and type of secondary structure
features (helices, multibranched-, internal-, and hairpin loops) are encountered in the same order when any 5 S RNA sequence is scanned from 5'- to 3'-
end, although the lengths of double-stranded and single stranded areas can
vary considerably. Among the secondary structure switches postulated in this
paper, some are universal, others are kingdom-specific, i.e. characteristic
for either eukaryotic or eubacterial 5 S RNA. As for the archaebacterial
5 S RNAs, their structure is not explicitly discussed in this paper because
relatively few sequences are published and hence the rules governing their
secondary structure are only beginning to emerge. Three structural types
have been distinguished, and their characteristics seem to consist of a
mosaic of eukaryote-like and eubacteria-like features (Willekens et al.,
1985).

There is little if any experimental evidence that can be cited at present to support the existence of any of the secondary structure switches that
have been postulated above. Some experimental support for the existence of
the universal switch in area I_1-C (structures C1 and C2 in Fig.2) has been
found by Christensen et al. (1985). However, it should be acknowledged that
the main argument for postulating the switches is comparative. A stronger
point could be made if some compensating substitutions were discovered that
support the existence of a switch. An example can be found in the universal
bulge-internal loop switch accounting for the C1 ⇌ C2 equilibrium (Fig.2

and 4). In <u>Bacillus subtilis</u>, as in many other species, the switch is made possible by the presence of the repeated sequence A C A C in the 5'-strand of area I_1-C, which is complementary to the G U in the opposite strand. In mammalian 5 S RNA, the corresponding positions show the complementarity G C G C / G C which can be considered as a (doubly) compensating substitution. This proof of the postulated switch is weakened somewhat by the fact that the repeated complementarity often has the structure R Y R Y / R Y and that the existence of G·U pairs in form C2 must be invoked sometimes. However, there are numerous examples where a quite different sequence is found in the corresponding area, yet the switch remains possible. As an example, halobacteria (Fig.4) show the complementarity A C U C /A A G A in form 1 (complementary sequences underlined), A C U C /A A G A in form 2. The basidiomycete <u>Agaricostilbum palmicolum</u> has complementarities A A C A A / A A U U U in form 1 and A A C A A / A A U U U in form 2. Other examples are found among the ascomycetes (Chen et al., 1984). In all these cases, the secondary structure of form C2 differs somewhat from the usual structure (Fig.2a), where a small internal loop of 4 bases is separated by 2 base pairs from loop I_1. Walker (1985) has raised the question whether this structural variability of form C2 weakens the evidence for its existence. However, form C1 has a variable structure as well. The usual shape is a bulge of two bases on helix C, separated from loop I_1 by two base pairs (Fig.2a,c), but variant forms are found, e.g. in chloroplasts (Fig.2b), yeasts (Fig.2d), and certain bacteria (Fig.4). In fact, structural variability is found all over the "static" secondary structure model, in the length of single-stranded and double-stranded areas, and in the distance separating bulges on helices C and E from the helix ends.

Many of the ambiguous complementarities that make possible the existence of switch structures in 5 S RNA occur in or near to internal loops. Since single-stranded areas tend to be more conserved in sequence than helix areas, it can be argued that the ambiguous complementarities may be due to chance, and that they are observed in all 5 S RNAs just because the local sequence is so conserved. While this possibility cannot be excluded, the argument can be turned around. If what is conventionally called an internal loop is in fact the site of a switch structure, then such sites may be very conserved in sequence because only triple compensating mutations allow a change in primary structure with conservation of function.

As stated in the introduction, no relation is proposed, in this paper, between the postulated secondary structure equilibria and the existence in certain 5 S RNAs of different conformers that are stable enough to be experimentally isolated. The switches described here are rather seen as reversible changes that occur rapidly enough to allow the 5 S RNA to participate in, or possibly to facilitate, transitions of the ribosome between multiple allosteric forms occurring during the stages of the translation process.

REFERENCES

Atmadja, J., Brimacombe, R., and Maden, B.E.H., 1984, Xenopus laevis 18 S ribosomal RNA : experimental determination of secondary structural elements, and location of methyl groups in the secondary structure model, <u>Nucl. Acids Res.</u>, 12:2649.
Aubert, M., Scott, J.F., Reynier, M., and Monier, R., 1968, Rearrangement of the conformation of Escherichia coli 5 S RNA, <u>Proc. Natl. Acad. Sci. USA</u>, 61:292.
Böhm, S., Fabian, H., and Welfle, H., 1982, Universal structural features of prokaryotic and eukaryotic ribosomal 5 S RNA derived from comparative analysis of their sequences, <u>Acta Biol. Med. Germ.</u>, 41:1.
Chen, M.W., Anné, J., Volcaert, G., Huysmans, E., Vandenberghe, A., and De Wachter, R., 1984, The nucleotide sequences of the 5 S rRNAs of seven molds and a yeast and their use in studying ascomycete phylogeny, <u>Nucl. Acids Res.</u>, 12:4881.

Christensen, A., Mathiesen, M., Peattie, D., and Garrett, R., 1985, Alternative conformers of 5 S ribosomal RNA and their biological relevance, Biochemistry, 24:2284.

Dams, E., Vandenberghe, A., and De Wachter, R., 1982, Nucleotide sequences of three poriferan 5 S ribosomal RNAs, Nucl. Acids Res., 10:5297.

Dams, E., Vandenberghe, A., and De Wachter, R., 1983a, Sequences of the 5 S rRNAs of Azotobacter vinelandii, Pseudomonas aeruginosa and Pseudomonas fluorescens with some notes on 5 S RNA secondary structure, Nucl. Acids Res., 11:1245.

Dams, E., Londei, P., Cammarano, P., Vandenberghe, A., and De Wachter, R., 1983b, Sequences of the 5 S rRNAs of the thermo-acidophilic archaebacterium Sulfolobus solfataricus (Caldariella acidophila) and the thermophilic eubacteria Bacillus acidocaldarius and Thermus aquaticus, Nucl. Acids Res., 11:4667.

Delihas, N., and Andersen, J., 1982, Generalized structures of the 5 S ribosomal RNAs, Nucl. Acids Res., 10:7323.

De Wachter, R., Chen, M.W., and Vandenberghe, A., 1982, Conservation of secondary structure in 5 S ribosomal RNA : a uniform model for eukaryotic, eubacterial, archaebacterial and organelle sequences is energetically favourable, Biochemie, 64:311.

De Wachter, R., Chen, M.W., and Vandenberghe, A., 1984, Equilibria in 5 S ribosomal RNA secondary structure. Bulges and interior loops in 5 S RNA secondary structure may serve as articulations for a flexible molecule, Eur. J. Biochem., 143:175.

Erdmann, V.A., Wolters, J., Huysmans, E., and De Wachter, R., 1985, Collection of published 5 S, 5.8 S and 4.5 S ribosomal RNA sequences, Nucl. Acids Res., 13:r105.

Fang, B.L., De Baere, R., Vandenberghe, A., and De Wachter, R., 1982, Sequences of three molluscan 5 S ribosomal RNAs confirm the validity of a dynamic secondary structure model, Nucl. Acids Res., 10:4679.

Fox, G.E., and Woese, C.R., 1975a, The architecture of 5 S rRNA and its relation to function, J. Mol. Evol., 6:61.

Fox, G.E., and Woese, C.R., 1975b, 5 S RNA secondary structure, Nature, 256:505.

Huysmans, E., Dams, E., Vandenberghe, A., and De Wachter, R., 1983, The nucleotide sequences of the 5 S rRNAs of four mushrooms and their use in studying the phylogenetic position of basidiomycetes among the eukaryotes, Nucl. Acids Res., 11:2871.

Luehrsen, K.R., and Fox, G.E., 1981, Secondary structure of eukaryotic cytoplasmic 5 S ribosomal RNA, Proc. Natl. Acad. Sci. USA, 78:2150.

MacKay, R.M., Spencer, D.F., Schnare, M.F., Doolittle, W.F., and Gray, M.W., 1982, Comparative sequence analysis as an approach to evaluating structure, function, and evolution of 5 S and 5.8 S ribosomal RNAs, Can. J. Biochem., 60:480.

Nelles, L., Fang, B.L., Volckaert, G., Vandenberghe, A., and De Wachter, R., 1984, Nucleotide sequence of a crustacean 18 S ribosomal RNA gene and secondary structure of eukaryotic small subunit ribosomal RNAs, Nucl. Acids Res., 12:8749.

Ninio, J., 1979, Prediction of pairing schemes in RNA molecules. Loop contributions and energy of wobble and non-wobble pairs, Biochimie, 61:1133.

Nishikawa, K., and Takemura, S., 1974, Structure and function of 5 S ribosomal ribonucleic acid from Torulopsis utilis. II. Partial digestion with ribonucleases and derivation of the complete sequence, J. Biochem., 76:935.

Nishikawa, K., and Takemura, S., 1978, Structure and function of 5 S ribosomal ribonucleic acid from Torulopsis utilis. IV. Detection of exposed guamine residues by chemical modification with kethoxel, J. Biochem., 84:259.

Noller, H.F., 1984, Structure of ribosomal RNA, Ann. Rev. Biochem., 53:119.

Ohlsen, G.J., McCarroll, R., and Sogin, M.L., 1983, Secondary structure of the Dictyostelium discoideum small subunit ribosomal RNA, Nucl. Acids Res., 11:8037.

Rabin, D., Kas, T.-H., and Crothers, D.M., 1983, A characterization of the low temperature structural transition of Escherichia coli 5 S RNA by enzymatic digestion, J. Biol. Chem., 258:10813.

Schwartz, R.M., and Dayhoff, M.D., 1978, Ribosomal and other RNAs, in: "Atlas of protein sequence and structure," M.O. Dayhoff, ed., National Biomedical Research Foundation, Washington.

Stahl, D.A., Luehrsen, K.R., Woese, C.R., and Pace, N.R., 1981, An unusual 5 S rRNA from Sulfolobus acidocaldarius, and its implication for a general 5 S rRNA structure, Nucl. Acids Res., 9:6129.

Studnicka, G.M., Eiserling, F.A., and Lake, J.A., 1981, A unique secondary folding pattern for 5 S RNA corresponds to the lowest energy homologous secondary structure in 17 different prokaryotes, Nucl. Acids Res., 2:1885.

Thompson, J.F., Wegnez, M.R., and Hearst, J.E., 1981, Determination of the secondary structure of Drosophila melanogaster 5 S RNA by hydroxymethyltrimethylpsoralen crosslinking, J. Mol. Biol., 147:417.

Toots, I., Misselwitz, R., Böhm, S., Welfle, H., Villems, R., and Saarma, M., 1982, Two distinct conformations of rat liver ribosomal 5 S RNA, Nucl. Acids Res., 10:3381.

Topal, M.D., and Fresco, J.R., 1976, Complementary base pairing and the origin of substitution mutations, Nature, 263:285.

Traub, W., and Sussmann, J.L., 1982, Adenine-guanine base pairing in ribosomal RNA, Nucl. Acids Res., 10:2701.

Trifonov, E.N., and Bolshoi, G., 1983, Open and closed 5 S ribosomal RNA, the only two universal structures encoded in the nucleotide sequences J. Mol. Biol., 169:1.

Vandenberghe, A., Chen, M.W., Dams, E., De Baere, R., De Roeck, E., Huysmans E., and De Wachter, R., 1984, The corrected nucleotide sequences of 5 S rRNAs from six angiosperms. With some notes on 5 S RNA secondary structure and molecular evolution, FEBS lett., 171:17.

Vandenberghe, A., Wassink, A., Raeymaekers, P., De Baere, R., Huysmans, E., and De Wachter, R., 1985, Nucleotide sequence, secondary structure and evolution of the 5 S ribosomal RNA from five bacterial species, Eur. J. Biochem., 149:537.

Walker, W.F., 1985, 5 S ribosomal RNA sequences from ascomycetes and evolutionary implications, System. Appl. Microbiol., 6:48.

Weidner, H., Yuan, R., and Crothers, D.M., 1977, Does 5 S RNA function by a switch between two secondary structures ?, Nature, 266:193.

Willekens, P., Huysmans, E., Vandenberghe, A., and De Wachter, R., 1985, Archaebacterial 5 S ribosomal RNAs : nucleotide sequence in two methanogen species, secondary structure models, and evolution, Syst. Appl. Microbiol., in press.

Woese, C.R., Gutell, R., Gupta, R., and Noller, H.F., 1983, Detailed analysi of the higher-order structure of 16 S-like ribosomal ribonucleic acids, Microbiol. Rev., 47:621.

A COMPARATIVE ANALYSIS OF STRUCTURAL DYNAMICS IN 5S rRNA

Martin Digweed, Tomas Pieler and Volker A. Erdmann

Institut für Biochemie
Freie Universität Berlin
D-1000 Berlin 33

INTRODUCTION

A natural consequence of the discovery that several bacterial 5S rRNAs could be incorporated into active Bacillus stearothermophilus 50S subunits (Wrede and Erdmann, 1973) was the search for a common secondary structure for all bacterial 5S rRNAs. Fox and Woese (1975) were the first to suggest a base-pairing scheme based on sequence comparison. This secondary structure has, in the meantime, been confirmed by a spectrum of physical and bio-chemical investigations: proof enough of the strength of the sequence comparison approach.

However, whilst the Fox and Woese model has been confirmed, most in-vestigators would insist that it can only represent the minimal extent of base pairing. A plethora of models based on extensions of Fox and Woese's has since been suggested (for a recent review see Pieler et al., 1983b).

Since the explosion in the number of available 5S rRNA sequences, thanks to the advent of rapid gel-sequencing methods (e.g. Peattie, 1979; Donis-Keller et al., 1977), others have applied the sequence comparison approach to the question of 5S rRNA secondary structure. Some have compared the sequences intuitively (DeWachter et al., 1982) whilst others have used sophisticated computer programs which allowed a simultaneous search for the energetically most favourable, homologous structure (Studnicka et al., 1981).

We have presented evidence previously for a tertiary interaction unique to eubacterial 5S rRNA which is not present in eukaryotic 5S rRNA (Pieler and Erdmann, 1982). As expected, 5S rRNA extracted from chloroplast ribo-somes displays eubacterial features (Pieler et al., 1983a), whilst archae-bacterial 5S rRNA shows both eukaryotic and eubacterial characteristics (Pieler et al., 1982).

A complication in the study of E. coli 5S rRNA structure has been the observation of structural heterogeneity. The occurrence of the structurally distinct A-form and B-form has been well documented (Aubert et al., 1968; Lecanidou and Richards, 1975). In addition, evidence is accumulating for a low-temperature melting transition in A-form 5S rRNA which occurs under phy-siological conditions and is therefore possibly of functional significance (Kao and Crothers, 1980; Kime and Moore, 1982). Finally, sequence heterogeneities have been found amongst the transcripts of the seven E. coli 5S rRNA genes (Jarry and Rosset, 1971). We were recently able to describe the dramatic effects which such sequence variation may have on 5S rRNA structure and function (Digweed et al., 1982). Such effects might have been predicted considering the data on stability of synthetic RNA duplexes accumulated by Tinoco and coworkers (Tinoco et al., 1971; 1973).

In other instances, these thermodynamic data have shown themselves to have little relevance for naturally occurring RNAs, and were unable to predict the 'clover-leaf' as the most stable conformation for tRNA sequences. This has now been remedied by the introduction of an empirical model for pairing-scheme evaluation by Ninio (1979). and Papanicolaou et al. (1984).

In this study, we have attempted to analyse the structure and stability of 5S rRNA from a range of organisms using two complementary techniques. Digestion of RNA with the single-strand specific nuclease S1 yields the location of loops and other single-stranded structures; carried out at increasing temperatures, it brings information on the melting behaviour of the molecule, which may then be related to its structure. The sodium bisulphite induced deamination of cytidines measures the accessibility of cytidines and hence the stability of base-pairs in which they are involved. When carried out for long periods of time under mild conditions, subtle dynamic features, such as "helix fraying", can be detected.

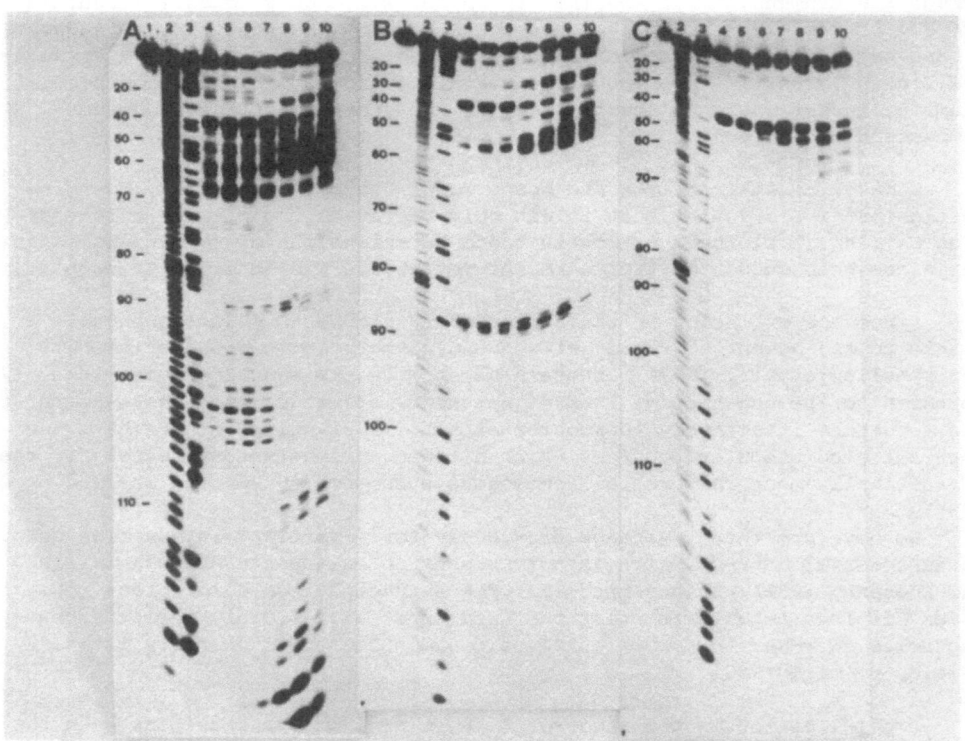

Fig. 1. S1 Nuclease digestion of eubacterial 5S rRNA. Sequencing gels of E. coli MRE 600 (A), B. stearothermophilus (B) and T. thermophilus (C) 5S rRNA digested with S1 nuclease at : 4) 3°C; 5) 23°C; 6) 30°C; 7) 40°C; 8) 50°C; 9) 55°C; 10) 60°C. Cleavage sites were located by comparison to the T1 RNase digest (3) and the alkaline hydrolysis (2). (1) control incubation.

We would like to show that the classical techniques of partial enzymatic digestion and chemical modification, when conducted under defined denaturing conditions can yield information on the dynamics of the 5S rRNA molecule. The use of 5S rRNAs from several organisms allows the role of the sequence itself in structural dynamics to be assessed. In particular, 5S rRNA variants which have point mutations have been examined to find the effects which such variation may have on structure and stability.

Figures 1 and 2 show the sequencing gels used to detect S1 nuclease cleavage sites and chemically modified cytidines in 5S rRNAs from several prokaryotic organisms. The results will be discussed in detail below, whilst the data are summarized in Tables 1 and 2.

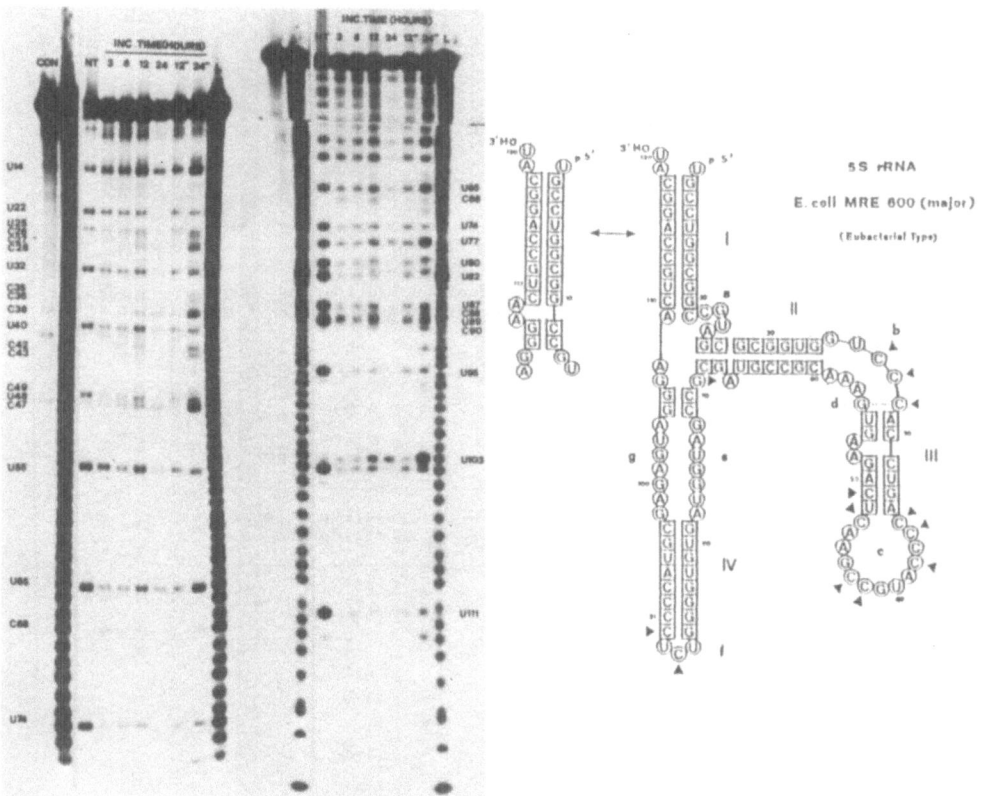

Fig. 2. Deaminated cytidines in E. coli 5S rRNA. The uridine specific sequencing reaction conducted on unmodified (NT) 5S rRNA and on 5S rRNA modified for 3 to 24 hours at 25°C (two samples in the absence of Mg^{++} ions: 12^- and 24^-). Extra bands arise from deaminated cytidines. The labelling on the left gives the positions of uridines and modifiable cytidines only. Electrophoresis for 4 hours and 16 hours on a 12% polyacrylamide, 7M urea gel at 1.5 kV. The secondary structure of E. coli 5S rRNA is shown with arrows indicating the modifiable cytidines.

TABLE 1 : Nucleotide accessibility in Eubacterial 5S rRNA[a]

Nucleotide	Structure[b]	Bacterium[c]	Temperature °C						
			3	23	30	40	50	55	60
20-24	II	E.c.	1	1	1	0	0	0	0
		B.s.	0	0	0	0	0	0	0
20		T.t.	1	2	2	2	2	1	1
30-33	III	E.c.	0	0	0	1	2	3	4
25-26		B.s.	0	0	1	1	2	2	2
36		T.t.	0	0	0	0	1	1	1
40-41	c	E.c.	4	4	4	4	4	2	2
39		B.s.	3	3	3	2	2	2	1
43		T.t.	3	3	3	3	3	3	2
45-46	c	E.c.	0	1	1	2	2	2	2
43-44		B.s.	0	0	1	1	2	2	2
48		T.t.	0	0	1	3	3	3	2
50-53	III	E.c.	1	1	2	2	3	3	3
51-52		B.s.	0	0	1	1	2	2	2
55-56		T.t.	0	0	0	0	1	2	2
57-60	d	E.c.	1	2	3	3	3	3	3
54-57		B.s.	1	2	2	3	3	3	3
60-61		T.t.	0	0	0	0	1	2	2
64-66	II	E.c.	3	3	3	3	2	1	1
		B.s.	0	0	0	0	0	0	0
		T.t.	0	0	0	0	0	0	0

[a]The extent of S1 cleavage at individual nucleotides as judged
from the intensities of bands on the sequencing gels in Fig. 1 is
given a value of 0 to 4.
[b]Structures are as shown in Fig. 2.
[c]E.c.: Escherichia coli; B.s.: Bacillus stearothermophilus; T.t.:
Thermus thermophilus

TABLE 2 : Cytidine modification in E. coli 5S rRNA[a]

Cytidine	Structure[b]	Relative Rate of Modification
90	IV	fast
88	f	fast
68	II	fast
49	III	fast
47	c	slow
43	c	slow
42	c	slow
38	c	fast
37	c	very slow
36	c	slow
35	c	slow
28	III	fast
27	b	slow
26	b	slow

[a]The modifiable cytidines in E. coli 5S rRNA were
detected on sequencing gels such as that of Fig. 2
and given relative rates of modification.
[b]Structures are as shown in Fig. 2.

TABLE 3 : Calculated Free Energy of 5S rRNA helices and loops.[a]

Organism Model[b]	Helix I	Helix II	Helix III	Helix IV	Helix V	Loop a/e/g	Loop b/d	Loop c	Loop f
E.coli									
thermodynamic	-29.6	-10.4	-11.3	-16.2	-	5.7	3.0	7.2	8.0
empirical	-23.4	-12.6	-10.7	-19.4	-	10.7	3.9	5.0	4.5
B.stearothermophilus									
thermodynamic	-12.4	-18.8	-14.1	-18.7	-	6.0	3.2	7.2	5.0
empirical	-18.4	-17.5	- 7.1	-18.3	-	10.6	4.2	5.0	5.0
T.thermophilus									
thermodynamic	-30.1	-19.2	-16.2	-13.1	-	6.0	3.0	7.2	5.0
empirical	-25.6	-14.4	-11.7	-18.4	-	10.9	3.9	5.0	5.0

[a]Values of ΔG are given in kcal. [b]Two models were used for the calculation, the thermodynamic model of Tinoco and coworkers (1971;1973) and the empirical model of Ninio (1979).

DISCUSSION

Helix III: A Dynamic Structure

Examination of the cleavage of double stranded regions can be expected to yield information on the dynamics of the RNA structure: It is striking that Helix III is susceptible in all three 5S rRNAs. As might have been expected, considering the origins of the three molecules, the cleavage takes place only at higher temperature in the thermophilic 5S rRNAs: 30°C in B. stearothermophilus, 50°C in T. thermophilus. Even at 3°C there is some cleavage in this helix in E. coli 5S rRNA. This is an indication of the labile nature of this helix, and is in agreement with the helix-fraying observed through modification experiments.

The susceptibility of Helix III to S1 digestion and bisulphite modification is not adequately explained by the lower calculated stability of this helix (see Table 3). Examination of the distribution of nearest-neighbour base-pairs in Table 4 shows several interesting points. Whilst the interaction GC/GC is favoured by most helices, in Helix III the mode CG/CG is the almost exclusive representative of this base-pair repeat. There is also a strong deficit of interactions containing only adenine and uridine nucleotides, although this helix has the highest proportion of A•U base-pairs of the four helices. The interaction AU/CG is overrepresented whereas it is underrepresented in the other helices. It is interesting to note that S1 cleavage was induced at A_{30} -U_{32} in E. coli 5S rRNA upon addition of urea (Pieler et al., manuscript submitted).

The analysis of G•U base-pairs in 5S rRNAs shown in Table 5 indicates that there are virtually no G•U pairs in Helix III (only one of the 25 sequences examined has a G•U pair). Clearly, Helix III has different basic properties than the other helices, which is also expressed by the fact that it is only the helix with a considerable amount of sequence conservation.

The function of 5S rRNA has yet to be determined. It is interesting to look for suspect regions of the molecule which may have a functional importance. Clearly, Helix III, with its labile nature and closing a highly structured loop which is intimately involved in maintenance of tertiary structure, is a good candidate.

In contrast to the behaviour of Helix III, Helix IV is not attacked in any of the molecules, even at 60°C. However, C_{90} $_{(E.c.)}$ is rapidly modified,

TABLE 4 : Occurrence of nearest-neighbour base-pair interactions in eubacterial 5S rRNA.[a]

Base-pair or interaction	Helix I Obs	Helix I Exp[b]	Helix II Obs	Helix II Exp	Helix III Obs	Helix III Exp	Helix IV Obs	Helix IV Exp
G·C	84	144	87	121	29	104	77	106
C·G	60		34		75		29	
A·U	11	54	36	54	31	55	11	18
U·A	43		18		24		7	
G·U	26	45	17	33	3	28	49	31
U·G	19		0		38		6	
$\frac{GG}{CC}$	16	39	29	16	3	14	5	16
$\frac{CC}{GG}$	14		17		14		6	
$\frac{GA}{UC}$	4	19	14	16	9	14	4	16
$\frac{AG}{CU}$	10		1		1		1	
$\frac{GU}{AC}$	6	19	3	15	11	15	1	5
$\frac{AG}{UG}$	3		2		22		3	
$\frac{GU}{CA}$	3	15	5	15	10	15	0	5
$\frac{AC}{UG}$	22		6		9		4	
$\frac{GU}{CU}$	12	15	2	15	11	15	0	5
$\frac{AG}{UC}$	3		20		0		4	
$\frac{AA}{UU}$	0	15	5	15	0	15	1	5
$\frac{AU}{UA}$	5	6	5	7	0	8	0	1
$\frac{AU}{UA}$	2	3	6	3	0	4	1	1
$\frac{AU}{UU}$	2	3	5	3	0	4	1	1

[a] The occurrence of given interactions in the helices of 25 species of eubacterial 5S rRNA is given.
[b] The expected values were calculated for random distribution as described
by Ninio (1979).

TABLE 5 : Occurrence of G·U base-pairs in the four helices of eubacterial 5S rRNA.[a]

Base-pair or interaction	Helix I		Helix II		Helix III		Helix IV		Helix terminals	Total	
	Obs	Exp[b]	Obs	Exp	Obs	Exp	Obs	Exp		Obs	Exp
5'G / U	13		10		0		23		12	46	
5'U / G	37		7		1		25		10	70	
↓GU/GC↑	6	13	0	6	1	0.4	13	12	0	20	28
↓GU/CG↑	27	13	5	6	0	0.4	9	12	3	41	28
↓UG/GC↑	15	13	14	6	0	0.4	6	12	0	35	28
↓UG/CG↑	5	13	0	6	0	0.4	2	12	3	8	28
↓GU/AU↑	2	5	6	2	0	0.1	2	2	1	10	11
↓GU/UA↑	5	5	4	2	0	0.1	6	2	3	15	11
↓UG/AU↑	1	5	1	2	0	0.1	1	2	0	3	11
↓UG/UA↑	7	5	1	2	0	0.1	1	2	3	9	11
UG/GU GU/UG	9	19	1	3	0	0	22	13	10	20	21

[a]The observed occurrence of nearest-neighbours containing G·U base-pairs in 25 eubacterial sequences is given. [b]The expected values were calculated for random distribution.

indicating that the 3 base loop of E. coli 5S rRNA is probably increased to 5 bases, the terminal base pair of Helix IV being dissolved. This instability of 3 membered loops has been previously found by Uhlenbeck et al. (1973) working with synthetic oligonucleotides.

Helix I in 5S rRNA from the thermophiles is also stable at this temperature, whilst in E. coli 5S rRNA there is distinct helix-fraying and the expected cuts around the looped-out adenines which are unique to this organism (A_{108} and A_{109}).

Helix II: looped out bases

Only E. coli 5S rRNA shows cleavage around the looped out adenine common to all eubacterial 5S rRNAs in this helix. An explanation for this difference may be obtained from an examination of the distribution of nearest-neighbour base-pair interactions in 5S rRNA.

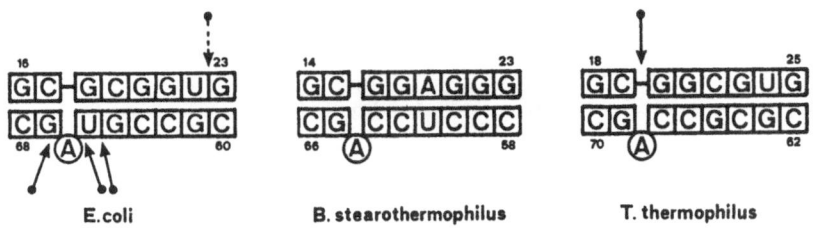

E.coli   B. stearothermophilus   T. thermophilus

An examination of the distribution of base-pair slices in a set of 25 eubacterial 5S rRNAs is presented in Table 4.

The susceptibility of the looped out adenine in Helix II of E. coli 5S rRNA may be caused by the presence of the base-pair G•U, although it is obviously in one of its favoured environments since the two slices GU/CG and CG/GU are represented normally and in excess, respectively, in Helix II of 5S rRNA (see Table 4). However, we have little knowledge of stacking interactions across bulge loops, perhaps a G•U pair in this position is particularly destabilizing.

Patel et al. (1982) have shown that an extra adenosine stacks into a strand of complementary deoxyribonucleotides in the neighbourhood of two G•C pairs. The equilibrium of the intercalated versus the looped out form may be strongly dependent on the neighbouring base-pairs, explaining why a bulge-loop in the neighbourhood of a 5'-G•U base pair is significantly more accessible than in the neighbourhood of a 5'-G•C base-pair.

No reactivity is found in the single stranded region e, in which two unreactive cytidines $C_{70+71 \ (E.C.)}$ are located, confirming the structured nature of this region. Single-stranded region g shows moderate susceptibility at low temperatures in E. coli 5S rRNA, and at elevated temperatures (23°C and above) in 5S rRNA from B. stearothermophilus and T. thermophilus.

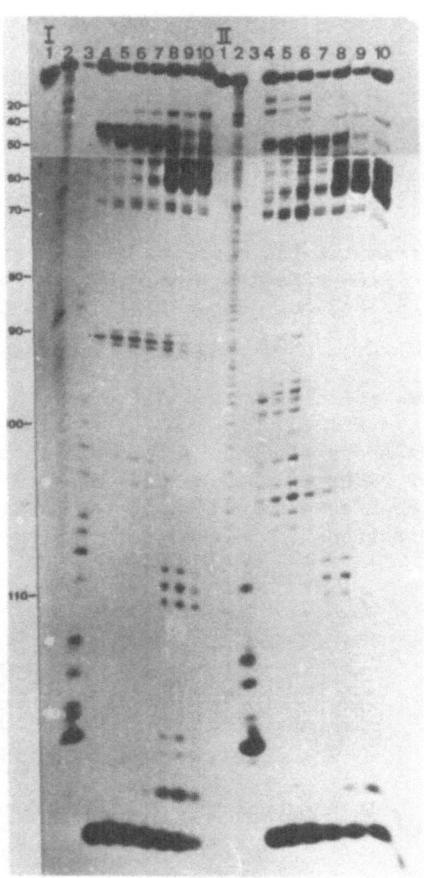

Fig. 3. S1 Nuclease digestion of E. coli A19 5S rRNA species. Sequencing gels of species 'AB' (I) and 'C' (II) digested with S1 nuclease at increasing temperatures. Details as for Fig. 1.

TABLE 6 : Nucleotide accessibility in E.coli A19 5S rRNAs.[a]

Nucleotide	Structure	5S species	Temperature °C						
			3	23	30	40	50	55	60
16-20	II	AB	0	0	0	0	0	0	0
		C	2	1 ·	2	0	0	0	0
21-24	II	AB	0	0	0	0	0	0	0
		C	2	1	2	0	0	0	0
30-33	III	AB	0	0	1	1	2	1	2
		C	0	0	0	0	1	1	1
40-41	c	AB	4	4	4	4	4	2	2
		C	3	3	3	3	3	1	1
45-46	c	AB	1	3	3	3	3	2	3
		C	1	1	1	1	2	1	1
70-71	e	AB	0	0	0	0	0	0	0
		C	0	1	1	0	0	0	0
87-89	f	AB	2	3	3	3	3	1	1
		C	1	1	1	1	1	0	0
94-97	IV	AB	0	0	0	0	0	0	0
		C	2	2	2	0	0	0	0
100-105	g	AB	0	0	0	0	2	2	1
		C	1	2	3	2	1	0	0
108-110	I	AB	0	0	0	0	2	2	1
		C	0	0	0	1	2	0	0

[a]The extent of S1 cleavage at nucleotides displaying differing
accessibility in the two 5S rRNA species is given a value of 0 to
4.

This region with its many purines is expected to stack particularly well
between the coaxial Helices I and IV. This region may also be involved in
formation of the fifth helix in eubacterial 5S rRNA as suggested by others.

5S rRNA with a Point Mutation

We have previously described a variant 5S rRNA species in E. coli A19
(Digweed et al., 1982) which contains the C_{92} to U transition first discovered by Jarry and Rosset (1971). This mutation enforces a G·U base-pair
in Helix IV. We were able to demonstrate the dramatic effects which this
transition has on the stability of the molecule. We have named this molecule
mutant species 'C', the remaining 5S rRNA population consisting of at least
two further species which we call 'A' and 'B'. Fig. 3 shows the autoradiogram of the sequencing gel used to examine S1 digestion of 5S rRNA 'C' and
'AB'. The differences in cleavage are collated in Table 6.

The main differences in cleavage are in Helix IV and the adjoining
single stranded region g. In species 'C', there is cleavage along Helix IV
towards the 3'-terminal from C_{93} past the helix end and into region g. This
is not the case in species 'AB'. The reactivity of these new areas in
species 'C' causes a displacement of the major cleavage in the hairpin-loop
f. We can conclude that in species 'C' the equilibrium between closed and

open forms of Helix IV has been pushed strongly in the open direction.

A-Form to B-Form Transition

We examined the A-form to B-form conversion of several E. coli 5S rRNA species. Fig. 4 shows the autoradiogram of a gel-electrophoresis separation of the two E. coli 5S rRNA conformations. In our procedures, all 5S rRNAs eluted from polyacrylamide gels after 3'-end labelling are exclusively in the A-form. Incubation at 55°C in the absence of Mg^{++} ions followed by rapid cooling yields the expected equilibrium of A-form and B-form for E. coli MRE 600 5S rRNA and two of the E. coli A19 subspecies, namely 'B' and 'C'. However, species 'A' from E. coli A19 shows negligible B-form formation.

In a thorough examination of the structural differences between E. coli 5S rRNA A-form and B-form, Göringer et al. (1984) found cleavage at U_{45}, G_{96}, G_{98} and A_{99}, with single-strand specific RNase only in the B-form. These results are similar to those found in E. coli A19 5S rRNA species 'C', suggesting a structure for this species similar to B-form, or a greater tendency to undergo the conformational change than species A or B of the same RNA.

Weidner et al. (1977) have suggested a secondary structure for B-form 5S rRNA which requires disruption of Helix IV and base-pairing of one side of this helix to bases of Helix III and loop C, as shown in Fig. 5. The cuts peculiar to 5S rRNA species 'C' are shown; clearly these agree well

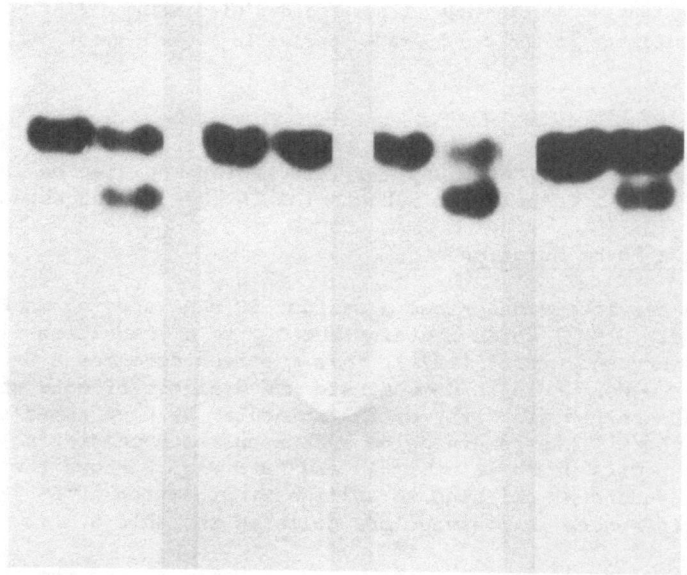

Fig. 4. A-form/B-form conformations in E. coli 5S rRNAs. 3'-[^{32}P] 5S rRNA from E. coli MRE 600 and three variant species (A, B & C) from E. coli A19 were loaded into a native 10% polyacrylamide gel NAT, native (no pretreatment); DEN, denatured. Electrophoresis at 3V cm^{-1} for 16 hours.

with this conformational rearrangement. However, the double-strand-specific RNase cuts at positions C_{92}, C_{93} and A_{94} found in the B-form examination of Göringer et al. (1984) suggest that other interactions must be occurring.

Our studies on the influence of urea on the secondary structure of E. coli MRE 600 5S rRNA have suggested a rearrangement of Helix IV (Pieler et al., manuscript submitted). A disturbed Helix IV, as in mutant species 'C', might facilitate this conformational switch.

Another interesting feature common to B-form 5S rRNA and mutant species 'C' is their protein binding capability. We have shown previously (Digweed et al., 1982 and unpublished results) that species 'C' is less efficient than species 'A' in ribosomal-protein-complex formation, and is incorporated into 50S ribosomal subunits to only 60% of the efficiency of species 'A'. Similarly, B-form 5S rRNA has been shown to be unable to form a specific protein complex with ribosomal proteins under reconstitution conditions (Bellemare et al., 1972).

Loop c and Tertiary Interactions

One particulary interesting region of 5S rRNA is the large hairpin-loop c, closed by Helix III. Several tertiary interactions have been proposed involving this region of the molecule. Böhm et al. (1981) have proposed a parallel base-pairing interaction between $C_{38}CAU_{40}$ and $G_{75}GUA$ whilst Jagadeeswaran and Cherayil (1980) and Faber and Cantor (1981) suggest an

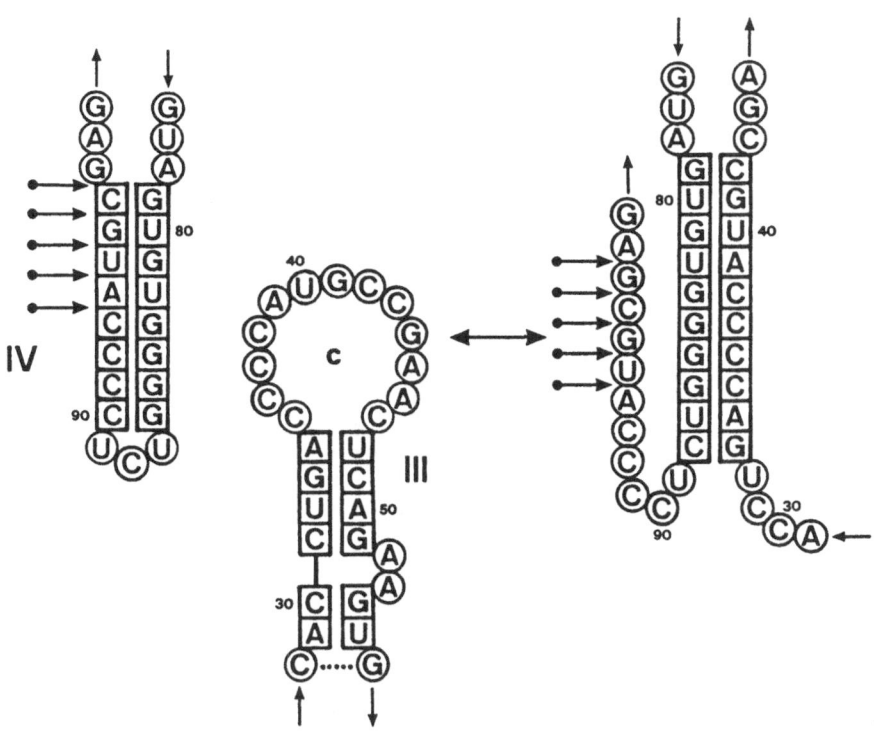

Fig. 5. Reorganisation of Helix IV, Helix III and Loop c in the B-form of E. coli 5S rRNA (after Weidner et al. (1977)). Arrows indicate the sites of S1 cleavage in E. coli A19 5S rRNA species 'C'.

antiparallel interaction between $A_{73} UGG_{76}$ and $U_{40} ACC_{37}$. We have previously suggested an interaction between $G_{41} CCG_{43}$ and $U_{74} GGU_{77}$ in 5S rRNA from <u>E. coli</u> which may be universal for eubacterial 5S rRNA since base-pairing is conserved throughout this group (Pieler and Erdmann, 1982). All these possibilities have found support in the crosslink formed between G_{41} and G_{72} by Hancock and Wagner (1982).

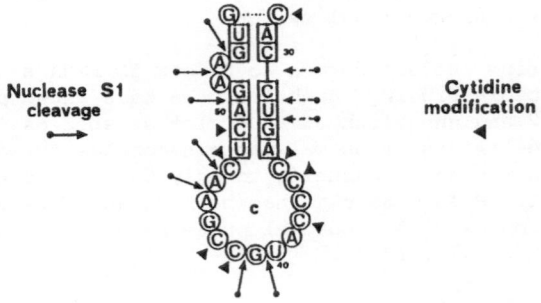

Characteristic for all three molecules is the primary cut in the large hairpin-loop c: U_{40} in <u>E. coli</u>, C_{39} in <u>B. stearothermophilus</u> and U_{43} in <u>T. thermophilus</u>. This cleavage dominates the pattern at 3°C and remains the most readily accessible site until over 50°C, at which point melting of the helices has increased the effective substrate concentration. Also typical for these eubacterial 5S rRNAs is a further cut in hairpin-loop c located 3 or 4 nucleotides further in the 3' direction. There is a gap in the cleavage pattern in all three molecules which may be due to the involvement of the intervening nucleotides in the tertiary interaction described earlier (Pieler and Erdmann, 1982).

The modification of cytidines in this loop in <u>E. coli</u> is unusual. Whilst C_{38} is deaminated rapidly, its neighbouring C_{37} is particularly resistant to modification. C_{35} and C_{36} are modified at normal rates. Clear-

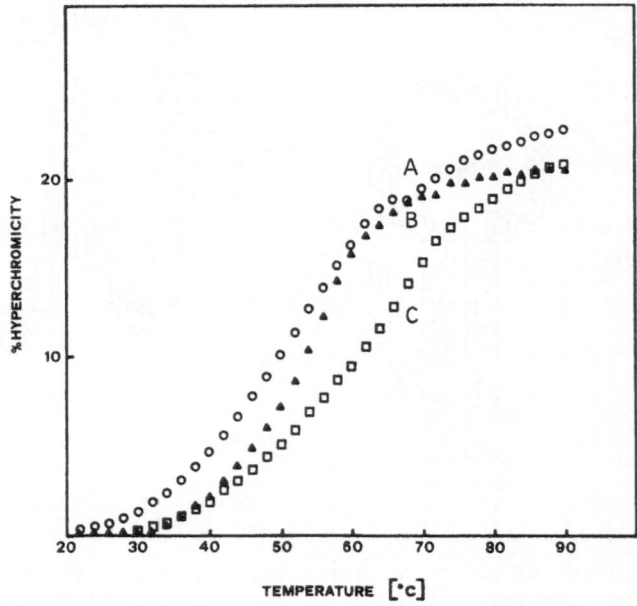

Fig. 6. Thermal melting profiles of eubacterial 5S rRNA. A- <u>E. coli</u> <u>MRE</u> <u>600</u>; B- <u>B. stearothermophilus</u>; C- <u>T. thermophilus</u> 5S rRNA.

TABLE 7 : Occurrence of nearest-neighbour base-pairs in
Eukaryotic and Eubacterial 5S rRNA.[a]

Base-pair or interaction	Eukaryotes		Eubacteria	
	Obs	Exp[b]	Obs	Exp
G̲C̲	309		277	
C̲G̲	344		198	
A̲U̲	140		89	
U̲A̲	148		92	
G̲G̲ / C̲C̲	76 / 100 → 176	183	95 / 63 → 158	132
G̲C̲	52	91	53	61
C̲G̲	65	91	51	61
G̲A̲ / U̲C̲	20 / 17 → 37	81	31 / 13 → 44	50
C̲U̲ / A̲G̲	66 / 76 → 142	81	21 / 30 → 51	50
G̲U̲ / A̲C̲	41 / 53 → 94	81	18 / 41 → 59	50
C̲A̲ / U̲G̲	46 / 23 → 69	81	25 / 27 → 51	50
A̲A̲ / U̲U̲	12 / 19 → 31	36	6 / 10 → 16	19
A̲U̲	7	18	9	9
U̲A̲	26	18	8	9

[a]The occurrence of given interactions in 25 species of eubacterial
and 30 species of eukaryotic 5S rRNA is given. [b]The expected values
were calculated for random distribution.

ly this loop is involved in some higher order structure.

Stability and Sequence

Although the general secondary and tertiary structural features of the
three eubacterial 5S rRNAs seems to be identical, their individual stabilities, as judged from the melting experiments, differ significantly. It is
apparent from the S1 nuclease studies that the thermophilic 5S rRNA is the
most thermostable (Table 1) which is in good agreement with the results
from the melting curve analysis (Fig. 6). This however, was unexpected from
the examination of the free enthalpy for the three secondary structures
following either the thermodynamic (Tinoco et al., 1971; 1973) or the empirical model (Ninio, 1979). This observation indicates that additional
tertiary interactions play a major part in the maintenance of RNA structure.

In mutant species 'C' the cytidine to uridine conversion at position 92
enforces a G·U pair which disrupts not only Helix IV but also single
stranded regions G and E. Romaniuk et al. (1979) have studied the melting
characteristics of synthetic duplexes containing G·U base pairs and have
established that such pairs are regions of local instability from which
melting can initiate. This is obviously the case in this mutant 5S rRNA
species.

In this context it is interesting to note that both the new slices formed by the C_{92} to U mutation in E. coli A19 5S rRNA species 'C', GU/GC and GC/GU, are unfavourable in 5S rRNA in general, and the latter in Helix IV in particular.

The analysis of slice distribution in Table 7 for a set of 30 eukaryotic 5S rRNAs indicates that several combinations have been conserved, whilst others may be classified as unfavourable. Thus, iterated G•C-pairs of either mode, much favoured by eubacterial 5S rRNAs, are 'normally' represented whilst the mixed slice GC/UA, in both modes, is greatly overrepresented. Mixed slice GC/AU is, on the other hand, in a deficit. The pair AU/UA appears to be similarly unfavourable.

In comparison, eubacterial 5S rRNA shows an almost random distribution of slices. Solely the excess of GC/GC and CG/CG interactions suggest that these positively influence stability. In the molecule as a whole there is a proclivity for 5'-G•C rather than 5'-C•G pairs. Within the individual helices the random slice distribution is no longer found, and the different helices appear to have adopted different favoured base-pair interactions. Either there are differences in stability contributions depending upon the helix's position in the structure as a whole, or there is a conservation of 'unfavourable' slices in some of the helices: this could be of functiona significance.

ACKNOWLEDGEMENT

The authors would like to thank A. Schreiber for drawing the figures and I. Brauer for typing the manuscript. The financial support of the Deutsche Forschungsgemeinschaft and the Fonds der Chemischen Industrie are acknowledged.

REFERENCES

Aubert, M., Scott, J.F., Reynier, M. and Monier, R. (1968) Proc. Natl. Acad Sci. USA 61, 292-299
Bellemare, G., Jordan, B.R., Rocca-Serra, J. and Monier, R. (1972) Biochimi 54, 1453-1466
Böhm, S., Fabian, H., Venyaminov, S.Y., Matvev, S.V., Lucius, H., Welfle, H. and Filimonov, V.V. (1981) FEBS Lett. 132, 357-361
Digweed, M., Kumagai, I., Pieler, T. and Erdmann, V.A. (1982) Eur. J. Biochem. 127, 531-537
DeWacher, R., Chen, M.-W. and Vandenberghe, A. (1982) Biochimie 64, 311-32
Donis-Keller, H., Maxam, A.M. and Gilbert, W. (1977) Nucl. Acids Res. 4, 2527-2538
Faber, N.N. and Cantor, C.R. (1981) J. Mol. Biol. 146, 241-257
Fox, G.E. and Woese, C.R. (1975) Nature (Lond) 256, 505-507
Göringer, H.U., Szymkowiak, C. and Wagner, R. (1984) Eur. J. Biochem. 144, 25-34
Hancock, J. and Wagner, R. (1982) Nucl. Acids Res. 10, 1257-1269
Jagdeswaran, P. and Cherayil, J.D. (1980) J. theor. Biol. 83, 369-375
Jarry, B. and Rosset, R. (1971) Mol. gen. Genet. 113, 43-50
Kao, T.H. and Crothers, D.M. (1980) Proc. Natl. Acad. Sci. USA 77, 3360-3364
Kime, M.J. and Moore, P.B. (1982) Nucl. Acids Res. 10, 4973-4983
Lecanidou, R. and Richards, E.G. (1975) Eur. J. Biochem. 57, 127-133
Ninio, J. (1979) Biochimie 61, 1133-1150
Papanicolaou, C., Gouy, M. and Ninio, J. (1984) Nucl. Acids Res. 12, 31-44
Patel, D.J., Kozlowski, S.A., Marky, L.A., Rice, J.A., Broka, C., Itakura, K. and Breslauer, K.J. (1982) Biochemistry 21, 445-451

Peattie, D.A. (1979) Proc. Natl. Acad. Sci. USA 76, 1760-1764

Peattie, D.A. and Gilbert, W. (1980) Proc. Natl. Acad. Sci. USA 77, 4679-4682

Pieler, T. and Erdmann, V.A. (1982) Proc. Natl. Acad. Sci. USA 79, 4599-4603

Pieler, T., Kumagai, I. and Erdmann, V.A. (1982) Zbl. Bakt. Hyg. I Abt. Orif. C3, 69-78

Pieler, T., Digweed, M., Bartsch, M. and Erdmann, V.A. (1983a) Nucl. Acids Res. 11, 591-604

Pieler, T., Digweed, M. and Erdmann, V.A. (1983b) Alfred Benzon Symposium 19, 353-376

Romaniuk, P.J., Hughes, D.W., Gregoire, R.J., Bell, R.A. and Neilson, T. (1979) Biochemistry 18, 5109-5116

Studnicka, G.M., Eiserling, F.A. and Lake, J.A. (1981) Nucl. Acids Res. 9, 1885-1904

Tinoco, I., Uhlenbeck, O.C. and Levine, M.D. (1971) Nature (Lond) 230, 362-367

Tinoco, I., Borer, P.N., Dengler, B., Levine, M.D., Uhlenbeck, O.C., Crothers, D.M. and Gralla, J. (1973) Nature New Biol. 246, 40-41

Uhlenbeck, O.C., Borer, P.N., Dengler, B. and Tinoco, I. (1973) J. mol. Biol. 73, 483-496

Weidner, H., Yuan, R. and Crothers, D.M. (1977) Nature (Lond) 266, 193-194

Wrede, P. and Erdmann, V.A. (1973) FEBS Lett. 33, 315-319

A DOMAIN OF 23S RIBOSOMAL RNA IN SEARCH OF A FUNCTION

Asser Andersen, Niels Larsen, Henrik Leffers,
Jørgen Kjems, and Roger Garrett

Biostructural Chemistry, Kemisk Institut
Aarhus Universitet, Aarhus, Denmark

SUMMARY

The 23S-like ribosomal RNAs (rRNA) consist of six main structural
domains that are stabilized by long range base pairing interactions. Domain
IV has remained one of the most highly conserved of these during evolution,
and this implies that it plays a crucial role at some stage of protein
biosynthesis. In this article, we examine the structure of domain IV of
Escherichia coli 23S RNA experimentally and consider its possible functional
roles.

We probed the accessible adenosines, and some cytidines, in this domain
using diethyl pyrocarbonate and dimethyl sulphate/hydrazine, respectively,
and we isolated homogeneous RNA fragments, for analysis, using a DNA hybridization technique (Van Stolk, B. and Noller, H.F. J. Mol. Biol., 180, 151,
1984). The results were compared with published secondary structural models
that are based primarily on phylogenetic sequence comparisons. Although
these models still contain uncertainties, the data were generally in good
agreement with recently revised versions. Most of the reactive adenosines
and cytidines were concentrated in the putative loop and interhelical regions. Only four moderately reactive adenosines, and no cytidines, were
found in putative double helices under "native" conditions (10 mM magnesium).
Two of these occurred in the putative helix 10 that appears to be disproven
both experimentally and by phylogenetic criteria. The experimental data
were compatible, however, with other helical segments, for which there is
no phylogenetic support because the region has been either too conserved or
too labile during evolution. Under semi-denatured conditions (1 mM EDTA)
increased reactivity was confined to relatively few nucleotides, situated
mainly in loop and interhelical regions, which suggests that the whole
domain exhibits a stable and intricate tertiary structure. This presumption
is supported further by the low reactivity of several nucleotides in "unstructured" regions under both native and semi-denatured conditions.

Three putative functional sites were identified on the basis of their
highly conserved sequences, post transcriptional modifications, intron
splicing sites, and location on the ribosomal surface. Possible functions
for each of these sites are considered and evaluated.

221

INTRODUCTION

Our current concept of protein biosynthesis, that the ribosomal RNAs play a dominating functional role, derives from the evolutionary argument that the first protein must have been produced on a "ribosome" consisting solely of RNA (Crick, 1968). This view has been strongly reinforced by increasing evidence of a direct role for the RNA in the mechanics of protein synthesis (reviewed by Noller, 1984) and of a secondary role for the proteins (reviewed by Garrett, 1983). Various sites have been implicated directly in function and they exhibit at least two of the following properties: (a) highly conserved sequences; (b) post transcriptional modification, and (c) location on the ribosomal surface. They include parts of the 3'-subdomain of 16S RNA that participate in mRNA and tRNA binding, and a site within domain V of 23S RNA that is involved in peptidyl transfer (Noller, 1984). Other potentially functional sites are located in domain IV because it contains several highly conserved sequences, some of which are accessible on the ribosomal surface, and it also exhibits some post transcriptionally modified nucleotides in *E. coli*.

Our insight into how the rRNAs function remains limited. They probably provide transient attachment sites for both mRNAs and tRNAs that involve base pairing, as well as binding sites for factors and ribosomal subunits; some of these may open and close in a reciprocating manner. Functional regions such as these are probably identifiable by the above mentioned criteria. The RNAs may also function in other ways, however, such as in catalysing peptidyl transfer (Garrett and Woolley, 1982), and in transmitting information allosterically through the ribosome (Garrett *et al.*, 1977); thi could occur along RNA helices, that may be co-axially stacked (Noller, 198₄ as a result, for example, of conformational changes (A ←→ A'-form) occurri within double helices.

In the present study, we have probed the accessibilities of the adenosines and some cytidine residues within domain IV and have correlated the results with phylogenetic analyses of the secondary structure. Sites that are of possible functional importance are identified and their possible roles are considered.

MATERIALS AND METHODS

Chemical Modification Procedure

Accessible adenosines and cytidines in *E. coli* 23S RNA were reacted with diethyl pyrocarbonate and dimethyl sulphate/hydrazine, respectively, under native (70 mM Na cacodylate, pH 7.2, 10 mM $MgCl_2$, 270 mM KCl) and semi-denatured (70 mM Na cacodylate, pH 7.2, 1mM EDTA) conditions (Peattie and Gilbert, 1980). The former RNA samples were renatured by heating in the CMK buffer for 10 min. at $45°C$ and cooling slowly; the latter were semi-denatured by precipitating from 0.3M Na acetate, 0.1 mM EDTA with ethanol and then redissolving in the modification buffer. Plasmid pJN 53 was prepared that carries the 23S RNA gene of the *rrnB* operon of *E. coli* (supplied by Prof. Harry Noller). It was derived from pKK 3535 (Brosius *et al.*, 1981) by digesting with *BamHI* and *BclI*, removing the rRNA promoters, and religating. 60 μg plasmid was restricted with *AluI* or *HhaI* and *HinfI* and the fragments were hybridized to 30 μg of modified 23S rRNA (3-4 times molar excess) in 20 μl of a buffer containing 89% (v/v) deionized formamid₍ 19 mM Hepes-OH, pH 7.0, 0.44 M NaCl, 1mM EDTA. The mixture was maintained ₍ $53°C$ for 1 h (Van Stolk and Noller, 1984). 300 μl of ice-cold 20 mM Tris-C pH 7.5, 0.3 M Na acetate was added, and non-hybridized RNA was digested with 1.5 units ribonuclease T_1 (RNAse - Sankyo) per μg RNA for 30 min. at

37°C. The reaction was stopped by extracting 3 times with 300 µl of phenol saturated with 0.3 M Na acetate followed by 3 ether extractions. The hybrids were fractionated on a Sephadex G-50 column (40 cm x 1 cm), and then dephosporylated by incubating with 0.5 units of *E. coli* alkaline phosphatase (Sigma) in 20 µl of 1 M Tris-Cl, pH 8.0, 1 mM $ZnCl_2$ at 37°C for 30 min. Another 0.5 units was then added and the incubation was continued for 30 min. before diluting with 300 µl of 0.3 M Na acetate and removing the phosphatase by extracting 5 times with phenol and chloroform.

The RNA fragments were 3'-end labelled with $[5'-^{32}P]$ pCp (Amersham) using RNA ligase (P.L. Biochemicals) (Bruce and Uhlenbeck, 1978) and fractionated on an 8% non-denaturing polyacrylamide gel (35 cm x 15 cm x 0.03 cm) containing 50 mM Tris-borate, pH 8.3, 1 mM EDTA at 8 W for 6 h. The hybrid bands were localized by autoradiography, excised, eluted by shaking for 2-4 hours at 4°C in 300 µl of 0.3 M Na acetate, 1 mM EDTA in the presence of carrier tRNA and precipitated with 2.5 vol. ethanol. The RNA fragments were heat denatured in 5 µl of 50 mM Tris-borate, pH 8.3, 1 mM EDTA, 8 M urea and electroforesed through a 5% denaturing polyacrylamide gel (40 cm x 40 cm x 0.04 cm) in the same buffer at 60-65 W. The RNA bands were eluted, as described for the hybrids, and precipitated with 2.5 vol. of cold ethanol. Strand scission was effected at the modified sites by aniline-acetic acid treatment and these sites were identified by running the samples on 8% sequencing gels alongside sequencing tracks (Peattie, 1979).

Phylogenetic Sequence Comparisons

We aligned the sequences and secondary structures of the 23S-like RNAs from 5 eukaryotes, 3 archaebacteria, 3 eubacteria, 2 chloroplasts and 7 mitochondria. Sequences derive from the following sources. Eukaryotes: Mouse (Hassouna *et al.*, 1984); *Xenopus laevis* (Clark *et al.*, 1984); *Saccharomyces cerevisiae* (Georgiev *et al.*, 1981); *Physarum polycephalum* (Otsuka *et al.*, 1983); *Dictyostelium discoideum* (Ozaki *et al.*, 1984). Archaebacteria: *Methanococcus vannielii* (Jarsch and Böck, 1985); *Halococcus morrhuae* and *Desulfurococcus mobilis* (Leffers *et al.*, 1985). Eubacteria: *Escherichia coli* (Noller *et al.*, 1981); *Bacillus stearothermophilus* (Kop *et al.*, 1984); *Anacystis nidulans* (Douglas and Doolittle, 1984). Chloroplasts: Tobacco (Takaiwa and Sugiura, 1982); *Zea mays* (Edwards and Kössel, 1981). Mitochondria: *Paramecium primaurelia* (Seilhamer *et al.*, 1984); *Aspergillus nidulans* (Köchel and Küntzel, 1982); *S. cerevisiae* (Sor and Fukuhara, 1983); mosquito (HsuChen *et al.*, 1984); mouse (Bibb *et al.*, 1981); bovine (Anderson *et al.*, 1982); human (Eperon *et al.*, 1980). In evaluating secondary structures only transitions between alternative Watson-Crick pairings were considered as support for a base pair. In studies on the trypanosome-like mitochondrial RNAs, we aligned and structured the sequences from *Trypanosoma brucei* (Eperon *et al.*, 1983); *Crithidia fasciculata* (Sloof *et al.*, 1985) and *Leishmania tarentolae* (de la Cruz *et al.*, 1985).

SECONDARY STRUCTURE OF DOMAIN IV

Several discrepancies existed amongst the earlier models for domain IV mainly because only two sequences were subjected to comparative analyses when deriving the Berlin-Freiburg (Glotz *et al.*, 1981) and the Strasbourg-Freiburg models (Branlant *et al.*, 1981); the three sequences examined in the Santa Cruz/Urbana study yielded a more lasting result. Double helices (numbered in our model) that have survived more comprehensive phylogenetic sequence comparisons (Maly and Brimacombe, 1983; Noller 1984) are as follows: Berlin-Freiburg model- all except helices 1a, 3c, 8a and b, 9 and 10; Strasburg model- all except helices 1a, 6b, 8c and base pairing within the internal loop of helix 8; Santa Cruz-Urbana model- all except for an extended version of helix 4 and 8a. Uncertainties remain, however, because some regions are not amenable to the sequence comparison approach for two reasons: (1) the sequence has been highly conserved, as is found in helices 5 and 10, and (2) the primary and secondary structure have evolved rapidly as has occurred in helices 3b and c, and 8c. We consider below these less certain regions in the light of more recent sequence data (see Materials and Methods) and our chemical modification results.

The chemical reactivities of the adenosines in domain IV, and some cytidines, were estimated under native conditions, after renaturing in the presence of 10 mM magnesium, and under semi-denatured conditions (1 mM EDTA) when some of the weaker tertiary and secondary structure had probably dissociated. Adenosines react when unstacked if the N-7 position is not involved in a tertiary interaction, and cytidines react at the N-3 position when not base paired; both reactions are primarily probes for secondary structure. The adenosine reactivities in the 5'-half of the domain are illustrated in the autoradiogram in Fig. 1 and the results are quantified in Table 1. Adenosine and some cytidine reactivities are also summarized on the putative secondary structure in Fig. 2.

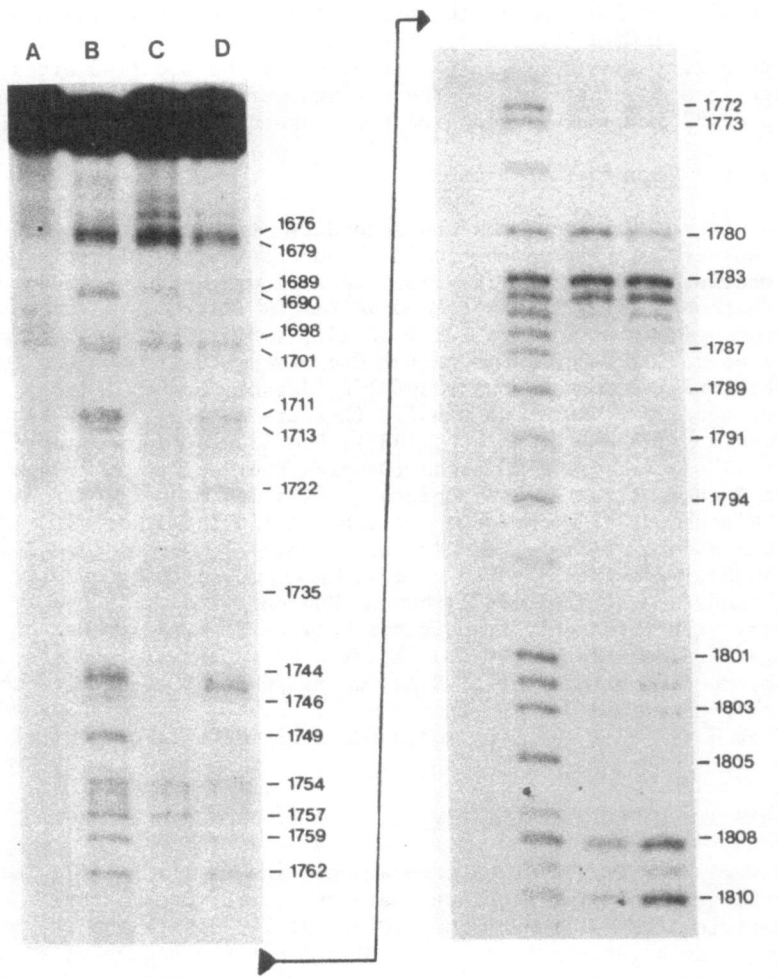

Fig. 1. Adenosine modification data for a fragment of 23S RNA isolated using a DNA restriction fragment extending between two *HhaI* sites at nucleotides 1643-1646 and 1867-1870. Track A-unmodified control sample; B - adenosine sequencing track; C - modification under native conditions, and D - modification under semi-denaturing conditions.

Table 1. Carbethoxylated Adenosines Within Domain IV of *E. coli* 23S RNA Under Native and Semi-Denatured Conditions at 30°C

Adenosine	10mM Mg^{2+}	1mM EDTA	Adenosine	10mM Mg^{2+}	1mM EDTA
1650	nd	nd	1810	++	+++
1652	nd	nd	1815	(+)	(+)
1654	nd	nd	1819	++	++
1655	nd	nd	1821	+	+
1664 } 1665	−	−	1829	−	−
			1847	+	+
1668 } 1669	(+)	(+)	1848	++	+++
			1853	(+)	+
1672	+	+	1854	−	−
1676 } 1677	++	++	1858	−	−
			1866	nd	nd
1678 } 1679	++	++	1871	nd	nd
			1872	nd	nd
1689 } 1690	+	+	1876	+	++
			1877	(+)	(+)
1698	−	−	1885	−	−
1700 } 1701	++	++	1889 } 1890	++	++
1705	−	−	1899	+	+
1711	−	++	1900	++	++
1713	(+)	(+)	1901	+	+
1717	−	−	1912	++	++
1722	−	+	1913	++	++
1735	−	−	1916	−	−
1739	−	−	1918	(+)	(+)
1744	−	+	1919	(+)	(+)
1745	−	++	1927	+	+
1746	−	+	1928	+	+
1749	−	+	1932	−	−
1754	+	+	1936	+	++
1755	+	+	1937	+	++
1757	++	+	1938	−	+
1759	+	+	1952	+++	++
1762	−	++	1953	+	+
1772	+	+	1960	−	−
1773	+	+	1966	(+)	++
1780	+++	++	1969	+	++
1783	+++	+++	1970	−	(+)
1784	+++	+++	1977	+	++
1785	+	++	1978	(+)	+
1786	−	+	1981	+++	++
1787	−	(+)	1987	(+)	(+)
1789	(+)	+	1998	−	−
1791	(+)	+	2003	−	−
1794	−	+	2005	+	+
1801	(+)	++	2009	(+)	(+)
1802	+	++	2013	+	+++
1803	+	++	2014	+	++
1805	−	(+)	2015	−	++
1808	++	+++	2019	+	+++
1809	(+)	+	2020	−	+

nd = not determined.

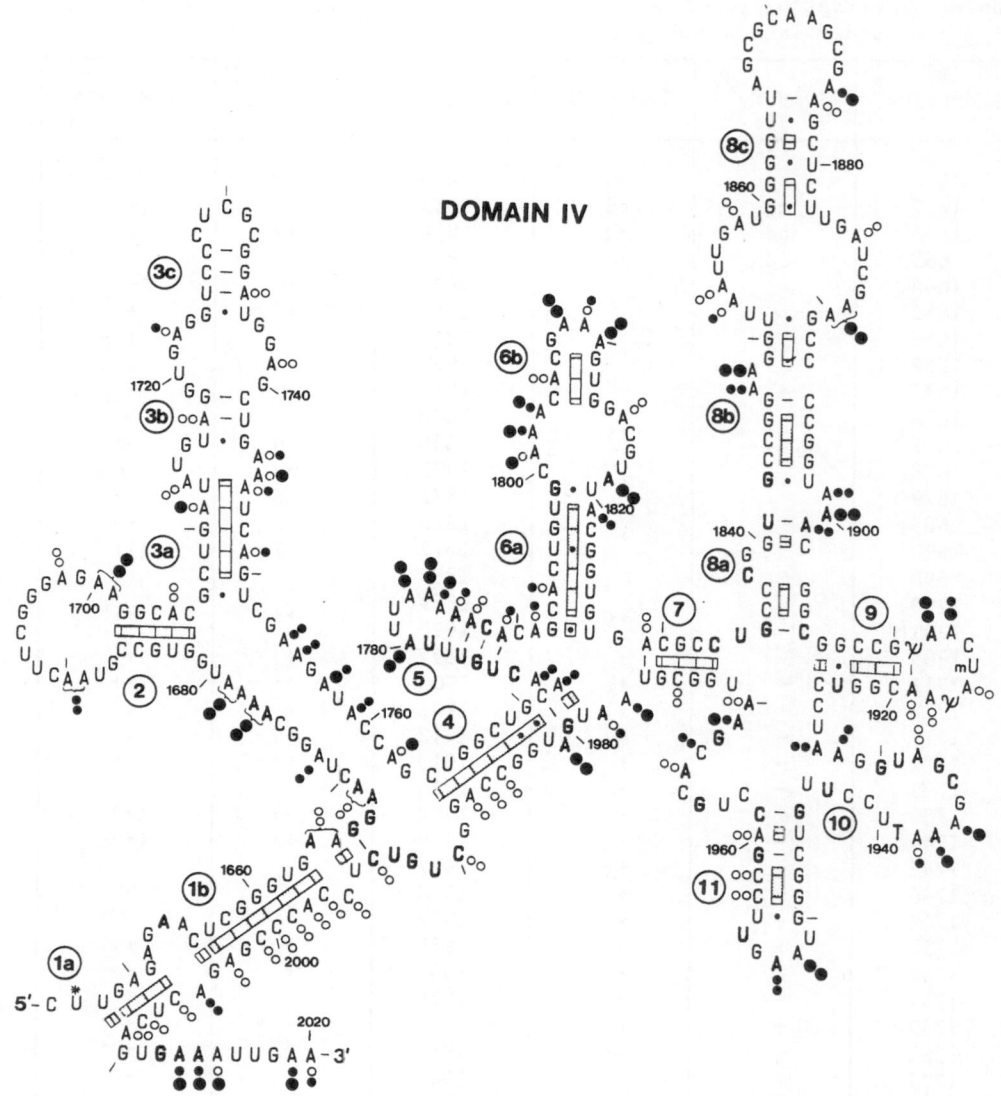

DOMAIN IV

Fig. 2. A secondary structural model of domain IV in *E. coli* 23S rRNA. It is based on the models that were derived primarily from phylogenetic sequence comparisons (Glotz *et al.*, 1981; Branlant *et al.*, 1981; Noller *et. al.*, 1981; Maly and Brimacombe, 1983; Noller, 1984). More extensive sequence data (see Materials and Methods) support many features of the most recent model (Noller, 1984). Nucleotides modified post transcriptionally are included (Fellner and Sanger, 1968) together with an unknown modification at U_{1647} (indicated by an asterisk)(Branlant *et al.*, 1979). Universally conserved nucleotides are drawn in darker type. Base pairs that can form universally and are supported by coordinated base changes are indicated with shaded squares. Those that are kingdom- or organelle-specific and supported by coordinated base changes are denoted by open squares. Chemically reactive adenosines or cytidines are represented as follows: Open circles - no modification; closed circles of increasing size indicate, respectively, increasing degrees of reactivity. The circle adjacent to the nucleotide and the second circle depict data obtained at 10 mM Mg^{2+} and 1 mM EDTA, respectively.

Helix 1: Is universal except for the terminal base pairs $U-A_{2009}$(1a) and $C-\overline{G_{2004}}$(1b) which are absent from some kingdoms/organelles. Moreover, another terminal base pair $G-C_{1994}$(1b) is conserved and, therefore, not supported by phylogenetic criteria. The carbethoxylation data reinforce the helix; each adenosine and cytidine that was probed was inaccessible except for the unpaired A_{2005}. The resistance of the juxtapositioned $A_{1664}-C_{1996}$ suggests that they stack into the helix.

Helix 2: Is conserved in the primary kingdoms and chloroplasts. *P. primaurelia* contains the only mitochondrial RNA from lower eukaryotes that generates a stable helix, and the helix is absent from mitochondrial RNAs of higher eukaryotes. A_{1705} within the helix was unreactive. The large terminal loop exhibits no detectable conserved secondary structure and, probably, four of the five adenosines were very reactive (Fig. 2).

Helix 3: Consists of three segments and has varied considerably during evolution even amongst the eubacteria; in *B. stearothermophilus* the RNA exhibits an extended version of helix 3a with no separate 3b and c, and in the RNAs of *An. nidulans* and chloroplasts there is only helix 3a. The three archaebacterial RNAs all exhibit extended versions of helix 3a and in the eukaryotic RNAs there are large additions of between 113 and 306 nucleotides with no recognizable equivalent of the *E. coli* RNA base-pairing (Fig. 2). Amongst the mitochondrial RNAs there are also large variations, ranging from *As. nidulans* RNA with a 67 nucleotide addition, rich in adenosines and uridines, to those of yeast and the higher eukaryotes where helix 3 and its 3'-flanking sequence are absent. The structures of helical segments 3b and 3c in the *E. coli* RNA (Fig. 2) remain speculative but they are supported by the chemical modification data; the latter also reveal structuring in the two internal loop regions.

Helix 4: Is universal and only the $C-G_{1980}$ pair is unproven phylogenetically because the G is conserved. The unpaired A_{1981} was typically hyperactive (Garrett *et al.*, 1984).

Helix 5: Constitutes a highly conserved region in which 9 of the 15 nucleotides are universal. There is no phylogenetic support for any base pair within this helix and there is a disproof of the two possible base pairs adjacent to the terminal loop in some archaebacterial and eukaryotic RNAs. The chemical modification data were compatible with the presence of base pairing in all except the terminal $U-A_{1785}$ that is left unpaired in Fig. 2. The loop region is very accessible in the RNA as judged by the high reactivity of the four adenosines.

Helix 6: Segment 6a is universal and 6b is present in all but the mitochondrial RNAs of higher eukaryotes. The experimental data strongly support the base pairing and all of the very reactive adenosines occur within the internal or terminal loops.

Helix 7: Exhibits some kingdom/organelle-specific effects that are summarized in Table 2. 5 base pairs cannot form in the RNAs of all kingdoms and organelles, although the modification data are compatible with the formation of a fifth base pair ($A-U_{1976}$) in the *E. coli* RNA.

Helix 8: Is very irregular with several unpaired nucleotides in helix 8a and a weak helix 8c. Its lability during evolution is illustrated in Fig. 3 where strong differences occur between the primary kingdoms and organelles. By phylogenetic criteria only segment 8a and a part of 8b occur universally; the illustrated version of 8c is proven only in archaebacteria, eubacteria/chloroplasts. The latter region constitutes one of only two in the 23S-like RNAs (the other is in domain V) where the eukaryotic RNAs exhibit fewer nucleotides than those of eubacteria, chloroplasts and the mitochondria of lower eukaryotes (Fig. 3). Helix 8 is reinforced by the location of the strongly reactive adenosines lying in unpaired regions. However, the presence of four resistant adenosines in the large internal loop indicates the presence of additional secondary and/or tertiary structural interactions.

Helix 9: Is universal by phylogenetic criteria. Two adenosines were strongly reactive within the terminal loop that also exhibited three post transcriptionally modified nucleotides.

Helix 10: As for helix 5 the region is highly conserved and the possible base pairing is not supported by phylogenetic criteria; although some base pairing can form in all RNAs frequent mismatches occur. The chemical modification data cast further doubt on this helix because the adenosines in two of the four base pairs A–U$_{1943}$ and A–U$_{1944}$ were both fairly reactive. We have intentionally omitted the base pairing in Fig. 2 because the helix appears to be disproven. The terminal loop exhibits two reactive and two resistant adenosines; the latter are indicative of extra structuring.

Helix 11: Is universal and strongly supported by the experimental data; the resistant adenosines and cytidines are localized in the helix and reactive adenosines are confined to the terminal loop.

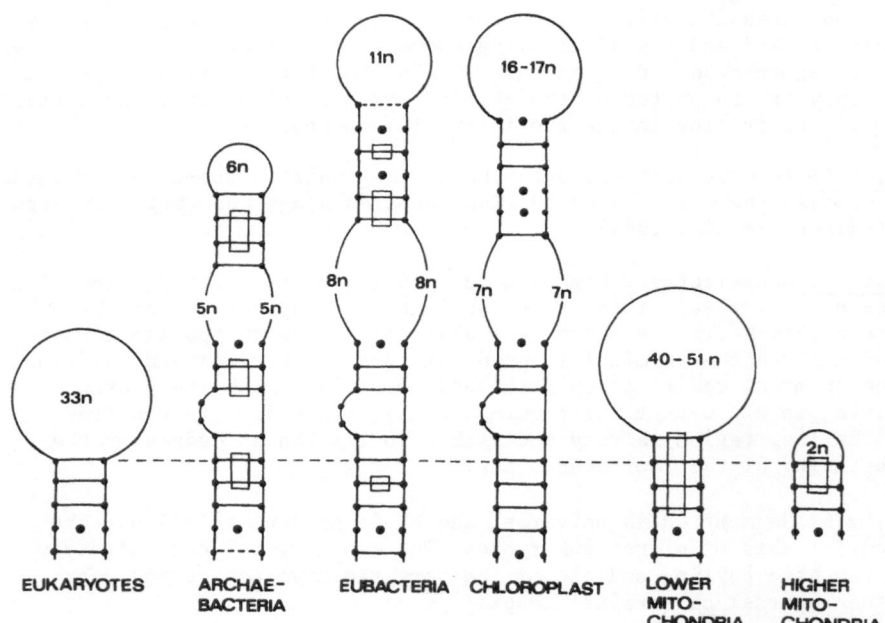

Fig. 3. Evolutionary divergence of helix 8b and c amongst the primary kingdoms and organelles. The structures illustrated are kingdom- or organelle-specific. Horizontal lines denote base pairs and closed circles between the bases depict G.U pairings; broken lines at the ends of helices reflect that mismatches sometimes occur. Base pairs that are confirmed by coordinated base changes within a kingdom or class of organelles are marked by open boxes. RNAs in which putative base pairing in the *E. coli* RNA was disproven by sequence comparison evidence, and for which no clear alternative exists, are drawn as open loops. The sequences were aligned at the points linked by the horizontal broken line.

Table 2. Kingdom- and Organelle-Specific Structural Features in Domain IV

Helix	Nucleotides	Eukaryotes higher/lower	Archae-bacteria	Eubacteria/chloroplasts	Mitochondria lower/higher
1	$A_{1665}-U_{1995}$	G-C	G-C	A-U	A-U
2 & 3	helices	see text			
3	$U_{1709}-A_{1749}$	C-A	C-G	U-A or C-G	absent
4	U_{1982}	G	G	Y/U	Y/G
	G_{1989}	A	A	G/A	R/R
5	loop	AAUU	AAUA	AYYA	$A(\frac{A}{C})$ UA/AYCA
6	G_{1817}	U	G	G	G
	C_{1820}	C	C	U	U
7	helix	3bp/2-3 bp	4-5 bp	5 bp	2-5 bp/3-5 bp
		no CBCs			
	C_{1832}	U	C	Ċ	Y/C
8		see figure 3			
9	$C_{1908}-G_{1922}$	G-C	G-C	C-G	C-G
	C_{1925}	U	U	C	C or A/C
10	helix	3 bp	4 bp	4 bp	4 bp/1-3 bp
11	$G_{1948}-C_{1958}$	A-U	G-C	G-C	non-conserved
	G_{1954}	U	G	G	non-conserved
	C_{1967}	U	U	C	Y/U

CBC = coordinated base changes. R = purine. Y = pyrimidine.

During the phylogenetic sequence analyses, several kingdom- or organelle-specific sequence and structural features were recognized. Those occurring within helices 2, 3, 7 and 8 were considered above but others that occur in loop or interhelical regions are summarized in Table 2. Of particular interest is the central column for the archaebacteria; they exhibit both eubacterial and eukaryotic features and are closer to both of these kingdoms than they are to one another.

Earlier Experimental Evidence

Various attempts have been made to probe the structure of domain IV within 50S subunits and 23S RNA of *E. coli*. All of the results pertain to the secondary structure but yield little detailed insight. Accessible RNA sites within native 50S subunits include the terminal loop of helix 11 that was sensitive to different single strand-specific probes, the kethoxal-reactive G_{1980} at the end of helix 4, and the putative helix 3c that was cut by the cobra venom nuclease (Noller *et al.*, 1981; Branlant *et al.*, 1981). Some insight has also been gained into base paired regions. A crosslink was induced between two large fragments of domain IV by ultraviolet irradiation of subunits; secondary analyses suggested an interaction between opposite strands of helical segments 8a and b ($CCCG_{1839}$ and $UAAACG_{1903}$) (Stiege *et al.*, 1983). Furthermore, partially deproteinized 23S RNA was digested with RNAse T_1, and then proteinase K, and co-migrating fragments were analysed (Glotz *et al.*, 1981) that are compatible with the presence of helices 1b, 3a and 8a and b. In the free 23S RNA unpaired cytidines were localized at the junction of helices 4 and 5 and in the terminal loop of helix 9 (Noller *et al.*, 1981). Several ribonuclease A, S_1 and T_1 cuts were also localized mainly in the terminal loops (Branlant *et al.*, 1981), but given the strength of the digestion conditions it is difficult to assess which were primary effects.

STRUCTURE-FUNCTION RELATIONSHIPS

The foregoing section emphasizes the high level of structural conservation of domain IV during evolution; it is equalled only by the adjacent domain V that has been strongly implicated in the peptidyl transferase reaction. This establishes that it is essential for the mechanics of protein synthesis and this inference is supported further by the results presented below.

Location of Splicing Sites

All of the splicing sites that have not been located in domain IV occur in one of two known functional centres: The peptidyl transferase centre within domain V and the α-sarcin cutting site in domain VI that is associated with elongation factor-dependent aminoacyl tRNA binding (Noller *et al.*, 1981). The other splicing sites, including one in the archaebacterium *D. mobilis* (Kjems and Garrett, 1985), all reside in domain IV within either highly, or universally conserved, sequences (Fig. 4). This observation indirectly reinforces the inference that these sequences are functionally important.

The reason for introns occurring in functionally important RNA regions remains unclear. It has been suggested that they regulate the function of ribosomes in that they are activated by the splicing reaction (Branlant *et al.*, 1981). Equally, they could fulfill a proof-reading role; this would require that the splicing reaction proceeds only if the correct ribosomal structure has formed in a critical functional site. They could also facilitate the accurate structuring of such critical regions during the relatively complex assembly of 50S subunits (R. Amils, *pers. comm.*) which might explain why all rRNA introns have been found, so far, in the 23S-like RNAs. There are clearly many possible reasons and it also remains likely that the introns are multifunctional.

Conserved Sequences, Post Transcriptional Modifications and Accessibility on the Ribosome

Further criteria for a functional role for domain IV are summarized schematically in Fig. 4. The universally conserved nucleotides are presented and they are clustered at three main sites denoted I, II and III. The post transcriptionally modified nucleotides that have so far been located

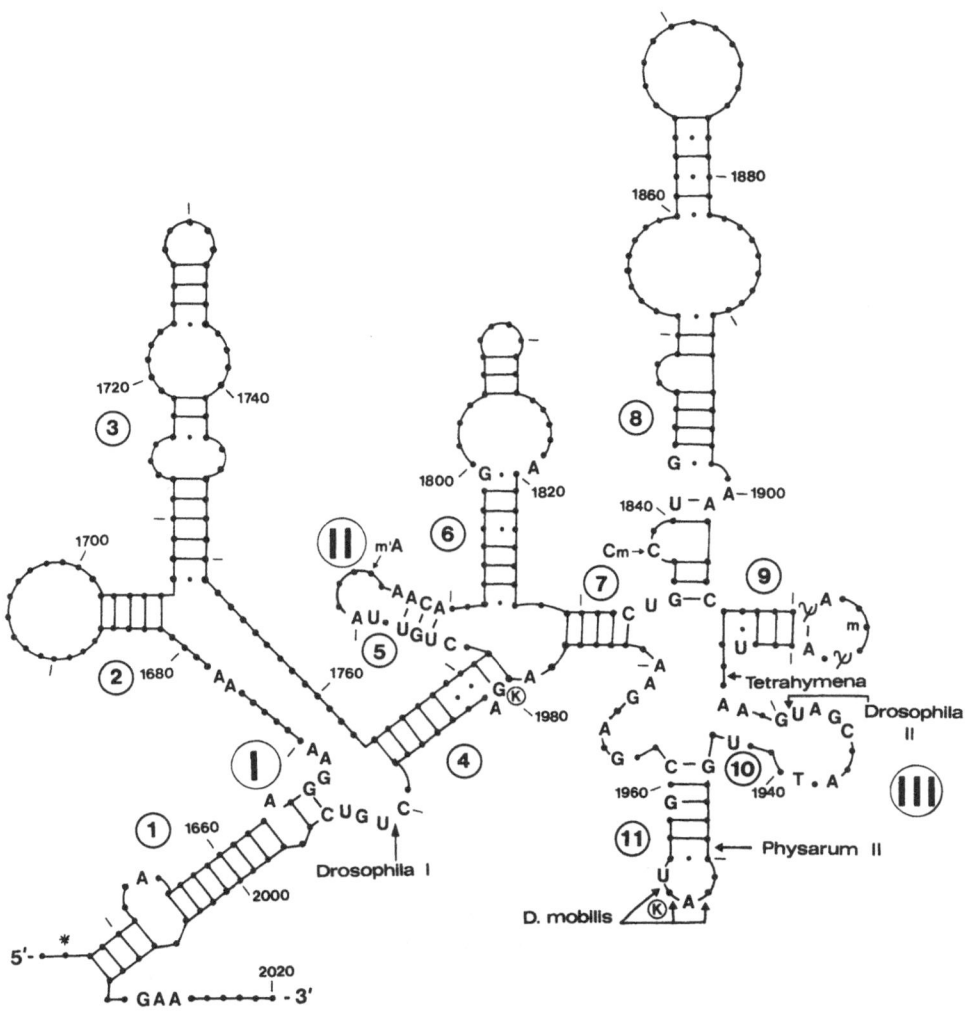

Fig. 4. Schematic representation of domain IV showing all universal nucleotides (sequence data from trypanosome mitochondrial RNAs are
excluded). Positions of the inserts and introns for eukaryotes
and an archaebacterium derive from the following sources.
Drosophila melanogaster I and II (Roiha *et al.*, 1981); *Tetrahymena
pigmentosa* (Wild and Sommer, 1980); *Physarum polycephalum* (Otsuka
et al., 1983); *Desulfurococcus mobilis* (Kjems and Garrett, 1985).
The inserts in the dipteran flies resemble transposable elements
rather than introns. The repeat sequence at the extremities of
the insert I of *Drosophila* leads to an ambiguous location of the
"splicing" site; the one shown is favoured over the alternative
position 2004 because it exhibits a characteristic uridine on its
3'-side. Post transcriptionally modified nucleotides in the *E.
coli* RNA are indicated; methylation sites that have been localized definitely in the *S. carlbergenesis* rRNA (Veldman *et al.*,
1981) are also denoted by arrows at positions 1784 and 1838. K's
represent kethoxal-reactive sites on the 50S subunit of *E. coli*.

in the 23S-like RNAs of *E. coli* (Fellner and Sanger, 1968) and yeast are shown; in the latter other modifications have been placed, tentatively, in the terminal loops of helices 9 and 10 (Veldman *et al.*, 1981). Two guanosines that were kethoxal-reactive in 50S subunits are located in conserved sequences (Noller *et al.*, 1981); the site at G_{1954} was also susceptible to RNAses S_1 and T_1 in 50S subunits (Branlant *et al.*, 1981). All of these result together with the locations of the introns, support functional roles for the three sites.

Trypanosome Mitochondrial RNAs

In order to establish the relative importance of the three putative functional sites we reasoned that only the most important would be conserved in the exceptionally small RNAs from trypanosome-like mitochondria. Previously, 23S-like rRNAs from *Trypanosoma brucei*, *Crithidia fasciculata* and *Leishmania tarentolae* have been aligned and structured mainly within

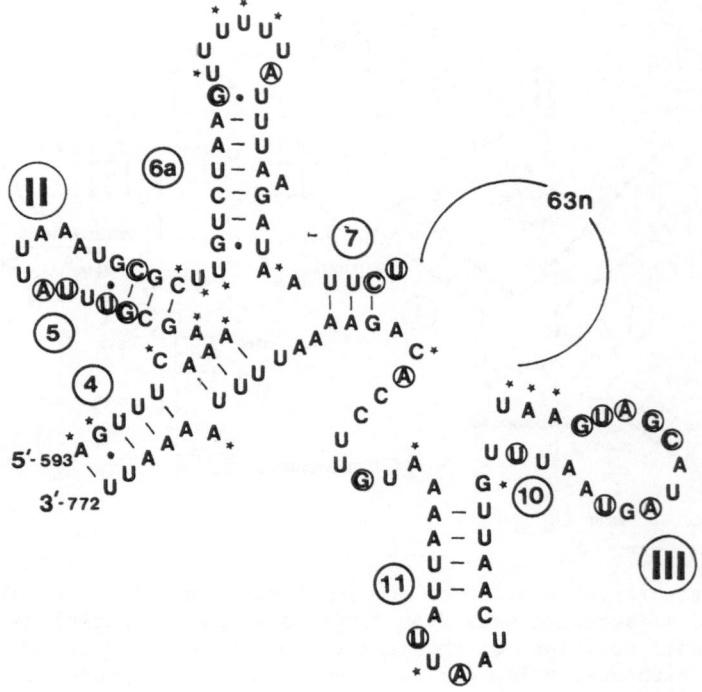

Fig. 5. Proposed secondary structure for part of domain IV of mitochondrial RNAs from the trypanosome *Crithidia fasciculata*, the same structure can be drawn for the partially homologous RNAs of *Trypanosoma brucei* and *Leishmania tarentolae*. Non conserved nucleotides amongst the three RNAs are denoted with asterisks. Universial nucleotides that are invariant in the primary kingdoms/organelles are encircled; they are concentrated in the putative functional sites II and III. We consider the sequence alignments and the secondary structure in the latter site to be the more reliable. The helices are numbered as for the *E. coli* 23S RNA (Fig. 2).

the functionally important domain V (Sloof *et al.*, 1985; de la Cruz *et al.*, 1985). In Fig. 5 we present our model for a part of domain IV of *C. fasciculata* that is shared by the other two RNAs. The conserved sequences at site I were absent but those of site II and most of site III could be aligned and structured. We conclude that the latter two sites are indispensible for the ribosome function.

POSSIBLE FUNCTIONS

At present, we have no definitive evidence for the functional roles of the three sites identified above; nevertheless we consider some possibilities below. In particular, we evaluate the weak evidence implicating site I in the initiation of translation and speculate on the possible involvement of site III in controlling mRNA movement through the ribosome.

Site I: A Role in Translational Initiation?

Domain IV of *E. coli* 23S RNA has been implicated in the initiation of translation by two lines of circumstantial evidence. First, Dahlberg *et al.*, (1978) detected a complex between initiator $tRNA_f^{Met}$ and 23S RNA after extracting *E. coli* cells at $100^{\circ}C$ in 0.5% Na dodecyl sulphate and 20 mM EDTA. It was inferred that a double helix formed between nucleotides 5 to 21 of $tRNA_f^{Met}$ and 1984 to 2001 of domain IV (helices 1 and 4). However, the base pairing is disproven by comparative sequence analyses for all kingdoms and organelles; it is probable, therefore, that the RNAs simply denatured and hybridized under the extreme extraction conditions.

The other evidence derives from studies on 5S RNA. While investigating the effects of 5S RNA carrying deletions at positions 42–46 or 42–52 on ribosome function, Zagorska *et al.*, (1984) found a large decrease in $tRNA_f^{Met}$ binding in the presence of an ApUpG template (although binding was normal when poly(A,U,G) or MS2 mRNA were used). Furthermore, Glotz *et al.*, (1981) reported that the protein L25 binding site on 5S RNA interacts with nucleotides 1754–1770 of domain IV in the ribosome. (Unfortunately, no data were presented and it is impossible to establish whether L25 was involved). Nevertheless, they concluded that an 8 base pair helix formed between nucleotides 69–77 of 5S RNA and 1759–1768 of domain IV (helix 4 and its 5'-flanking region). Although this is clearly disproven by sequence comparisons for archaebacterial, eubacterial and chloroplast RNAs, some other interaction may occur.

The close proximity within domain IV of the two abovementioned sequences prompted site-directed mutagenesis experiments on a plasmid-encoded 23S RNA gene from *E. coli* to examine the effect of small sequence deletions. Elimination of C_{1985} from the putative $tRNA_f^{Met}$ site impaired assembly of 50S subunits (Stark *et al.*, 1982), while three small overlapping deletions within the putative 5S RNA-binding region (nucleotides 1775–1761) had no detectable effect on 5S RNA assembly into 50S subunits (Zwieb and Dahlberg, 1984). However, the former, and two of the latter mutations, effected slower growth rates. Therefore, although these experiments do not implicate either RNA sequence directly in function, they emphasize the deleterious effects on the cell of minor changes in RNA structure.

Site II: A Metastable Helix?

Potentially this is a very interesting site. No introns have been localized here but it does exhibit an m'A at position 1784 in yeast (Fig. 4). There is no phylogenetic evidence for a helix, although its existence and unstable nature are supported by our experimental data. If it is a helix then it is one of the few in the 23S RNA that does not become more stable in the extreme thermophile *D. mobilis* (Leffers *et al.*, 1985). Potentially the structure here could reciprocate between conformers and play a dynamic role in protein synthesis; it may also be relevant that the neighbouring end of helix 4 is unstable. Preliminary evidence was reported for a cross-link in 50S subunits, induced by ultraviolet irradiation, between this region and the peptidyl transferase centre in domain V (Stiege *et al.*, 1983) but this result has not yet been confirmed.

Site III: Facilitating mRNA Movement?

The structure and location of the intron in the 23S RNA of the archaebacterium *D. mobilis* (Kjems and Garrett, 1985) suggests a possible function for site III. The intron occurs within the terminal loop of helix 11 in a G-A repeat at the exon-intron junction (Fig. 4). While all eukaryotic introns can be classified into 3 classes, (1) rRNA, (2) mRNA and (3) tRNA, on the basis of their structure and splicing mechanisms (Cech, 1983), the *D. mobilis* intron exhibits properties of classes 1 and 3. It contains the two conserved sequences that are essential for correct splicing of rRNA introns and other characteristics of nuclear tRNA introns that include a putative 12-13 base pair helix forming between the ends of the intron, the lack of an internal core structure, and the presence of any nucleotide at the 3'-end of the 5'-exon (Waring and Davies, 1984). The tRNA introns lie in the anti-codon loops of tRNAs and it may be relevant that helices 10 and 11, flanking the splicing site in the 23S RNA, could be co-structural with the D-arm and anticodon-arm, respectively, that generate one branch of the L-shaped tertiary structure of tRNA. The *D. mobilis* intron is then also located within an "anticodon" loop. The 23S RNA region shares some of the invariant features of the tRNA structure, including the 7 nucleotide terminal loop of the anticodon arm with a 5 base pair stem, the highly conserved U_{1951}, 5' to the anticodon, and the universal $A-G_{1933}$ in the D-loop. One possible function of such a tRNA-like structure, within the 23S-like rRNAs, could be to stabilize the mRNA, intermittently, during its passage through the ribosome.

ACKNOWLEDGEMENTS

Roger Garrett thanks Barbara Van Stolk and Harry Noller for their hospitality and counsel while they were developing the described method in Santa Cruz, and NATO for providing grants that made the visits possible. We are grateful to Jane Hjerrild and Anni Christensen for their excellent artistic help. Henrik Leffers was supported by the Carlsberg Foundation. The research was supported by grants from the Danish Natural Research Council and the Aarhus University Fund.

REFERENCES

Anderson, S., de Bruijn, M.H.L., Coulson, A.R., Eperon, I.C., Sanger, F. and Young, I.G., 1982, Complete sequence of bovine mitochondrial DNA. Conserved features of the mammalian mitochondrial genome, J. Mol. Biol. 156:683.

Bibb, M.J., Van Etten, R.A., Wright, C.T., Walberg, M.W. and Clayton, D.A., 1981, Sequence and gene organization of mouse mitochondrial DNA, Cell 26:167.

Branlant, C., Krol, A., Machatt, M.A., and Ebel, J.P., 1979, Structural study of 23S rRNA from *E. coli*, FEBS Lett. 107:177.

Branlant, C., Krol, A., Machatt, M.A., Pouyet, J., Ebel, J.P., Edwards, K. & Kössel, H., 1981, Primary and secondary structures of *E. coli* MRE 600 23S rRNA, Nucl. Acids Res. 9:4303.

Brosius, J., Ullrich, A., Raker, M.A., Gray, A., Dull, T.J., Gutell, R.R. and Noller, H.F., 1981, Construction and fine mapping of recombinant plasmids containing the *rrnB* rRNA operon of *E. coli*, Plasmid 6:112.

Bruce, A.G. and Uhlenbeck, O.C., 1978, Reactions at the termini of tRNA with T4 RNA ligase, Nucl. Acids Res. 5:3665.

Cech, T.R., 1983, RNA splicing: Three themes with variations, Cell 34:713.

Clark, C.G., Tague, B.W., Ware, V.C. and Gerbi, S.A., 1984, *Xenopus laevis* 28S rRNA; a secondary structure model and its evolution and functional implications, Nucl. Acids Res. 12:6197.

Crick, F.H.C., 1968, The origin of the genetic code, J. Mol. Biol. 38:367.

Dahlberg, J., Kintner, C., and Lund. E., 1978, Specific binding of tRNA$_f^{Met}$ to 23S RNA of *E. coli*, Proc. Natl. Acad. Sci. USA, 75:1071.

De la Cruz, V.F., Simpson, A.M., Lake, J.A., Simpson, L., 1985, Primary sequence and partial secondary structure of the 12S kinetoplast (mitochondrial) RNA from *Leishmania tarentolae*: Conservation of peptidyl-transferase structural elements, Nucl. Acids Res. 13:2337.

Douglas, S.E. and Doolittle, W.F., 1984, Complete nucleotide sequence of the 23S rRNA gene of the cyanobacterium, *Anacystis nidulans*. Nucl. Acids Res. 12:3373.

Eperon, I.C., Anderson, S. and Nierlich, D.P., 1980, Distinctive sequence of human mitochondrial rRNA genes, Nature 286:460.

Eperon, I.C., Janssen, J.W.G., Hoeijmakers, J.H.J. and Borst, P., 1983, The major transcripts of the kinetoplast DNA of *Trypanosoma brucei* are very small RNAs, Nucl. Acids Res. 11:105.

Edwards, K. and Kössel, H., 1981, The rRNA operon from *Zea mays* chloroplasts: Nucleotide sequence of 23S rDNA and its homology with *E. coli* 23S rDNA, Nucl. Acids Res. 9:2853.

Fellner, P. and Sanger, F., 1968, Sequence analysis of specific areas of the 16S and 23S rRNAs, Nature 219:236.

Garrett, R.A., 1983, Structure and role of eubacterial ribosomal proteins, in: "Genes: Structure and Expression", A. Kroon, ed. J. Wiley and Sons, London.

Garrett, R.A., Ungewickell, E., Newberry, V., Hunter, J. and Wagner, R., 1977, An RNA core in the 30S subunit of *E. coli* and its structural and functional significance, Cell Biol. Internat. Reps. 1:487.

Garrett, R.A., Vester, B., Leffers, H., Sørensen, P.M., Kjems, J., Olesen, S.O., Christensen, A., Christiansen, J. and Douthwaite, S., 1984, Mechanisms of protein-RNA recognition and assembly in ribosomes, in: "Gene Expression", ed. Clark, B.F.C. and Petersen, H.U., Munksgaard, Copenhagen, 1984.

Garrett, R.A., and Woolley, P., 1982, Identifying the peptidyl transferase centre, Trends Biochem. Sci. 7:385.

Georgiev, O.I., Nikolaev, N., Hardjiolov, A.A., Skryabin, K.G., Zakharyev, V.M. and Bayev, A.A., 1981, Complete sequence of the 23S rRNA gene from *S. cerevisiae*, Nucl. Acids Res. 9:6953.

Glotz, C., Zweib, C., Brimacombe, R., Edwards, K. and Kössel, H., 1981, Secondary structure of the large subunit rRNA from *E. coli*, Zea mays chloroplast and human and mouse mitochondrial ribosomes, Nucl. Acids Res. 9:3287.

Hassouna, N., Michot, B. and Bachellerie, J.P., 1984, The complete nucleotide sequence of mouse 28S rRNA gene. Implications for the process of size increase of the large subunit rRNA in higher eukaryotes, Nucl. Acids Res. 12:3563.

HsuChen, C.-C., Kotin, R.M. and Dubin, D.T., 1984, Sequences of the coding and flanking regions of the large subunit rRNA gene of mosquito mitochondria, Nucl. Acids Res. 12:7771.

Jarsch, M. and Böck, A., 1985, Sequence of the 23S rRNA from the archae-
    bacterium *Methanococcus vannielii*: Evolutionary and functional im-
    plications, Mol. Gen Genet. 200:305.
Kjems, J. and Garrett, R.A., 1985, An intron in the 23S rRNA gene of the
    archaebacterium *Desulfurococcus mobilis*, Nature, in press.
Köchel, H.G. and Küntzel, H., 1982, Mitochondrial L-rRNA from *Aspergillus
    nidulans*: Potential secondary structure and evolution, Nucl. Acids
    Res. 10:4795.
Kop, J., Wheaton, V., Gupta, R., Woese, C.R. and Noller, H.F., 1984, Com-
    plete nucleotide sequence of a 23S rRNA gene from *B, stearothermo-
    philius*, DNA 3:5.
Leffers, H., Kjems, J. and Garrett, R.A., 1985, The gene sequences of the
    23S RNAs of *Halococcus morrhuae* and *Desulfurococcus mobilis*: A
    comparative study, manuscript in preparation.
Maly, P. and Brimacombe, R., 1983, Refined secondary structure models for
    the 16S and 23S rRNAs of *E. coli*, Nucl. Acids Res. 11:7263.
Noller, H.F., 1984, Structure of rRNA, Ann. Rev. Biochem., 53:119.
Noller, H.F., Kop, J., Wheaton, V., Brosius, J., Gutell, R.R., Kopylov,
    A.M., Dohme, F., Herr, W., Stahl, D.A., Gupta, R. and Woese, C.R.,
    1981, Secondary structure model of 23S rRNA, Nucl. Acids Res. 9:6167.
Otsuka, T., Nomiyama, H., Yoshida, H., Kukita, T., Kuhara, S. and Sakaki,
    Y., 1983, Complete nucleotide sequence of the 26S rRNA gene of
    Physarum polycephalum: Its significance in gene evolution, Proc.
    Natl. Acad. Sci. USA 80:3163.
Ozaki, T., Hoshikawa, Y., Ida, and Iwabuchi, M., 1984, Sequence analysis
    of the transcribed and 5'-non transcribed regions of the rRNA gene
    in *Dictyostelium discoideum*, Nucl. Acids Res. 12:4171.
Peattie, D.A., 1979, Direct chemical probes for sequencing RNA, Proc. Natl.
    Acad. Sci. USA, 76:1760.
Peattie, D.A., and Gilbert, W., 1980, Chemical probes for higher order
    structure in RNA, Proc. Natl. Acad. Sci. USA 77:4679.
Roiha, H., Miller, J.R., Woods, L.C. and Glover, D.M., 1981, Arrangements
    and rearrangements of sequences flanking the two types of rRNA in-
    sertions in *D. melanogaster*, Nature 290:749.
Seilhamer, J.J., Gutell, R.R., Cummings, D.J., 1984, *Paramecium* mitochon-
    drial genes. II. Large subunit rRNA gene sequence and microevolu-
    tion, J. Biol. Chem. 259:5167.
Sloof, P., Van Den Burg, J., Voogd, A., Benne, R., Agostinelli, M., Borst
    P., Gutell, R. and Noller, H.F., 1985. Further characterization of
    the extremely small rRNAs from trypanosomes: A detailed comparison
    of the 9S and 12S RNAs from *Crithidia fasciculata* and *Trypanosoma
    brucei* with rRNAs from other organisms, Nucl. Acids Res., 13:4171.
Sor, F. and Fukuhara, H., 1983, Complete DNA sequences coding for the lar-
    ger rRNA of yeast mitochondria, Nucl. Acids Res. 11:339.
Stark, M.J.R., Gourse, R.L. and Dahlberg, A.E., 1982, Site-directed muta-
    genesis of rRNA: Analysis of rRNA deletion mutants using maxicells,
    J. Mol. Biol. 159:417.
Stiege, W., Glotz, C., and Brimacombe, R., 1983, Localization of a series
    of intra-RNA cross-links in the secondary and tertiary structure of
    23S RNA induced by ultraviolet irradiation of *E. coli* 50S subunits,
    Nucl. Acids Res. 11:1687.
Takaiwa, F. and Sugiura, M., 1982, The complete nucleotide sequence of a
    23S rRNA gene from tobacco chloroplasts, Eur. J. Biochem. 124:13
Van Stolk, B.J. and Noller, H.F., 1984, Chemical probing of conformation
    in large RNA molecules, J. Mol. Biol. 180:151.
Veldman, G.M., Klootwijk, J., de Regt, V.C.H.F., Planta, R.J., Branlant,
    C., Krol, A. and Ebel, J.P., 1981, The primary and secondary struc-
    ture of yeast 26S rRNA, Nucl. Acids Res. 9:6935.
Waring, R.B. and Davies, R.W., 1984, Assessment of a model for intron RNA
    secondary structure relevant to RNA self-splicing - a review, Gene
    28:277.

Wild, M.A. and Sommer, R., 1980, Sequence of a ribosomal RNA gene intron from *Tetrahymena*, <u>Nature</u> 283:693.

Zagorska, L., Van Duin, J., Noller, H.F., Pace, B., Johnson, K.D. and Pace, N.R., 1984, The conserved 5S rRNA complement to tRNA is not required for translation of natural mRNA, <u>J. Biol. Chem.</u> 259:2798.

Zwieb, C. and Dahlberg, A.E., 1984, Structural and functional analysis of *E. coli* ribosomes containing small deletions around position 1760 in 23S rRNA, <u>Nucl. Acids Res.</u>, 12:7135.

THE THREE-DIMENSIONAL ORGANIZATION OF ESCHERICHIA COLI RIBOSOMAL RNA

Richard Brimacombe

Max-Planck-Institut für Molekulare Genetik
Abteilung Wittmann
Berlin-Dahlem, Germany

INTRODUCTION

The structural analysis of a large RNA molecule proceeds in the four obvious stages of determining first the primary, then the secondary, then the tertiary structure, and finally - if the molecule is part of a multi-component system - the quaternary structure. In the case of the ribosomal RNA molecules of Escherichia coli, current research is focussed on the last two of these stages. The primary structures of the 16S and 23S RNA have been known for some time, and secondary structures for both molecules have been derived by a combination of experimental approaches together with phylo-genetic comparisons with the corresponding RNA sequences from a variety of other species. These structural derivations, as well as the high degree of structural conservation among different organisms, have been the subject of a number of detailed recent reviews (e.g. 1-4). The secondary structures are still not complete (particularly in regions where the primary sequence is most highly conserved), and there are still a few discrepancies (reviewed in refs. 5,6) between the different versions that have been proposed (2,4), but nevertheless the current models for both 16S and 23S RNA provide a reliable framework into which data concerning the tertiary and quaternary arrangement of these molecules can be incorporated.

In the absence of crystallographic data, there are essentially two approaches for studying the tertiary folding of the large ribosomal RNA molecules. The first of these - usually referred to as "higher order struc-ture analysis" - involves the detailed determination of the accessibility of individual residues in the RNA to single- or double-strand specific nucleases or to various chemical modifying agents. This technique yields a vast amount of very precise data (see e.g. ref. 5 for review), but the results can un-fortunately only be interpreted if the structure of the molecule is already known, as was the case in the elegant analysis of tRNA by Peattie & Gilbert (7). For RNA molecules whose structure is not known, the data from this type of approach become progressively less susceptible to interpretation in terms of tertiary folding as the size of the molecule under consideration in-creases.

The second approach to the analysis of tertiary structure involves the application of intra-RNA cross-linking techniques. Two types of cross-link can be distinguished, namely "secondary cross-links" (lying within the sec-ondary structure) and "tertiary cross-links" (connecting separate secondary structural elements). The cross-linking method does not give such precise information as the higher order structure approach, since the sites of cross-linking cannot always be pinpointed to single nucleotide residues, and the detailed chemistry of the cross-linking reaction is often not known. Furthermore, the spanning distance of a flexible cross-linking molecule is not a constant parameter. Thus the cross-linking approach can only be expected to give relatively crude information concerning "neighbourhoods" between various regions of the RNA molecule being studied, but on the other hand it has two obvious advantages over the higher order structure approach. First, the data obtained are directly interpretable in terms of tertiary folding of the RNA, and, secondly, the interpretability does not decrease with larger RNA molecules; on the contrary, as the RNA under consideration becomes larger, so the imprecision of the data becomes progressively less important, in relation to the overall features of the structure.

In my laboratory we have opted to use the cross-linking approach to investigate the tertiary structure of the 16S and 23S ribosomal RNA from E. coli, and we also use cross-linking techniques - namely cross-linking between RNA and ribosomal proteins - for the study of the quaternary structure of the ribosomal subunits. In this article I summarize the current status of our experiments, and compare our experimental strategy critically with that used by other workers in the field, concentrating on the data available for the 30S subunit. I describe how the cross-link results can be combined with other topographical information to give preliminary models of the three-dimensional organization of the RNA in situ in the ribosome, and also briefly discuss the question of possible alternative conformations or "dynamic changes" in the RNA structure.

INTRA-RNA CROSS-LINKING

In planning a cross-linking study of ribosomal RNA, there are three important parameters to consider, viz. the choice of substrate, the choice of reagent, and the choice of method for the analysis of the cross-links. In our experiments (8-12) the substrates are either intact 30S or 50S subunits, 70S ribosomes, or, in the case of intra-RNA cross-links induced by direct ultraviolet irradiation (see below), growing E. coli cells. We use these substrates because only RNA in situ in the ribosomal subunits can be expected to display the correct tertiary structure.

In contrast, most other authors, for technical reasons, use isolated 16S or 23S RNA as a substrate for cross-linking studies (e.g. 13-17), and justify their choice by arguing that the conformation of the isolated RNA reflects its conformation in situ. While this may be reasonable for the secondary structure (which can "snap back" after a denaturing isolation procedure), it is a highly questionable assumption for the tertiary structure. It is well known that the conformation of isolated RNA is very dependent on the salt conditions (e.g. 18), and that a certain sub-set of ribosomal proteins must be present before the RNA takes up the morphology observed in the intact subunits (19). Further, the physical properties of isolated RNA, such as radius of gyration or hypochromicity, are only similar but not identical to those in situ in the subunits (20). Even more serious from the point of view of a cross-linking study is the likelihood that the isolated RNA, no longer constrained by the ribosomal proteins and denatured by (e.g.) phenol extraction, is heterogeneous in its tertiary structure. The potentially disastrous consequences of such heterogeneity can easily be seen by considering a simplified example. Suppose that the RNA consists of three similar secondary structural domains, arranged in a "Y" shape with arms A and B, and stalk C. Now suppose that in situ the arms A and B are closed together in an upwards direction, so that the tip of domain A can be cross-linked to the tip of domain B. When the RNA is in the isolated state, however, arm A could swing downwards in a sub-set of the molecules towards the stalk C, so that the tip of domain A can also now cross-link to the tip of domain C. These two different conformers would be physicochemically indistinguishable from one another, and, being unaware of the heterogeneity, one would conclude that the tip of domain C is close to that of B (since both are observed cross-linked to the tip of A), although in both conformers these two points are the remotest parts of the structure.

In fact, the cross-linking data that have been published for isolated 16S RNA (13-16) strongly suggest that such a situation is indeed occurring. Apart from those cross-links lying within the secondary structure, the majority of the tertiary structural cross-links observed (13,15) involve the potentially flexible 3'-terminal sub-domain of the 16S RNA (bases ca. 1400-1540, cf. Fig. 1, below) linked to a wide variety of locations distributed across the rest of the 16S molecule. At this stage it is impossible to tell how many (if any) of these cross-links are relevant to the in situ structure of the RNA.

The second parameter, namely the choice of a suitable cross-linking agent, is primarily a matter of deciding whether to opt for a reversible or an irreversible cross-linker. We use the latter approach, and the intra-RNA cross-links are generated either by direct ultraviolet irradiation (8,9,11, 12), or by treatment with bis-(2-chloroethyl)-methylamine, "nitrogen mustard" (10). Other similar reagents, such as diepoxybutane or chlorambucil can also be applied (W. Stiege, unpublished results), all of them leading to the formation of relatively short cross-link "bridges". The point of choosing an irreversible cross-linker is that the cross-link remains intact right through the subsequent analytical procedures. In contrast, if a reversible reagent is used, then as soon as the reaction has been reversed the cross-link site becomes in general either "invisible" or else indistinguishable from a site where the reagent has formed a monoaddition product. Sites of monoaddition will under these circumstances lead to an incorrect cross-link analysis, and in cases where the reversed cross-link is "invisible" the site of cross-linking has to be deduced from the expected specificity of the cross-linker (e.g. 15). This can however also be misleading. It is already known for instance that nitrogen mustard, which would normally be expected to react primarily with G-residues, can in addition generate A-G cross-links (see below), and the intra-RNA cross-links induced by ultraviolet irradiation tend to involve purine residues more often than – as might be expected – pyrimidines (11,12). Similarly, in experiments with psoralen, Turner & Noller (17) found that the most predominant intra-RNA cross-link in 23S RNA involved a cytidine residue rather than the expected uridine.

Thus, particular micro-environments in a complex particle such as the ribosome can cause reactions to occur which contradict the normal or expected specificity of the cross-linking method being used, and it is obviously important to be able to take direct account of such aberrations. This can only be done if the identification of the cross-linked nucleotides is made either with the cross-link still intact (i.e. with an irreversible cross-linker), or if the cross-link can be reversed in such a way as to leave no ambiguity. (An example of the latter possibility is the ingenious system devised by Ehresmann & Ofengand (21) for the analysis of photoreversible cross-links).

The third parameter in planning a cross-linking study is the choice of method for the analysis of the cross-links. This inevitably involves at some stage a partial nuclease digestion procedure in order to release cross-linked complexes of a suitable size for sequence determination, and it has already become clear that the nature of the partial digestion exerts an enormous influence on the spectrum of cross-links observed. Ribonuclease T_1 digestion for instance favours the production of fragments containing stable hairpin loops, and using this enzyme we were only able to find cross-links in such loops (8,9), i.e. cross-links within the secondary structure of 16S or 23S RNA. The more interesting tertiary structural cross-links (cf. Fig. 1, below were first found when the partial digestion was made with the double-strand specific cobra venom nuclease (10,11), or with the latter enzyme in combination with ribonuclease H and specific deoxyoligonucleotides (11,12). The selectivity means that no single digestion procedure will allow all the cross-links in a particular cross-linked ribosome preparation to be observed simultaneously, and this same selectivity makes it difficult if not impossible to estimate the yield in which individual RNA cross-links were formed in the initial reaction step. For this reason, in our experiments the levels of cross-linking are deliberately kept low, in order both to simplify the patterns of digestion products as well as to maximize the probability that each observed cross-link is the first cross-linking event in any individual molecule.

After the partial digestion procedure, the cross-linked material is separated by two dimensional gel electrophoresis (10,11). Each cross-linked complex that we isolate from the gels comprises either two distinct RNA fragments joined together by the cross-link, or a single uninterrupted RNA fragment within which the cross-link forms a covalent loop. Since the cross-links are irreversible, conventional end-labelling sequencing techniques can not be applied in either case; the first type of complex would give rise to two labelled ends, and the second type could not be sequenced within the closed loop. We have tested a number of possible modifications to existing end-labelling methods with a view to overcoming this problem, but so far without finding a satisfactory solution. In our routine analyses we therefor continue to use ribosomes that are uniformly ^{32}P-labelled, and the individua isolated cross-linked complexes are analysed by RNA "fingerprinting" techniques (22). The cross-link site is characterized by the absence of two oligonucleotide spots from the fingerprint, together with the appearance of a new spot corresponding to a covalent complex between these two missing oligonucleotides. The method has the disadvantage that if one or both components of the cross-link site occur at unfavourable positions in the RNA sequence, then we are unable to locate them, and as a result a number of potentially important cross-links have escaped analysis. In general, however the method yields unambiguous identifications in which the sites of cross-linking are located to within two or three nucleotides.

Figure 1 shows the latest status of the intra-RNA cross-links that we have found in the E. coli 30S subunit, generated by reaction with nitrogen mustard (J. Atmadja, unpublished results) or by direct irradiation with ultraviolet light (8,12). In the latter instance several of the cross-links have recently been confirmed in experiments where the cross-linking was induced in vivo by irradiation of growing E. coli cells (W. Stiege & J. Atmadja, unpublished data), in order to minimize still further the chance that the results represent conformational or isolational artefacts. The data in Fig. 1 include both "secondary" and "tertiary" cross-links, and the incorporation of this information into a three-dimensional model of the RNA structure will be discussed below. A similar data set (9-11) is also available for the 50S subunit, which has already been summarized elsewhere (6).

RNA-PROTEIN CROSS-LINKING

RNA-protein cross-linking represents the "missing link" in the determination of the quaternary structure of ribosomal RNA. On the one hand, the tertiary structure of the RNA itself is beginning to emerge from the type of data discussed in the preceding section, and in addition the orientation of the RNA within the ribosomal subunits has been defined to a certain extent b immune electron microscopical localization of specific sites in the RNA mole cule, such as the 5'- and 3'-ends, or the modified nucleotides (see ref. 23 for summary, and see below). On the other hand, there is a wealth of information concerning the arrangement of the ribosomal proteins, including in particular protein-protein cross-linking, immune electron microscopical localization of individual proteins, and triangulation of protein mass centres by neutron scattering (see ref. 24 for review). However, data on the interactions between ribosomal RNA and proteins and their mutual orientations are relatively scarce. The classical "binding site" approach to this problem is of limited applicability, and among those proteins able to bind specifically to 16S RNA, only protein S8 (25) has yielded a binding site (ca. 70 nucleotides) which is of a useful size in this context. While information is available on the binding of other single proteins or groups of proteins to larger regions of the RNA (such as S4 (26), or S6, S8, S15, S18 (27), or S7, S9, S10, S14, S19 (28)), it is not possible to derive the precise locations of the individual proteins relative to the RNA from such data. The RNA-protein cross-linking approach offers a way out of this dilemma, the objective being the determination of a series of sites of cross-linking on the RNA, so as to define a three-dimensional pocket for each ribosomal protein in the 16S or 23S RNA structure.

Just as with the intra-RNA cross-linking described above, our RNA-protein cross-linking experiments use intact ribosomal subunits as the substrate, irreversible reagents as the cross-linkers, and partial nuclease digestion coupled with RNA fingerprint analysis to locate the sites of cross-linking. The cross-links are induced either by direct ultraviolet irradiation (29,30), or by reaction with nitrogen mustard (cf. ref. 31), azidophenylacetimidate (32) or 2-iminothiolane (33,34). In the first two cases the RNA-protein cross-linking reaction occurs concomitantly with the intra-RNA cross-linking already described above, whereas the last two compounds react specifically with lysine residues in the proteins, and the subsequent coupling to RNA is achieved by a very mild ultraviolet irradiation step. After partial digestion, the cross-linked complexes are separated by two-dimensional gel electrophoresis, in this case in a system involving both non-ionic and ionic detergents (30), which has the property of separating free RNA fragments from those attached to protein. Each isolated RNA-protein complex is analysed for both its RNA and protein content, the identity of the protein being established by gel electrophoresis and confirmed by Ouchterlony tests with antibodies to the individual ribosomal proteins.

The two-dimensional gel patterns of partial digestion products are extremely complex. Most of the ribosomal proteins are cross-linked at least to some extent by the bifunctional reagents, and many of them are linked to more than one site on the RNA. In addition, each individual RNA-protein complex usually appears in several spots on the gel which differ only in the length of the attached RNA fragment. As a result of this complexity, it is obviously essential that both the protein and RNA analyses are carried out on one and the same complex isolated from the two-dimensional gel. By using irreversible cross-linkers and uniformly-labelled RNA we can be certain that our cross-link assignments are correct, any ambiguities or cross-contamination with other complexes being immediately apparent from the analysis. In contrast, in RNA-protein cross-link site determinations described by other authors, where the cross-link was reversed (35) or the cross-linked protein digested with proteinase (36) prior to end-labelling procedures for RNA sequence determination, there is no guarantee that the RNA fragment finally analysed was in fact the one attached to the protein. The complexity of the digestion patterns can be significantly reduced by making a total digestion with ribonuclease T1, as some authors have done (37,38), but this has the result that the cross-linked complexes are left with attached oligonucleotides that are mostly too short to allow an unambiguous location in the 16S or 23S RNA sequence. In these cases also there is therefore a serious danger that the cross-link site determination is erroneous.

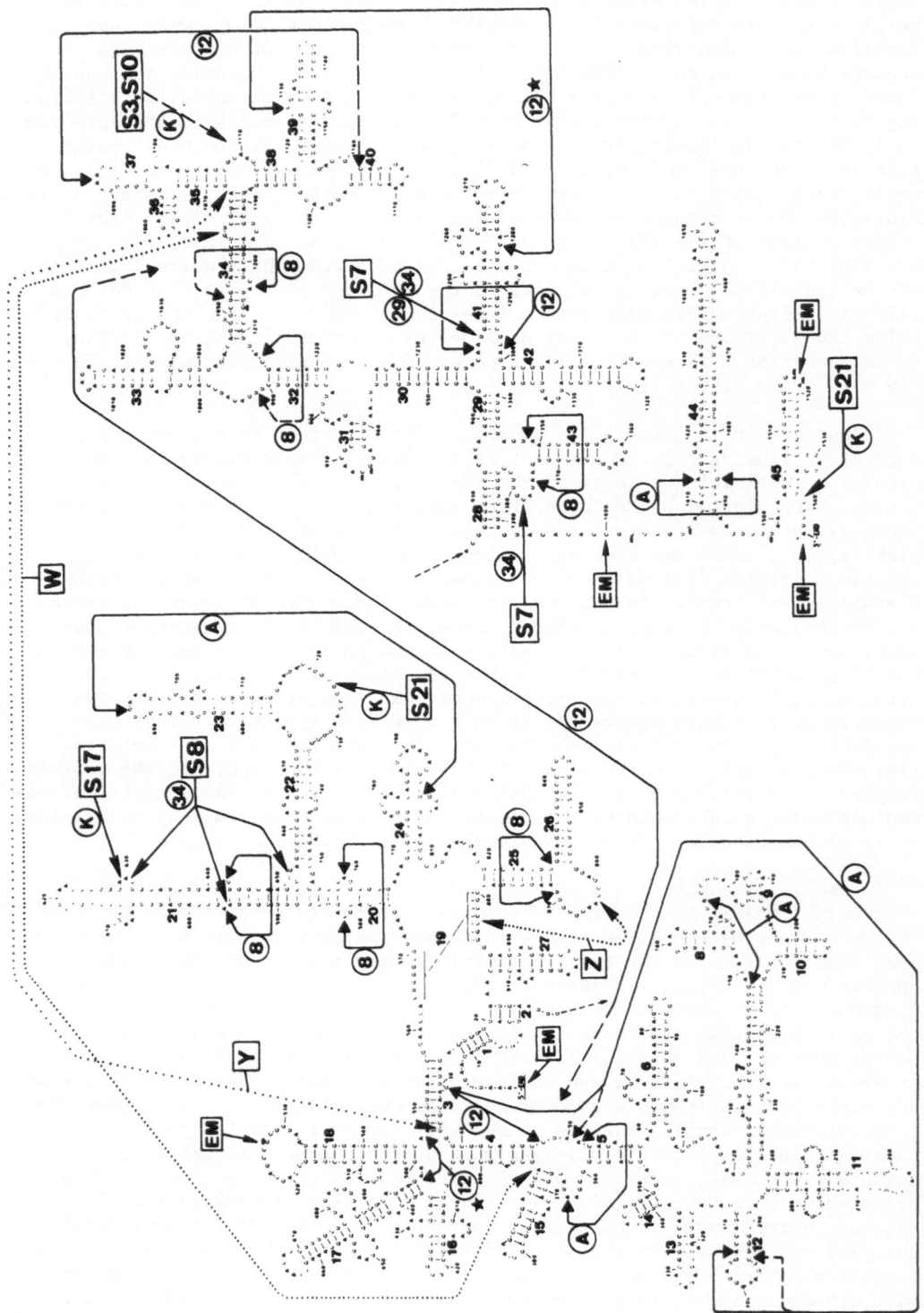

As with the intra-RNA cross-linking, the spectrum of cross-linked products observed in a partial digest is highly dependent on the nuclease used. We have worked with various different partial digestion procedures, and all of the RNA-protein cross-link sites that we have so far identified in the 30S subunit (29,34, and A. Kyriatsoulis, unpublished data) are included in Figure 1. In addition, for each of the cross-linking agents there is a large amount of partially analysed or un-confirmed data from which it is clear that many more cross-link sites remain to be elucidated. The corresponding data for the 50S subunit (30,33) are summarized in ref. 6.

THREE-DIMENSIONAL MODEL-BUILDING

Figure 1 shows the secondary structure of 16S RNA (2), and incorporates the data that are relevant to its tertiary and quaternary structure. These data include the intra-RNA and RNA-protein cross-links described in the preceding sections, the positions in the RNA sequence that have been located on the 30S subunit surface by electron microscopy, and the sequences that have been shown to be involved in conformational "switching" (see following section). As will be evident from the foregoing discussion, Figure 1 does not include intra-RNA cross-links generated in isolated RNA, or RNA-protein cross-links where we consider the cross-link site determination to be suspect; it is clear that the incorporation of an erroneous cross-link in the early stages of model-building would be disastrous.

We have constructed models of the 16S and 23S RNA from plastic-covered wire, at a scale of 2 mm = 1 Å. The wire represents the phosphate backbone, and is marked off at intervals of 1.5 cm to indicate the individual nucleotides. The secondary structure is built by fixing the wire around suitable short sections of plexiglass (perspex) tubing (4 cm diameter) to generate the double-helical segments (cf. Fig. 1), and the first step in packing these raw secondary structure models into three dimensions is to build in the tertiary intra-RNA cross-links. The data for 16S RNA (Fig. 1) include several new cross-links which have enabled considerable refinements to be made to the 16S model, and the following section is therefore confined to a brief discussion of the small subunit, the status of the large subunit being essentially unchanged from that described in ref. 6. The models of the 30S and 50S subunits themselves will be published elsewhere.

The most important cross-link in the 5'-region of the molecule is the

Figure 1: (See facing page). The secondary structure of E. coli 16S RNA, including sites of intra-RNA and RNA-protein cross-linking, RNA "switches", and positions that have been localised on the 30S subunits by electron microscopy. The structure is divided into two domains and the helices numbered as in ref. 2. The arrows connected by lines denote intra-RNA cross-links, broken lines indicating an uncertainty in the cross-link site analysis. The encircled number by each cross-link gives the appropriate literature reference, "A" denoting unpublished nitrogen mustard data of J. Atmadja. Cross-links marked with an asterisk have also been observed in in vivo cross-linking experiments (see text). RNA-protein cross-links are indicated by an arrow to the corresponding protein (boxed), again with numbers giving the literature reference; "K" denotes unpublished data of A. Kyriatsoulis. The precision of determination of the individual cross-link sites varies from one case to another (cf. the original literature). Sequences involved in RNA switches (6,39 and see text) are joined by dotted lines and denoted by the letters W, Z and Y (tentative). Electron microscopically located nucleotides (23) are marked "EM".

ultraviolet-induced link between residues G-31 and C-48 (12, cf. ref. 6). At
first sight this cross-link might appear to be sterically impossible between
helices 3 and 4 (Fig. 1), but a detailed Nicholson model of this region of
the RNA shows that if helices 3 and 4 are doubled back into an antiparallel
conformation and then tilted so that their helix axes are roughly at 45° to
one another, then G-31 and C-48 do indeed come into close proximity
(J. Atmadja, unpublished data). Furthermore, in this orientation the looped-
out residue A-397 in helix 4 points into the major groove of helix 3, and
the two proposed "switching" sequences (39,6) (bases 35-37 and 388-395)
lie in favourable orientations with regard to the sequential order
of their potential partners in helix 34 (bases 1067-1069 and 1058-1066, respectively). The finding of a further ultraviolet-induced cross-link between
regions 1-72 and 1020-1095 (although this cross-link could not be localised
(12)) provides further circumstantial evidence that these regions of the
RNA, namely helices 3-4 and helix 34, are neighboured in the 30S subunit.

The neighbourhood between G-31 and C-48 is supported by cross-linking
data with nitrogen mustard. Here a cross-link has been established between
G-31 and A-306, but the cross-link is sometimes observed to be heterogeneous
(J. Atmadja, unpublished data), involving a nucleotide between G-46 and G-57
instead of G-31. If G-31 and C-48 are close neighbours, then this is entirely reasonable. Another heterogeneity is also observed with the 3'-component
of the same cross-link, in this case involving a nucleotide in the region
of G-293 to G-297 instead of A-306 (Fig. 1). This type of "local" heterogeneity could reflect either the flexibility of the cross-linking reagent
or the flexibility of the 16S structure itself, or it could indicate some
minor distortion of the structure as a result of the cross-linking reaction.
In terms of the RNA topography, however, these data are all consistent with
one another, at least to a first approximation, and support the concept
that the whole 5'-domain of the 16S RNA has a "tight" tertiary structure
(cf. the studies on the binding site of protein S4 (26)). Also noteworthy
in this region of the RNA is the nitrogen mustard cross-link between G-148
and A-174; in the secondary structure model of Noller (4), this cross-links
an A-G base-pair in a double-helical segment.

The cross-link between G-31 and C-48 has a further important consequence. Other authors (40) have proposed "rules" for the coaxial stacking
of helices in ribosomal RNA, based on the coaxiality observed in the established structure of tRNA. According to these criteria, helices 3 and 4 are
potential candidates for coaxiality (40), but as just discussed, the G-31/
C-48 cross-link has the consequence that the two helices are in fact arranged
in an almost antiparallel configuration. This demonstrates that, although
coaxial stacking of helices undoubtedly plays an important role in the structure of 16S and 23S RNA, it is entirely premature to use hypothetical rules
for such stacking as a model-building criterion.

In the central region of the 16S molecule there is a new high-yield
nitrogen mustard cross-link between G-693 and G-799, which places a precise
constraint on the orientations of helices 23 and 24. This helps to some
extent to unravel the orientation of the crucial "ring" in the central
domain of the structure, formed by the junctions of helices 1, 2, 3, 20,
24, 25 and 27 (Fig. 1). The configuration adopted by this "ring" is complicated by several factors. The discrepancy between our secondary structure model (2) and that of Noller (4) concerning the existence of helix 19
is obviously important in this context, as is also the switch structure "Z"
(Fig. 1), in which helix 25 can be replaced by an interaction between bases
862-869 and 880-886 (41). Further, the status of helices 1 and 2, which
also form part of the "ring", is not clear. Although a Nicholson model has
shown that these two helices can in principle co-exist simultaneously
(P. Maly, unpublished data, and cf. ref. 42), it is not yet known whether

this is the case in practice. Helix 2, being a double-helix formed within
the loop-end of helix 1, is effectively a tertiary structural interaction,
and is particularly important as it brings the whole 3'-domain of the 16S
RNA (defined by the region enclosed by helix 28) into juxtaposition with
helix 1, the 5'-terminus of the molecule. The configuration of the 3'-do-
main, in addition to the cross-link already mentioned between bases 1-72
and 1020-1095, begins to take shape as a result of the two ultraviolet-
induced cross-links between helices 37 and 40, and helices 39 and 41 (12).
The latter is a high-yield cross-link which is also seen in 30S subunits
isolated from in vivo irradiated E. coli cells (W. Stiege & J. Atmadja,
unpublished data).

Having incorporated the intra-RNA cross-link data, the next phase in
model-building is to orient the RNA in relation to the known features of
the 30S subunit. The positions in the 16S sequence that have been located
directly on the 30S subunit by electron microscopy (see ref. 23 for review)
are the 5'- and 3'-ends, the methylated G-residue at position 527, the
methylated A-residues at positions 1518-1519, and the C-residue at position
1400, which has been cross-linked to the anti-codon loop of tRNA. Other
locations are inferred by correlating the RNA-protein cross-link data with
the positions of the ribosomal proteins in the 30S subunit, as determined
by neutron scattering (43) and immunoelectron microscopy (44,45). The
result is a coherent model of the 30S subunit, in which the general locations of the majority of the 16S helices - and the precise relative locations of several of them - are established. Some of the more interesting
features of the model are summarized in Fig. 2.

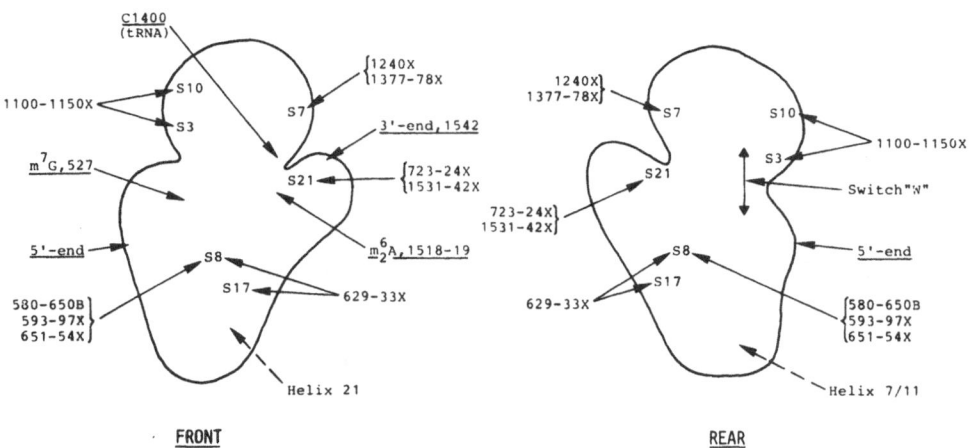

FRONT   REAR

Figure 2: Sketches of the electron microscopically derived model of
the E. coli 30S subunit (cf. 44), showing the locations of
specific regions of the 16S RNA (cf. Fig. 1). The sites
directly localised by immune electron microscopy are underlined, whereas those deduced from their proximity to a
protein position are distinguished by the suffix "X"
(deduced from the RNA-protein cross-link data) or "B"
(deduced from the protein S8 binding site (25) on the RNA).
Protein locations are either from immune electron microscopic (44,45) or neutron scattering (43,46) data. The
position of the switch structure W is indicated and the
probable locations of sequence regions where large insertions occur in eukaryotic 18S RNA are indicated by the
appropriate helix number (cf. Fig. 1).

The multiple cross-links to protein S8 are in agreement with the known binding site of the protein (25), but, since all three cross-links are on the same side of helix 21 (34), the cross-link data give the additional information as to where the protein must lie in relation to the helix. Protein S17 also cross-links to the same oligonucleotide as one of the S8 cross-links (Fig. 1, positions 629-633, A. Kyriatsoulis, unpublished data), which agrees well with the recent finding by neutron scattering that S17 and S8 are very close together in the lower part of the 30S subunit (46). The binding site of S17 is however in the 5'-region of the molecule (26), so that the 5'-region as well as helix 21 must extend down into the lower part of the subunit. The principal insertions in eukaryotic 18S RNA occur at the junction of helices 7 and 11 in the 5'-domain, and in the region of helix 21 (41), so both these inserted regions are likely to be situated at the "base" of the subunit (see Fig. 2 and cf. ref. 45). The corresponding inserted regions in the 50S subunit probably lie at the "back" of the subunit (6), so that the interface region of both subunits contains the most highly conserved regions of the secondary structure.

Further useful constraints are placed on the 30S structure by the multiple cross-links to proteins S21 and S7 (Fig. 1). S21 is cross-linked to two positions in the 16S RNA (nucleotides 723-724 and the 3'-terminal oligonucleotide 1531-1542, A. Kyriatsoulis, unpublished data, and cf. ref. 47). Since S21 is a very small protein, this information represents virtually an "intra-RNA cross-link", and together with the RNA cross-link between helices 23 and 24 serves to locate the latter helices, as well as helix 45, in the region of the large protuberance or "platform" of the 30S subunit (Fig. 2). The two S7 cross-links, on the other hand, create a compact "pocket" for S7 between helices 41 and 43. Taken together with the intra-RNA cross-link between helices 39 and 41, as well as partially localised cross-links to S3 and S10 (between positions 1100 and 1150, A. Kyriatsoulis, unpublished data), this leads to the consequence that the "head" region, encompassing helices 28 to 43 (1), is topographically now the most well-defined part of the 30S subunit.

The most difficult helix to place in the structure is helix 44. Because of its length and probable rigidity, it has been reasonably proposed (40) that the helix extends down towards the base of the subunit, but so far there is no positive evidence in favour of this arrangement, and here is an obvious case where the cross-linking approach should eventually give the answer. The data as a whole have reached the stage where each new cross-link helps to refine the model step by step, and, as indicated in the preceding sections, it is clear that many cross-links still remain to be elucidated with the reagents we are currently using. Thus far no irreconcilable contradictions in the results have arisen, and hopefully this will remain so!

DYNAMICS OF RIBOSOMAL RNA

It is appropriate to a meeting entitled "Three-dimensional structure and dynamics of RNA" to end with some comments on the existence of alternative conformations or "switches" in the ribosomal RNA structures. This is a controversial subject, and many ribosomologists, while tacitly accepting the concept that the secondary structure of mRNA has to be entirely unfolded loop by loop during the translation process, show a surprising resistance to the idea that secondary structural changes can occur in the ribosomal RNA during the same process. The usual objection is that the energy involved would be too great for such changes to take place under physiological conditions. In fact, however, nucleic acids are well able to circumvent this energy problem, as evidenced by the case of mRNA just cited (not to mention the replication of DNA!), and they presumably do this, not by "brute-force" melting of double helices as in a laboratory hyperchromicity measurement, but by "rolling" substitutions of one structural element for another, so that complete melting of a structure as such never occurs and the energy niveau remains roughly constant. Furthermore, secondary structural re-arrangements of ribosomal RNA can be observed under very mild conditions. In our experiments on the analysis of base-paired RNA fragments by two-dimensional gel electrophoresis (39), mutually exclusive secondary structures for certain segments of the RNA were found in one and the same experiment. Thus the energy argument is untenable, and the question is not whether secondary structural re-arrangements can occur in ribosomal RNA, but rather whether or not the observed re-arrangements are experimental artefacts.

Just as with the derivation of the secondary structures (1-4), the phylogenetic test can be applied to a potential switch or structural re-arrangement, and in recent articles (6,41) we have shown that the experimentally observed switches "W" and "Z" depicted in Figure 1 (between bases 388-395 and 1058-1066, and between bases 862-869 and 880-886, respectively) are indeed phylogenetically conserved. A further interaction "Y" (between bases 35-37 and 1067-1069) remains tentative, although consistent with the 16S model (cf. the foregoing section). The point was also made (6) that the alternative structures were only observed experimentally because in the system concerned in each case an artefactual extension to the base-pairing was possible. Thus it is highly probable that many more such structural switches remain to be discovered, where the interactions are too weak to be detected by this type of experimental procedure.

The most likely ribosomal function to require a cycle of conformational switches is the translocation process (39,48), where the entire complex of mRNA, tRNA and nascent peptide has to be moved efficiently and precisely by one codon's length across the ribosome. This universal process is certain to involve the conserved "core" (1,49) of the secondary structures of 16S and 23S RNA. Furthermore, the evolutionary pressure for conservation of an RNA sequence taking part in more than one base-paired interaction will be enormously high, so that any conformational switches will tend to be in regions of the RNA where also the primary sequence is highly conserved (in some of the double-helices (cf. ref. 2), or possibly in some of the supposedly "single-stranded" regions of the structure (Fig. 1)). In such regions, the phylogenetic method cannot be effectively applied to test the validity of the alternative structures (cf. ref. 49), and it will therefore be difficult to gather convincing evidence for them; compensating base changes in other organisms in support of the switch structure "W" (Fig. 1) are for instance quite rare (6). A conformational switch must also be topographically feasible (cf. the discussion of switches W and Y in the foregoing section), and therefore the whole question of switches should perhaps best be shelved until considerably more accurate models for the tertiary structure of the RNA than those described in this article have been developed. There is of course no guarantee that we will be able to unfathom the mechanical movements occurring during the ribosomal RNA function when such detailed structures finally become available, but on the other hand it is equally clear that without a detailed structural knowledge our chances of understanding the function are zero.

ACKNOWLEDGEMENTS

I am indebted to my co-workers, past and present, who have contributed to this work, most notably Christian Zwieb, Peter Maly, Jacek and Iwona Wower, Apostolos Kyriatsoulis, Wolfgang Stiege and Johannes Atmadja. I am also grateful to Drs. Georg and Marina Stöffler for making antibodies available, to Drs. Hans Bäumert and Hugo Fasold for designing and synthesizing cross-linking reagents, and to Drs. Ronald Frank and Helmut Blöcker for synthesizing deoxyoligonucleotides. Finally, I thank Dr. H. G. Wittmann for his continual support.

REFERENCES

1. Brimacombe, R., Maly, P. and Zwieb, C. (1983) Progr. Nucleic Acids Res. 28, 1-48.
2. Maly, P. and Brimacombe, R. (1983) Nucleic Acids Res. 11, 7263-7286.
3. Woese, C.R., Gutell, R., Gupta, R. and Noller, H.F. (1983) Microbiol. Rev. 47, 621-669.
4. Noller, H.F. (1984) Ann. Rev. Biochem. 53, 119-162.
5. Brimacombe, R. and Stiege, W. (1985) Biochem. J. 229, 1-17.
6. Brimacombe, R., Atmadja, J., Kyriatsoulis, A. and Stiege, W. (1986) Texas Ribosome Conference, Springer-Verlag, in press.
7. Peattie, D.A. and Gilbert, W. (1980) Proc. Natl. Acad. Sci. USA 77, 4679-4682.
8. Zwieb, C. and Brimacombe, R. (1980) Nucleic Acids Res. 8, 2397-2411.
9. Glotz, C., Zwieb, C., Brimacombe, R., Edwards, K. and Kössel, H. (1981) Nucleic Acids Res. 9, 3287-3306.
10. Stiege, W., Zwieb, C. and Brimacombe, R. (1982) Nucleic Acids Res. 10, 7211-7229.
11. Stiege, W., Glotz, C. and Brimacombe, R. (1983) Nucleic Acids Res. 11, 1687-1706.
12. Atmadja, J., Brimacombe, R., Blöcker, H. and Frank, R. (1985) submitted for publication.
13. Wollenzien, P., Hearst, J.E., Thammana, P. and Cantor, C.R. (1979) J. Mol. Biol. 135, 255-269.
14. Turner, S., Thompson, J.F., Hearst, J.E. and Noller, H.F. (1982) Nucleic Acids Res. 10, 2839-2849.
15. Thompson, J.F. and Hearst, J.E. (1983) Cell 32, 1355-1365.
16. Expert-Bezançon, A., Milet, M. and Carbon, P. (1983) Eur. J. Biochem. 136, 267-274.
17. Turner, S. and Noller, H.F. (1983) Biochemistry 22, 4159-4164.
18. Boublik, M., Oostergetel, G.T., Wall, J.S., Hainfeld, J.F., Radermacher, M., Wagenknecht, T., Verschoor, A. and Frank, J. (1986) Texas Ribosome Conference, Springer-Verlag, in press.
19. Serdyuk, I.N., Agalarov, S.C., Sedelnikova, S.E., Spirin, A.S. and May, R.P. (1983) J. Mol. Biol. 169, 409-425.
20. Vasiliev, V.D., Serdyuk, I.N., Gudkov, A.T. and Spirin, A.S. (1986) Texas Ribosome Conference, Springer-Verlag, in press.
21. Ehresmann, C. and Ofengand, J. (1984) Biochemistry 23, 438-445.
22. Volckaert, G. and Fiers, W. (1977) Analyt. Biochem. 83, 228-239.
23. Gornicki, P., Nurse, K., Hellmann, W., Boublik, M. and Ofengand, J. (1984) J. Biol. Chem. 259, 10493-10498.
24. Wittmann, H.G. (1983) Ann. Rev. Biochem. 52, 35-65.
25. Thurlow, D.L., Ehresmann, C. and Ehresmann, B. (1983) Nucleic Acids Res. 11, 6787-6802.
26. Zimmermann, R.A. (1980) in "Ribosomes" (Chambliss, G., Craven, G.R., Davies, J., Davis, K., Kahan, L. and Nomura, M., eds.) pp. 135-169, University Park Press, Baltimore.
27. Gregory, R.J., Zeller, M.L., Thurlow, D.L., Gourse, R.L., Stark, M.J.R. Dahlberg, A.E. and Zimmermann, R.A. (1984) J. Mol. Biol. 178, 287-302.
28. Yuki, A. and Brimacombe, R. (1975) Eur. J. Biochem. 56, 23-34.
29. Zwieb, C. and Brimacombe, R. (1979) Nucleic Acids Res. 6, 1775-1790.
30. Maly, P., Rinke, J., Ulmer, E., Zwieb, C. and Brimacombe, R. (1980) Biochemistry 19, 4179-4188.
31. Ulmer, E., Meinke, M., Ross, A., Fink, G. and Brimacombe, R. (1978) Mol. Gen. Genet. 160, 183-193.
32. Rinke, J., Meinke, M., Brimacombe, R., Fink, G., Rommel, W. and Fasold, H. (1980) J. Mol. Biol. 137, 301-314.

33. Wower, I., Wower, J., Meinke, M. and Brimacombe, R. (1981) Nucleic Acids Res. 9, 4285-4302.
34. Wower, I. and Brimacombe, R. (1983) Nucleic Acids Res. 11, 1419-1437.
35. Sköld, S.E. (1983) Nucleic Acids Res. 11, 4923-4932.
36. Golinska, B., Millon, R., Backendorf, C., Olomucki, M., Ebel, J.P. and Ehresmann, B. (1981) Eur. J. Biochem. 115, 479-484.
37. Ehresmann, B., Backendorf, C., Ehresmann, C., Millon, R. and Ebel, J.P. (1980) Eur. J. Biochem. 104, 255-262.
38. Chiaruttini, C., Expert-Bezançon, A., Hayes, D. and Ehresmann, B. (1982) Nucleic Acids Res. 10, 7657-7676.
39. Glotz, C. and Brimacombe, R. (1980) Nucleic Acids Res. 8, 2377-2395.
40. Noller, H.F. and Lake, J.A. (1984) in "Membrane Structure and Function, Vol. 6" (Bittar, E.E., ed.) pp. 217-297, John Wiley and Sons.
41. Atmadja, J., Brimacombe, R. and Maden, B.E.H. (1984) Nucleic Acids Res. 12, 2649-2667.
42. Pleij, C.W.A., Rietveld, K. and Bosch, L. (1985) Nucleic Acids Res. 13, 1717-1731.
43. Ramakrishnan, V., Capel, M., Kjeldgaard, M., Engelmann, D.M. and Moore, P.B. (1984) J. Mol. Biol. 174, 265-284.
44. Breitenreuter, G., Lotti, M., Stöffler-Meilicke, M. and Stöffler, G. (1984) Mol. Gen. Genet. 197, 189-195.
45. Lake, J.A. (1983) Prog. Nucleic Acid Res. 30, 163-194
46. Moore, P.B., Capel, M., Kjeldgaard, M. and Engelmann, D.M. (1986) Texas Ribosome Conference, Springer-Verlag, in press.
47. Czernilofsky, A.P., Kurland, C.G. and Stöffler, G. (1975) FEBS Lett. 58, 281-284.
48. Thompson, J.F. and Hearst, J.E. (1983) Cell 33, 19-24.
49. Brimacombe, R. (1984) Trends in Biochem. Sciences 9, 273-277.

THE EXTREMELY SMALL MITOCHONDRIAL RIBOSOMAL RNAs FROM TRYPANOSOMES

P. Sloof, R. Benne and B. F. De Vries

Section for Molecular Biology/AMC, Laboratory of
Biochemistry, University of Amsterdam, Academic Medical
Center, Meibergdreef 15, 1105 AZ Amsterdam, The
Netherlands

SUMMARY

The mitochondrial 9S and 12S RNAs from trypanosomes are the smallest
known ribosomal RNAs (about 600 and 1150 nucleotides in length,
respectively) and they exhibit the most extreme A+U content (80%). The
nucleotide sequences of the genes for the 9S and 12S RNAs from three
trypanosome species, Trypanosoma brucei, Crithidia fasciculata and
Leishmania tarentolae have recently been determined. In this paper we
identify primary and secondary stuctures in the trypanosomal 9S and 12S
RNA molecules which are highly conserved among ribosomal RNAs from
members of the three Primary Kingdoms. The presence of these conserved
structural elements in the otherwise highly diverged trypanosomal
mitochondrial ribosomal RNAs, indicates their fundamental role in
ribosome function.

INTRODUCTION

The variation in size of ribosomal RNAs (rRNAs) among organisms and
organelles is extremely high: they range in length from 950 and 1550
nucleotides for mammalian mitochondrial small and large subunit rRNA to
1800 and 5100 nucleotides for the corresponding RNAs from eukaryotic
cytoplasmic ribosomes. However, inspection of the nucleotide sequences of
rRNAs from phylogenetically diverse sources shows that some regions are
universally conserved. Moreover, model building studies showed that all
sequenced rRNAs can be accommodated in the models proposed for the E.coli
16S and 23S rRNAs, which are usually taken as reference models (1,2).
Size differences in the various rRNAs can be attributed to site-specific
insertions or deletions within the E.coli rRNA models.

The high structural homology of rRNAs from various organisms and
organelles suggests that all ribosomes are related through a common
ancestral ribosome. The sections in the primary and secondary structures
which remained constant in all rRNAs are therefore considered essential
for ribosome function. In these investigations the mitochondrial rRNAs
are of special comparative value because their short sizes indicate that
they must represent the stripped-down versions of rRNAs, having retained

features essential to function in translation. In this context the analysis of the 9S and 12S mitochondrial rRNAs of trypanosomes is particularly interesting since they are the smallest rRNAs characterized thusfar (about 600 and 1150 nucleotides in length).

Recently, the nucleotide sequences of the genes for the 9S and 12S RNAs in three trypanosome species, Trypanosoma brucei, Crithidia fasciculata and Leishmania tarentolae have been determined (3-6). In this report we summarize the structural data on trypanosomal mitochondrial rRNAs and relate these to the current models for the rRNAs of E.coli, in order to identify primary- and secondary-structure elements that are still present in these extremely small rRNAs and which may therefore be considered as absolutely indispensable for ribosome function.

MODEL BUILDING

Although the construction of secondary structure models is straight forward for the majority of rRNAs as soon as their nucleotide sequences are known, this does not hold for the trypanosomal mitochondrial rRNAs for three reasons:

(i) The degree of sequence homology between trypanosome mitochondrial rRNAs and the corresponding E.coli rRNAs, is extremely low. This is rather unfortunate because conserved sequences can be used as landmarks for folding flanking sequences into secondary structures using the models for the E.coli rRNAs (7,8) as reference secondary structure models.

(ii) The A+U % of trypanosome mitochondrial rRNAs is about 80% and the A/U ratio is approximately 1 (3-6). Therefore, the potential to form many alternative helical regions in either 9S or 12S RNA is extremely high.

(iii) The level of sequence homology among the corresponding mitochondrial rRNAs of the three trypanosome species is rather high (about 75%). This aspect complicates the assessment of helical regions by the phylogenetic approach, because the number of compensating base replacements in proposed helical regions in corresponding trypanosomal mitochondrial rRNAs is limited.

Another important consideration deals with the special features of mitochondrial rRNAs which combine their small size and low GC content to a higher frequency of irregularities in helices in comparison to other rRNAs. If this can be extrapolated to the trypanosomal mitochondrial rRNAs, which are the smallest rRNAs in nature with the lowest GC content, the number of irregularities or mispairs in helical regions would be the highest. Consequently, most of the helices in the reference structures would be obscured from detection in the trypanosome mitochondrial rRNAs.

In view of the above considerations we have restricted ourselves to those regions of the trypanosome mitochondrial rRNAs for which secondary structures can be derived that are in agreement with the universal aspects of the reference model and in which the structural and phylogenetic criteria are met with respect to the existence of proposed helices (1). Our strategy to establish such structural relationships has been two fold.

a) We searched in the trypanosome mitochondrial rRNAs for universally conserved sequences. Mitochondrial rRNAs show, however, significant variation in these sequences (1). Therefore, we derived consensus sequences from corresponding "universally conserved" regions in mitochondrial rRNAs from a large number of organisms including mammals, fungi, ciliate protozoa, insects and plants. Because of the abnormally high A/U bias observed in trypanosome mitochondrial rRNAs , we carried out a search for these consensus sequences after generalizing G, A, C and U to either purines or pyrimidines. This approach led to the identification of a limited number of conserved sequences in trypanosome mitochondrial rRNAs which are summarized in ref.3. These have been used as landmarks for attempts to fold flanking sequences into secondary structures as present in the reference models.

b) We also searched a matrix plot of all possible base pairing interactions for a particular secondary structure element which is constant
both in configuration and in location in all rRNAs.

These approaches led to secondary structure models for a number of
regions of the trypanosomal mitochondrial rRNAs. Although the basic
features of these models have been recognized previously in secondary
structure models by ourselves (3) and others (4,5), the assessment of
interacting RNA regions within the secondary structures is, in some of
the models presented below, completely different. This is due to a more
reliable application of the phylogenetic approach when corresponding
sequences from three trypanosome species are considered, instead of two
(3-5).

TRYPANOSOMAL 9S RNA

Among prokaryotic small subunit rRNA, 29 universally conserved
sequences have been identified (9). Only 11 of these are present in mammalian mitochondrial small subunit rRNAs (10,11). The most strongly conserved sequences in the small subunit rRNAs of prokaryotes and eukaryotes, including both cytoplasmic and mitohondrial rRNAs, are found in
three proposed single stranded regions around positions 530, 1400 and
1500 in E. coli (1). Only these three highly conserved regions were identified in the trypanosome 9S RNAs by the strategy outlined above. These
are shown in Fig.1 (boxed regions a, b and c) in which invariant nucleotides, occurring in all small subunit RNAs, are indicated by shading.
Regions a, b and c are constant among the three trypanosome species.

The most conserved sequence is region a. Sequences flanking region a
can readily be adapted to the secondary structure of the corresponding
region in the reference model, with all the conserved structural characteristics:
(i) the presence of two adjacent A's in a series of bulged residues at
the 5'-side of the hairpin structure containing region a in the apex
loop,
(ii) the presence of a helical stem at the 3'-side of the region a containing hairpin structure bearing one irregularity at a conserved location. In a previously proposed model for L.tarentolae 9S RNA (4) this
helix is composed from other parts of the 9S RNA sequence and shows several irregularities not found in other rRNAs. This interaction can therefore be considered as ambiguous.

Region b (Fig.1) corresponds to a highly conserved region which is
located in E.coli at the subunit interface (1). Furthermore, the anticodon of a P-site bound tRNA has been cross-linked to C_{1400} of
this region in E.coli , indicating that codon-anticodon recognition
occurs in close proximity to this position (12). At the corresponding
position in trypanosome 9S mitochondrial rRNA and in the mitochondrial
rRNAs of two Paramecium subspecies (13) a C-U transition has been found
(as indicated by an asterisk in Fig.1). If this universal sequence is
involved in the codon-anticodon recognition process, the 3'-half will be
the crucial part since a large number of base replacements have accumulated in the 5'-half of the corresponding trypanosome sequence.

Region c is highly variable. Like region b, region c is localized at
the subunit interface in E.coli (1). Sequences flanking regions b and c
can be accomodated to a secondary structure resembling the model proposed
for the 3'-minor domain of E.coli 16S RNA (Fig.1).

The helical stem between conserved sequences b and c contains one
compensating base replacement and a transitional substitution changing an
A-U into a G-U basepair when the three trypanosome species are compared.
The helix at the 3'-end of trypanosomal 9S RNA contains 2 compensating
base replacements. Furthermore, an interruption of this helical stem in
C.fasciculata 9S RNA is eliminated by an A-U transversion in both
T.brucei and L.tarentolae. The nucleotide sequence in the loop of this

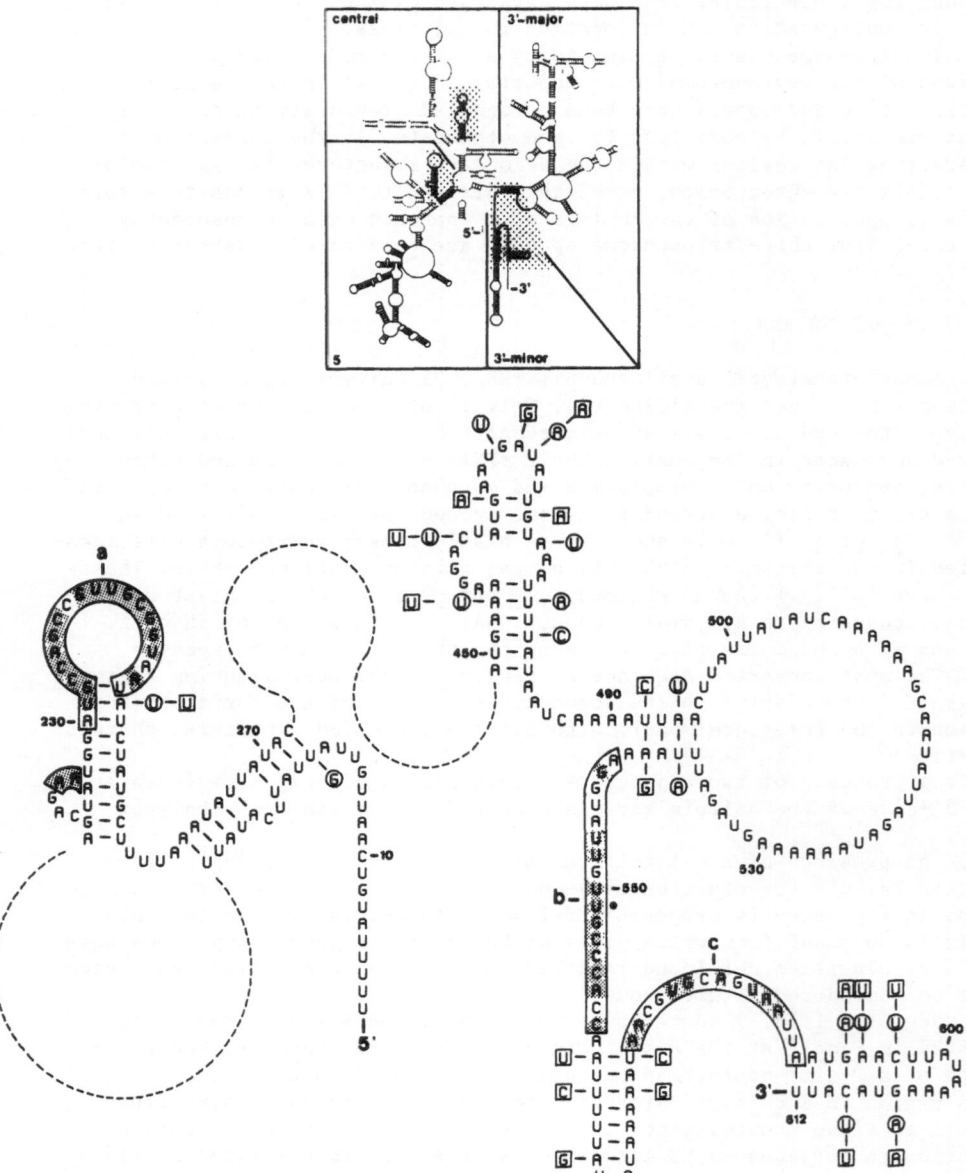

Fig.1: Conserved secondary structure elements in trypanosome 9S mtrRNA.
The sequences of relevant C. fasciculata 9S RNA regions are given
(3) in which the nucleotide numbering is such that the 5'-terminal
residue is position 1. Conserved sequences a, b and c are boxed
and constant nucleotides are shaded. Asterisk in conserved
sequence b indicates a residue discussed in the text. Differences
with corresponding T.brucei and L.tarentolae 9S RNA regions are
indicated by respectively encircled and boxed nucleotides. Dashed
lines indicate areas of trypanosome 9S RNA which could not
reliably be structured because of reasons mentioned in the text.
In the insert a schematic drawing of the secondary structure of
16S rRNA of E.coli is shown (7) in which the domains are
indicated. Regions of this model, conserved in trypanosome 9S
mtrRNA, are marked by shading.

3'-terminal hairpin is highly conserved in small subunit rRNAs from both
nuclear origin (UGAA) and from mitochondrial-, chloroplast- and bacterial
origin (GGAA) (14); the two adjacent A-residues are dimetylated in a
number of organisms. In trypanosomes the sequence is AUAA, but no
information on the state of methylation of the two adjacent A's is
available. Because of their AUAA sequence at this position, trypanosomes
do not confirm the scheme of possible evolutionary links between
bacteria, eukaryotic cytoplasms, mitochondria and chloroplasts (14) which
suggests an independent evolution of trypanosome mitochondrial rRNAs (see
below).

The difference between our model (Fig.1) and a previously proposed
model for this region in L.tarentolae 9S RNA (4) concerns the 9S RNA
segments involved in the long-range RNA interaction at the 5'-side of
conserved sequence b. In contrast to the L.tarentolae 9S RNA model (4),
the interaction proposed here fulfils phylogenetic criteria by the
presence of two compensating base replacements. No complementarity could
be identified between the 3'-terminal sequence of trypanosome 9S rRNA and
sequences in the vicinity of the initiator codon of the trypanosomal
mitochondrial protein genes, indicating that "Shine-Dalgarno" type
interactions do not play a role in initiation of trypanosomal
mitochondrial protein synthesis. The same observation was made for plant
mitochondria (15)

Our second strategy, the search for conserved secondary structure
elements in trypanosome mitochondrial rRNA without the aid of universally
conserved primary sequences, has yielded one such element. It corresponds
to the E.coli 769-810 region which is constant in all small subunit
rRNAs. However, in mitochondrial small subunit rRNAs, this region shows
some variation from which a consensus can be derived (see Fig.2). The
only combination of base paired- and single stranded regions in
trypanosome 9S RNA which fulfils the criteria of the consensus structure,
is region 449-483 (Figs.1 and 2). Comparison of the helical stems d and e
(Fig.2) in three trypanosome species shows a number of base replacements,
all but one (U-C transition in T.brucei at position 481, see Fig.1)
leaving the base pairing scheme intact. This provides strong phylogenetic
evidence for the existence in vivo of this structure. In the previously
proposed L.tarentolae 9S RNA model (4) these helices have been composed
from different 9S RNA sequence elements. However, multiple nucleotide
substitutions, which disrupt the proposed helices, exist in the
corresponding T.brucei and C.fasciculata sequences. The L.tarentolae
derived structure is therefore ambiguous. The sizes of the
single-stranded regions in the structure proposed here (Fig.1) fit the
criteria formulated in the consensus structure (Fig.2). The location of
this secondary structure element is as expected for the central domain:
between the conserved sequences in regions a and b.

The dashed areas in the 9S RNA model (Fig.1) are not structured
since they lack conserved sequences and can be folded in many alternative
ways, none of them bearing on phylogenetic evidence. The four structural
domains, composing the E.coli small subunit rRNA are not equally reduced
in size in trypanosomes. Relative to E.coli 16S rRNA the 5'-domain and
the central domain are each for 60% reduced in size. The 3'-major domain
shows the most drastic reduction in size (about 90%) whereas the 3'-minor
domain is also 60% smaller than the corresponding E.coli domain. The
areas in the secondary structure scheme of E.coli 16S rRNA, which are
conserved in trypanosome 9S mitochondrial rRNA, are indicated by shading
in the insert of Fig.1.

TRYPANOSOMAL 12S RNA

The functionally important 3'-area of large subunit rRNA (domains V
and VI) is highly conserved. Many conserved sequences can also be
identified in the 3'-terminal one third of 12S RNA from the three

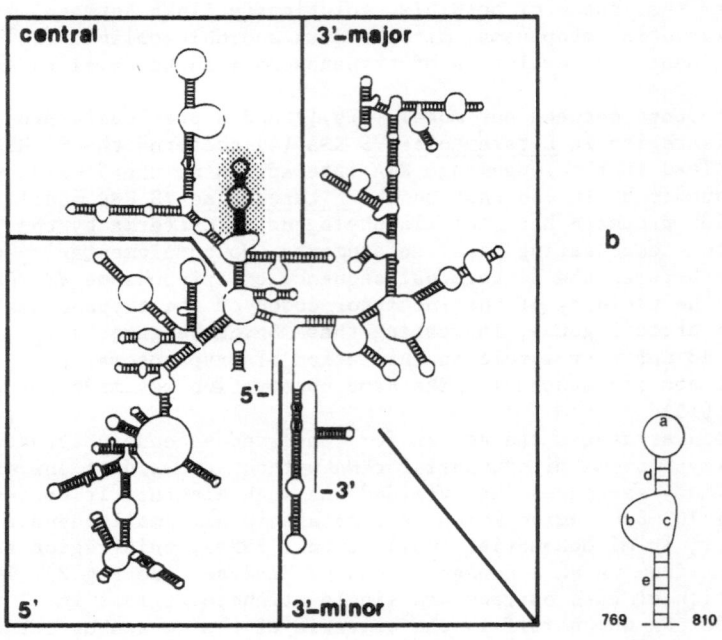

c

SMALL SUBUNIT rRNA	NUCLEOTIDES IN			BASEPAIRS IN	
	a	b	c	d	e
POSITION					
E.COLI 769-810	9	7	4	4	7
mt CONSENSUS	9	6-8	3-5	3-4	6-8
TRYPANOSOMEmt 449-483	9-8	6	3	3	6

Fig.2: Conserved secondary structure elements in the central domain of
small subunit ribosomal RNA.
(a) Secondary structure scheme of 16S rRNA of E.coli in which the
769-810 area is shaded.
(b) Schematic drawing of the 769-810 area of small subunit rRNA in
which single stranded regions a, b and c, and duplex regions d and
e are indicated.
(c) Comparison of the 769-810 region of E.coli with corresponding
regions in small subunit rRNAs of other mitochondria (see ref.3)
and trypanosomes with respect to the number of nucleotides and
base pairs in the single stranded- and in the double helical
conformation, respectively.

trypanosome species (3,5). Region a (see Fig.3) is a highly conserved 12-nucleotide sequence containing the alpha-sarcin cleavage site (16,17) which is implicated in the energy requiring steps in protein synthesis namely the EF1 dependent binding of amino-acylated tRNA and the EF2 and ribosome coupled hydrolysis of GTP (18). This region can be folded in a secondary structure (see Fig.3), exposing the alpha-sarcin cleavage site in the loop as has been done for E.coli and Bacillus stearothermophilus (19). This is the only structure of domain VI which has been retained in trypanosome 12S mitochondrial rRNA.

The central part of domain V of large subunit rRNA is involved in peptidyl transferase activity. Regions b and c are highly homologous to corresponding regions in E.coli to which either the aminoacyl end of a P-site tRNA (region b) or puromycin derivatives (region c) can be cross-linked to U-residues (indicated by asterisks in Fig.3; 20,21). Also regions which are involved in binding of peptidyl transferasae inhibiting antibiotics or which have acquired base substitutions in resistant mutants, are clearly recognizable in trypanosome 12S RNA. The regions specified in d, are involved in chloramphenicol (CAP) sensitivity of pro-tein synthesis. In trypanosomes many differences with corresponding regions in other organisms are found as indicated by the low number of shaded nucleotides (Fig.3). From CAP resistant mitochondrial mutants of yeast, mouse and human is known that base replacements at five positions occur in this region (22-25). Two of these coincide with substitutions in the corresponding trypanosome region (indicated by asterisks) and it might be expected that the trypanosome mitoribosome shows CAP resistance. As in CAP resistant mutants, resistance to erythromycin (ERY) results from one base substitution in an otherwise highly conserved region (26). The position of this difference coincides with one of the 5 substitutions in the corresponding trypanosome region e (indicated by an asterisk in Fig.3). The sequence data on regions d and e are consistent with the observation that mitochondrial protein synthesis in trypanosomes could not be demonstrated with the aid of chloramphenicol or erythromycin (27). Three other sequences have been identified in 12S RNA which show a high level of homology with the consensus sequences (regions f, g and h).

Using these conserved sequences as landmarks, a secondary structure could be devised for the 3'-terminal region of trypanosomal 12S RNA which is very similar to domains V and VI of the reference model (see insert Fig.3). In contrast to our previous model, based on C.fasciculata and T.brucei 12S RNA data (3), the sequence characterized by cross-linked puromycin (region c) is now located in the loop of an irregular helix, which is highly conserved (21). This structure was recognized, however, in the recently published model for the corresponding region of L.taren-tolae 12S RNA (5). Domain V has been reduced by 50% in trypanosomal 12S rRNA relative to E.coli 23S rRNA. The central part of domain V has been well conserved, however protruding helices are reduced considerably in size or have been eliminated. This trend has also been observed in mamma-lian mitochondrial large subunit rRNAs (28), although less extensive than in trypanosomes. Domain VI has been reduced severely (90% relative to E.coli). The areas in the secondary structure scheme of domains V and VI of E.coli 23S rRNA which are conserved in trypanosome 12S rRNA are indi-cated by shading in the insert of Fig.3. The remaining 5'-two thirds of the 12S RNA has not been structured because many alternative ways of fol-ding are possible. None of these can be proven by phylogenetic evidence. The functionally important UAGCUGGUU sequence, found at position 811 in E.coli 23S rRNA and which is conserved in rRNA from various phylogenetic sources, cannot be identified in trypanosome 12S rRNA. It has been suggested that the UGG of this region interacts with the CCA-3' of tRNAs (20). Absence of this region in trypanosomes suggests that such a mode of binding does not occur and that trypanosomal mitochondrial tRNAs will turn out to lack CCA at their 3'-termini. Since precise positioning of

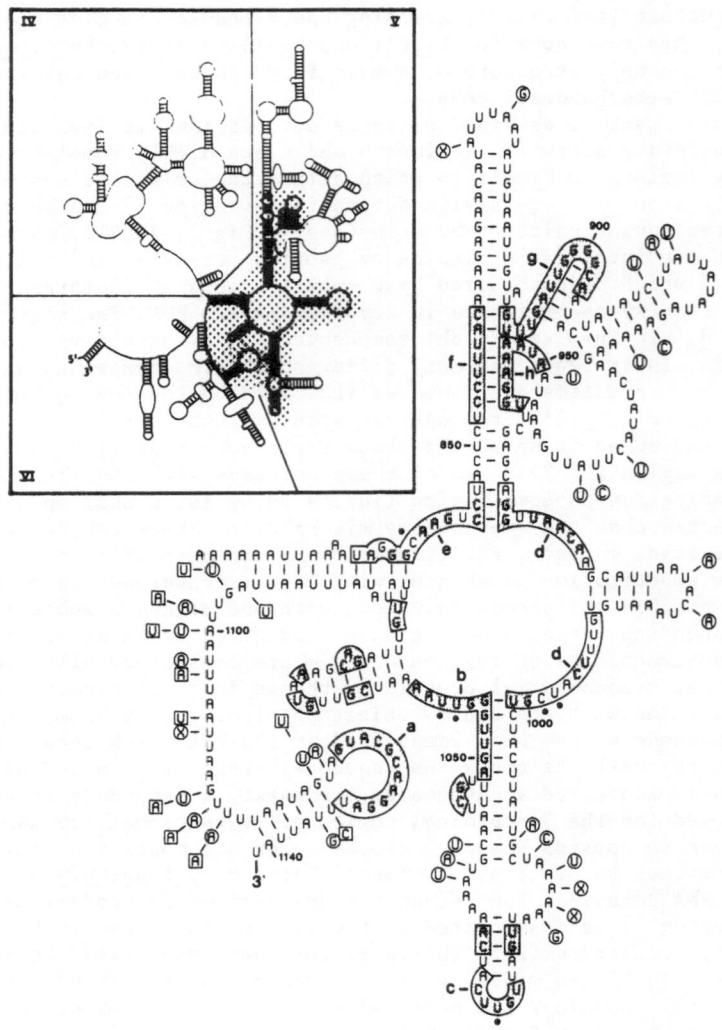

Fig.3: Secondary structure of the 3'-terminal region of trypanosome 12S
mtrRNA.
The sequence of C.fasciculata 12S RNA is given in which the
5'-terminal nucleotide is position 1. Conserved sequences a, b, c,
d, e, f, g and h are boxed. Shaded residues represent constant
nucleotides. Asterisks in conserved sequences indicate nucleotides
discussed in the text. Differences with the corresponding sequence
in T.brucei and L.tarentolae are indicated by encircled and boxed
nucleotides respectively. In the insert a schematic drawing of the
secondary structure of the corresponding region of E.coli 23S rRNA
is given. Domains are indicated and regions conserved in
trypanosome 12S mtrRNA are marked by shading.

the 3'-ends of tRNAs on the ribosome is obviously crucial for the peptidyl transferase reaction, it is possible that in trypanosome mitochondria
this is accomplished in some other way.

CONCLUSIONS AND PERSPECTIVES

The mitochondrial 9S and 12S RNAs from trypanosomes are rRNAs which are
unique in many ways (reviewed in 29,30). They (i) are the smallest known
rRNAs, (ii) possess the lowest GC content, (iii) are encoded in tandem on
maxicircle DNA, the trypanosomal equivalent of mtDNA in other organisms,
but in contrast to all known rRNAs the direction of transcription is from
large to small rRNA. Furthermore, the above analysis shows that only a
limited number of the secondary structures and sequences which are universally conserved in rRNA from various phylogenetic lines, are retained
in the trypanosome mitochondrial rRNAs. This suggests that these structural elements (summarized in Figs.1 and 3) are absolutely indispensible for
ribosome function.
    Variation from the norm can also be expected for other elements of
the protein synthesizing machinery in trypanosome mitochondria. First of
all mitoribosomes have thusfar resisted purification by standard techniques. This suggests that they are highly unstable or have other extraordinary properties by which they escape detection. There is a trend in
mitochondria, in which a smaller size and a lower GC content of the rRNA
is accompanied by a high protein/RNA ratio of the ribosomes. If this can
be extrapolated to trypanosomes, an extraordinary high protein/RNA ratio
is expected for their mitoribosomes. Second, the existence of mitochon-
drially encoded tRNAs has not been demonstrated yet (27), nor have tRNA
genes been detected on maxicircle DNA sequences of T. brucei (6,31,32)
and L. tarentolae (33). The possibility arises, however, that mitochon-
drially encoded tRNAs have an unusual structure, which allows them to go
undetected in hybridization analysis or in computer searches for classical cloverleaf structures.
    Considering the current state of the art in trypanosome mitochon-
dria, the presence of freak rRNAs and the lack of demonstration thusfar
of ribosomes and tRNAs, one may question whether protein synthesis does
occur in trypanosome mitochondria or not. In the latter case, the 9S and
12S RNAs are consequently products of pseudogenes. However, this view
cannot be correct since the corresponding mitochondrial rRNAs of the
three trypanosome species show a high level of overall sequence homology
(75%) which is locally significantly higher (c.f. the 3'-terminal region
of 12S rRNA is more than 90% conserved in the three trypanosome species)
notwithstanding their separate evolution for about 80 million years
(3,33). Furthermore, as we show in this paper, certain structural
elements of 9S and 12S RNAs are highly homologous to universally
conserved structures in rRNAs from the three primary kingdoms. It is
therefore hard to envisage that these major maxicircle DNA transcripts
are non-functional. In addition, mitochondrial protein genes identified
on maxicircle DNA of T. brucei and L. tarentolae exhibit the UGA =
tryptophan codon assignment (30) which is also found in other
mitochondrial systems. This underlines the existence of a separate
compartment for mitochondrial protein synthesis in trypanosomes, since
UGA specifies a translational stop in the cytoplasm.
    Why are the mitochondrial rRNAs and possibly other components of the
mitochondrial protein synthesizing machinery in trypanosomes so different
from their counterparts in other systems? We attribute this divergence to
a different evolutionary constraint which has been exerted on trypanosomal mtDNA in comparison with mtDNA from other organisms. This is illustrated both by its unique structure, consisting of a complex network of
thousands of catenated circular DNA molecules and by the low degree of
nucleotide sequence homology with other mtDNAs. Our future research
focuses on the characterization of the components of the trypanosomal

protein-synthesizing machinery, (c.f. mitoribosomes and tRNAs) and its
products. Since these promise to be highly exceptional, they might
provide some insight in the essential, functional elements of both the
translational apparatus and the enzymes of the respiratory chain.

ACKNOWLEDGEMENTS

We thank Drs L.A.Grivell and H.F.Tabak for critically reading the
manuscript and Drs H.F.Noller and R.Gutell, University of California at
Santa Cruz, for stimulating discussions.

REFERENCES

1. Noller, H.F., Structure of ribosomal RNA, Ann.Rev.Biochem., 53:119--
   162 (1984).
2. Maly, P.R. and Brimacombe, R., Refined secondary structure models
   for the 16S and 23S ribosomal RNA of Escherichia coli, Nucl.Acids
   Res., 11:7263-7286 (1983).
3. Sloof, P., Van den Burg, J., Voogd, A., Benne, R., Agostinelli, M.,
   Borst, P., Gutell, R. and Noller, H.F., Further characterization of
   the extremely small mitochondrial ribosomal RNAs from trypanosomes:
   a detailed comparison of the 9S and 12S RNAs from Crithidia fasciculata and Trypanosoma brucei with rRNAs from other organisms, Nucl.-
   Acids Res., 13: 4171-4190 (1985).
4. De la Cruz, V.F., Lake, J.A., Simpson, A.M., and Simpson, L., A
   minimal ribosomal RNA: Sequence and secondary structure of the 9S
   kinetoplast ribosomal RNA from Leishmania tarentolae, Proc.Natl.-
   Acad.Sci.USA, 82:1401-1405 (1985).
5. De la Cruz, V.F., Simpson, A.M., Lake, J.A., Simpson, L., Primary
   sequence and partial secondary structure of the 12S kinetoplast
   (mitochondrial) ribosomal RNA from Leishmania tarentolae: conservation of peptidyl-transferase structural elements, Nucl.Acids Res.,
   13:2337-2356 (1985)
6. Eperon, I.C., Janssen, J.W.G., Hoeijmakers, J.H.J. and Borst, P.,
   The major transcripts of the kinetoplast DNA of Trypanosoma brucei
   are very small ribosomal RNAs, Nucl.Acids Res., 11:105-125 (1983).
7. Woese, C.R., Gutell, R., Guptas, R. and Noller, H.F., Detailed analysis of the higher-order structure of 16S-like ribosomal ribonucleic acids, Microbiol.Rev. 47:621-669 (1983).
8. Noller, H.F., Secondary structure model for 23S ribosomal RNA,
   Nucl.Acids Res., 9:6167-6189 (1981).
9. Woese, C.R., Fox, G.E., Zabler, L., Uchida, T., Bonen, L., Pechman,
   K., Lewis, B.J. and Stahl, D., Conservation of primary structure in
   16S ribosomal RNA, Nature 254:83-86 (1975).
10. Eperon, I.C., Anderson, S. and Nierlich, D.P., Distinctive sequence
    of human mitochondrial ribosomal RNA genes, Nature 288:60-63
    (1980).
11. Anderson, S., De Bruijn, M.H.L., Coulson, A.R., Eperon, I.C.,
    Sanger, F. and Young, I.G., Complete sequence of bovine mitochon-
    drial DNA. Conserved features of the mammalian mitochondrial genome,
    J.Mol.Biol., 156:683-717 (1982).
12. Prince, J.B., Taylor, B.H., Thurlow, A.L., Ofengand, J. and
    Zimmerman, R.A., Covalent crosslinking of tRNA^{Val} to 16S
    RNA at the ribosomal P site: Identification of crosslinked residues,
    Proc.Natl.Acad.Sci.USA 79:5450-5454 (1982).
13. Seilhamer, J.J., Olsen, G.J. and Cummings, D.J., Paramecium mito-
    chondrial genes I: small subunit rRNA gene sequence and micro-evolu-
    tion, J.Biol.Chem., 259: 5167-5172 (1984).

14. Van Knippenberg, P.H., Van Kimmenade, J.M.A. and Heus, H.A., Phylogeny of the conserved 3' terminal structure of the RNA of small ribosomal subunits, Nucl.Acids Res., 12:2592-2604 (1984).
15. Boer, P.H., McIntosh, J.E., Gray, M.W. and Bonen, L., The wheat mitochondrial gene for apocytochrome b: absence of a prokaryotic ribosome binding site, Nucl.Acids Res., 13:2281-2292 (1985).
16. Schindler, D.G. and Davies, J.E., Specific cleavage of ribosomal RNA caused by alpha sarcin, Nucl.Acids Res., 4:1097-1110 (1977).
17. Chan, Y.L., Endo, Y. and Wool, I.G., The sequence of the nucleotides at the alpha-sarcin cleavage site in rat 28S ribosomal ribonucleic acid, J.Biol.Chem., 258: 12768-12770 (1983).
18. Fernandez-Puentes, L. and Vasquez, D., Effects of some proteins that inactivate the eukaryotic ribosome, FEBS Lett., 78:143-146 (1977).
19. Garrett, R.A., Christensen, A. and Douthwaite, S., Higher order structure in the 3'-terminal domain VI of the 23S ribosomal RNAs from Escherichia coli and Bacillus stearothermophilus, J.Mol.Biol., 179: 689-712 (1984).
20. Barta, A., Steiner, G., Brosius, J., Noller, H.F. and Kuechler, E., Identification of a site on 23S ribosomal RNA located at the peptidyl transferase center, Proc.Natl.Acad.Sci.USA 81:3607-3611 (1984).
21. Branlant, C., Krol, A., Machatt, M.A., Pouyet, J. and Ebel, J.P., Primary and secondary structures of E.coli MRE600 23S ribosomal RNA. Comparison with models of secondary structure for maize chloroplast 23S rRNA and for large portions of mouse and human 16S mitochondrial rRNAs, Nucl.Acids Res., 9:4303-4324 (1981).
22. Dujon, B., Sequence of the intron and flanking exons of the mito-chondrial 23S rRNA gene of yeast strains having different alleles at the rib-1 loci, Cell 20:185-197 (1980).
23. Blanc, H., Adams, C.W and Wallace, D.C., Different nucleotide changes in the large rRNA gene of the mitochondrial DNA confer chloramphenicol resistance on two human cell lines, Nucl.Acids Res., 9: 5785-5795 (1981).
24. Blanc, H., Wright, C.T., Bibb, M.J., Wallace, D.C. and Clayton, D.A. Mitochondrial DNA of chloramphenicol-resistant mouse cells contain a single nucleotide change in the region encoding the 3' end of the large ribosomal RNA, Proc.Natl.Acad.Sci.USA 78:3789-3793 (1981).
25. Kearsey, S.E. and Craig, I.W., Altered ribosomal RNA genes in mito-chondria from mammalian cells with chloramphenicol resistance, Nature 290:607-608 (1981).
26. Sor, F. and Fukuhara, H., Identification of two erythromycin resis-tance mutations in the mitochondrial gene coding for the large ribo-somal RNA in yeast, Nucl.Acids Res., 10:6571-6577 (1982).
27. Benne, R., Agostinelli, M., De Vries, B.F., Van den Burg, J., Klaver, B. and Borst, P., Gene expression and organization in trypa-nosome mitochondria, in: Mitochondria 1983: Nucleo-Mitochondrial Interactions (R.J.Schweyen, K.Wolf and F.Kaudewitz, eds), De Gruyter, Berlin, pp. 285-302 (1983).
28. Glotz, C., Zwieb, C., Brimacombe, R., Edwards, K. and Kössel, H., Secondary structure of the large subunit ribosomal RNA from E.coli, Z.mays chloroplast and human and mouse mitochondrial ribosomes, Nucl.Acids Res., 9: 3287-3306 (1981).
29. Borst, P. and Hoeijmakers, J.H.J., Kinetoplast DNA, Plasmid 2:20-40 (1979).
30. Benne, R., Mitochondrial genes in trypanosomes, Trends in Genetics 1:117-121 (1985).
31. Benne, R., De Vries, B.F., Van den Burg, J. and Klaver, B., The nucleotide sequence of a segment of Trypanosoma brucei mitochondrial maxi-circle DNA that contains the gene for apocytochrome b and some unusual unassigned reading frames, Nucl.Acids Res., 11, 6926-6941 (1982).

32. Hensgens, L.A.M., Brakenhoff, J., De Vries, B.F., Sloof, P., Tromp, M.C., Van Boom, J.H. and Benne, R., The sequence of the gene for cytochrome c oxidase subunit I, a frameshift containing gene for cytochrome c oxidase subunit II and seven unassigned reading frames of Trypanosoma brucei mitochondrial maxi-circle DNA, Nucl.Acids Res., 12:7327-7344 (1984).

33. De la Cruz, V.F., Neckelmann, N. and Simpson, L., Sequences of six genes and several open reading frames in the kinetoplast maxicircle DNA of Leishmania tarentolae, J.Biol.Chem., 259: 15130-15147 (1984).

MUTAGENESIS OF RIBOSOMAL RNA AS A METHOD TO

INVESTIGATE INTERACTIONS BETWEEN rRNA, mRNA and tRNA

A. Dahlberg, W. Jacob, M. Santer*, C. Zwieb** and D. Jemiolo

Brown University
Providence, RI 02912, USA

*Haverford College
Haverford, PA 19041, USA

**European Molecular Biological Laboratory
Heidelberg, Federal Republic of Germany

INTRODUCTION

Our understanding of the three dimensional structure of ribosomal RNA is just beginning to unfold. Information has come from a variety of approaches including crosslinking experiments of rRNA (1-3) and neutron and proton magnetic resonance studies (4). Comparative phylogenetic studies, so useful for determining the secondary structure of ribosomal RNA (5), may now also be applied to the study of tertiary structure (Noller, personal communication and 6). Another new approach to the study of rRNA structure involves the application of genetic techniques to introduce mutations in the rRNA (7). This latter approach, while in its infancy, provides the means to confirm results obtained by other methods, as well as explore new areas.

We have been interested in studying the relationship between structure and function in E. coli rRNA. Using mutagenic techniques now available it is possible to introduce mutations at any site in rRNA. While we have obtained only limited information on the tertiary structure of rRNA, we have begun to explore the quartenary structure, or intermolecular interactions at one particular area, the 3' end of 16S rRNA. It has been known for some time that the 3' end of 16S rRNA is intimately involved in initiation of protein synthesis, including interactions with mRNA and tRNA (8). In this article we will describe some recent studies from our laboratory in which we have investigated these RNA-RNA interactions by site directed mutagenesis of the rRNA.

The two mutants we shall describe here have a number of similarities. Both mutations are at or near sites thought to be involved in interactions with other RNA molecules. One mutation is at C1538, in the Shine-Dalgarno region, the reported mRNA binding site (9). The second mutation is at C1402, near C1400, the base crosslinked to the wobble base of tRNA in the P site on the ribosome (10). Both mutations involve a single base transition of C to U. Both sites are near the 3' end of 16S rRNA, in single stranded,

highly conserved regions (see Fig. 1). The mutant rRNA transcripts of both
mutant plasmids are processed (17S to 16S) and assembled into 30S subunits

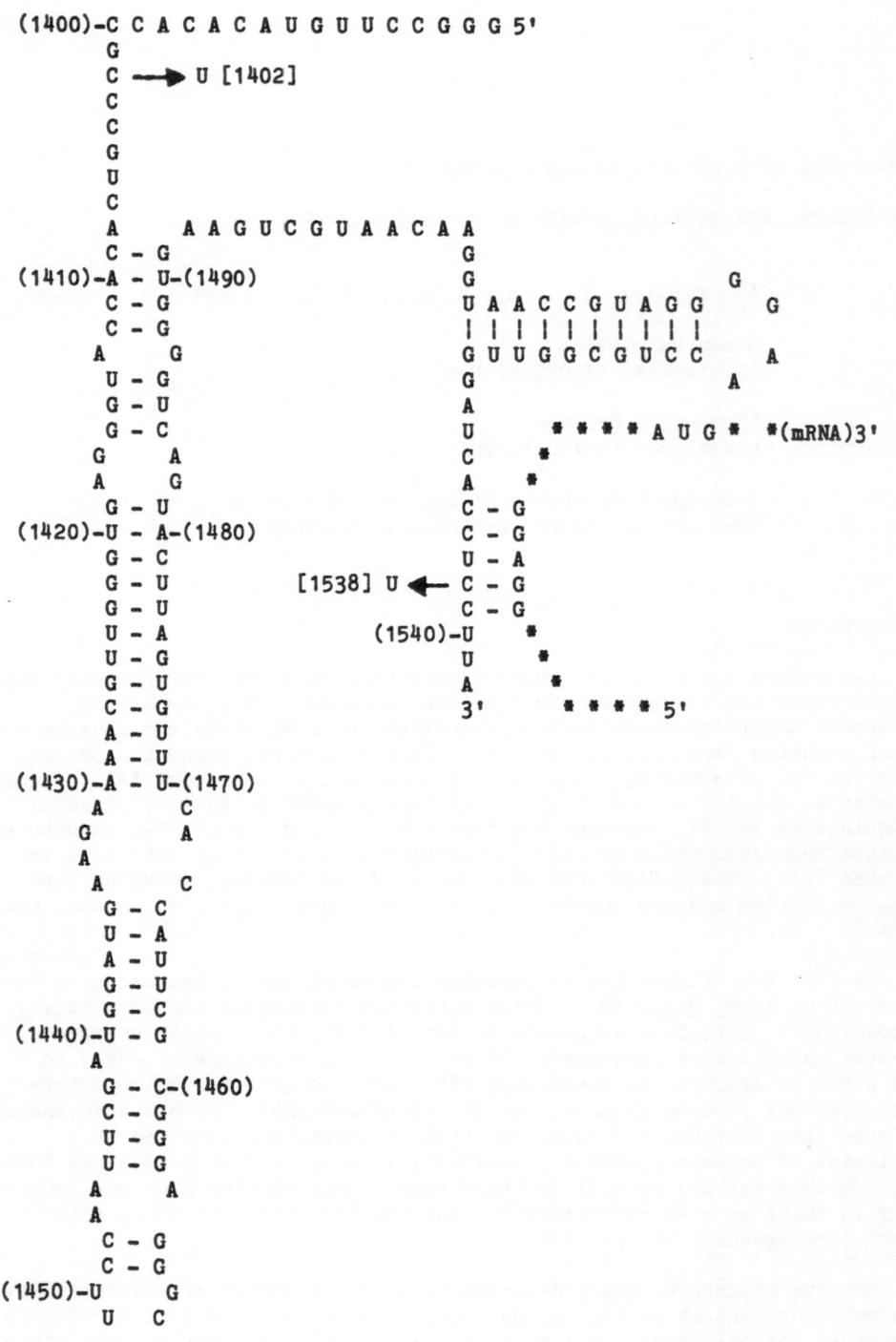

```
(1400)-C C A C A C A U G U U C C G G G 5'
       G
       C ──► U [1402]
       C
       C
       G
       U
       C
       A       A A G U C G U A A C A A
       C - G                         G
(1410)-A - U-(1490)                  G                          G
       C - G                         U A A C C G U A G G      G
       C - G                         | | | | | | | | | |
       A       G                     G U U G G C G U C C        A
       U - G                         G                    A
       G - U                         A
       G - C                         U     * * * * A U G *   *(mRNA)3'
       G       A                     C   *
       A       G                     A  *
       G - U                         C - G
(1420)-U - A-(1480)                  C - G
       G - C                         U - A
       G - U           [1538] U ◄─ C - G
       G - U                         C - G
       U - A                (1540)-U    *
       U - G                         U   *
       G - U                         A    *
       C - G                         3'     * * * * 5'
       A - U
       A - U
(1430)-A - U-(1470)
       A       C
       G       A
       A
       A       C
       G - C
       U - A
       A - U
       G - U
       G - C
(1440)-U - G
       A
       G - C-(1460)
       C - G
       U - G
       U - G
       A       A
       A
       C - G
       C - G
(1450)-U    G
       U  C
```

Fig. 1. The secondary structure model for the 3' minor domain of
E. coli 16S rRNA. Transitions at C1402 and C1538 are
labeled with arrows.

which appear to function in translation. The mutants differ in that C1402 → U alters translational accuracy and is compatible with cell growth while C1538 → U affects translational initiation and is lethal to the cells. The two mutants were also produced by different mutagenic techniques and it is appropriate at this point to describe them separately.

MUTAGENESIS OF THE SHINE-DALGARNO REGION OF 16S rRNA

The site of the mutation in the Shine-Dalgarno (SD) region and the base to be substituted were selected after considering what intramolecular base-pairing might be necessary for processing of the mutant precursor 17S rRNA. To avoid introducing too great of a change into the SD region a C → U transition at residue 1538 was chosen, which changed the sequence from (5')CCUCC(3') to (5')CCUUC(3'). The C → U transition was constructed at base 1538 using a synthetic oligonucleotide and the M13mp11 cloning vector (11). A restriction fragment of DNA containing this mutation was inserted into pKK3535, a multicopy plasmid containing the entire rrnB operon of E. coli (12). The mutant plasmid was introduced into the recA host, HB101. To our surprise, plasmids with this single base change could not be isolated. All of the transformants that were isolated contained plasmids with second site mutations in the 16S rRNA which suppressed the SD muta- tion. This indicated that the single base change at position 1538 was lethal to the transformed cells. It was necessary, therefore, to insert the SD mutant into a plasmid with a repressible promoter replacing the two ribosomal promoters (P1 and P2) normally present in the rrnB operon. Just such a system (pNO2680) was constructed by R. Gourse and provided to us. He had substituted the lambda pL promoter-operator for P1 and P2 (13). Transcription from the pL promoter can be regulated by introducing the plasmid into cells containing the temperature sensitive repressor CI857. Expression of the rDNA operon is induced by shifting the temperature from 30°C to 42°C. At 30°C there is no transcription of the plasmid-coded rRNA genes. A small degree of expression is noted at 37° and this increases to a maximum at 42°.

A restriction fragment of rDNA in M13, containing the SD mutation, was cloned into pNO2680, producing plasmid pWFJ1538. Cells containing the temperature sensitive repressor CI857 were transformed with pWFJ1538. Two cell strains were used. One contained a single copy of CI857 inserted in the host chromosome (N4830) and the other (HB101) carried CI857 on a multi- copy plasmid (pCI857). Growth rates of transformed cells were measured at the restrictive (30°) and the non-restrictive (42°) temperatures. Cells containing mutant and wild type plasmids had identical rates at 30°. The rates at 42° were similar for the first two hours, after which the growth of cells containing pWFJ1538 leveled off considerably. Cells with the wild type plasmid also ceased growing after several more hours as reported (13). To our surprise when HB101 cells without the pCI857-containing plasmid were transformed with pWFJ1538 we detected a large number of very slow growing transformants after three or four days at 30°. This was in contrast to the result when HB101 was transformed with the SD mutant in pKK3535, where no slow growing colonies were detected. Presumably the expression of the SD mutant rRNA is sufficiently less with the pL promoter than with the natural P1 and P2 promoters in pKK3535 to permit the cells to survive.

In order to study expression of the SD rRNA mutant on the plasmid independent of host-coded rRNA synthesis, we used a modification of the maxicell procedure originally described by Sancar et al. (14), and adapted by M. Stark (15). The original maxicells were derived from strains of E. coli unable to repair ultraviolet light-damaged DNA. Upon irradiation with u.v. light, strains such as CSR603 (recA1, uvrA6, phr-1) cease DNA synthesis and most of the chromosomal DNA becomes degraded over several hours. However, plasmids contained in such strains largely escape u.v. induced damage due to their small target size in comparison with the host chromosome. They continue to replicate within the "maxicells," which are therefore active in the expression of plasmid-coded genes only. (It is also possible to use strain HB101 (recA) if the cells are kept in the dark after u.v. treatment.) The modified maxicell system provided a means to study expression of the plasmid-coded rRNA in the complete absence of host-coded rRNA synthesis.

Under the conditions specified by Stark the maxicells continued to synthesize proteins from the host chromosome including ribosomal proteins[*]. The availability of ribosomal proteins and the presence of the specific rRNA processing enzymes permitted the mutant rRNAs transcribed from the plasmids to be processed and assembled into 30S ribosomal subunits. Electrophoretic analysis by composite gels (18) showed that all of the SD mutant rRNA was processed to the 16S form and assembled into 30S subunits incorporated into 70S ribosomes. This alleviated an initial concern that the mutant at C1538 might affect processing of the 17S rRNA. Earlier studies had shown that the 17S rRNA processing enzymes are exceedingly stringent, that assembly of 30S subunits is particularly sensitive to perturbations of the rRNA structure, and that 16S rRNA maturation and 30S subunit assembly are tightly coupled (15).

Having shown that the SD mutant rRNA was indeed processed correctly and assembled into 30S subunits, we then sought to determine whether the mutant ribosomes were actually functional. We reasoned that ribosomes with a SD mutation in the 16S rRNA might translate mRNA with a complementary SD mutation more efficiently than would wild type ribosomes. We obtained a plasmid containing the appropriate SD mutation (ZEM72 of T4 rIIb) cloned in front of the lac Z gene from L. Gold. The synthesis of β-galactosidase from this plasmid containing the SD mutant (rIIB/lac Z fusion) was measured in cells carrying either pNO2680 or pWFJ1538. Cells containing both the SD mutant ribosomes and the SD mutant mRNA produced 40% more β-galactosidase after 3 hours at 42° than did cells containing the wild type ribosomes and SD mutant mRNA. This level was reproducible and provided the first direct evidence that rRNA is involved in base-pairing with mRNA as proposed by Shine and Dalgarno (9). Similar experiments have also been performed by Herman DeBoer who has constructed two quite different SD mutants (GGAGG and CACAC) by procedures essentially the same as described here.

The relatively modest increase (40%) in β-galactosidase production in cells with the SD mutant plasmid can be explained in part by the fact that wild type ribosomes can also translate the mutant rRNA. Gold and co-workers found that the SD mutant mRNA was translated 85% as efficiently as the wild type mRNA by wild type ribosomes at 25°. At 37° the efficiency was reduced to 63% (L. Gold, personal communication). Wild type ribosomes recognize a wide range of mRNA initiation sequences. One might also expect that a single base change in the SD region of rRNA allows these mutant ribosomes to recognize a wide, overlapping range of mRNAs. If so, this might create a global effect on mRNA utilization in cells harboring the SD mutant ribosomes. Some mRNAs may be bound more efficiently, resulting in

---

[*]The highly specific synthesis of plasmid-coded rRNA under conditions where maxicells appear to continue to transcribe mRNA could be due to prefer-ential relaxation (unwinding rather than degradation) of the host chromo-some as the result of low u.v. influence. The transcription of rDNA in vitro and in vivo is particularly sensitive to inhibition of DNA gyrase (16, 17) indicating that superhelicity of rDNA is required for efficient expression. This relaxation of host-coded rRNA could account for the specificity of rRNA transcription of the plasmid-coded rRNA while main-taining the ability to synthesize host-coded proteins.

an increased frequency of translation, while other mRNAs may be bound less efficiently, giving a reduced level of translation. There are many facets of mRNA structure which determine its affinity for and translation by ribosomes. These may regulate the relative levels of the many different proteins in the cell. Disrupting this balance should have a catastrophic effect on cells, as indeed it appears did happen when the SD mutant was introduced into pKK3535 (lethal) or the pL plasmid, pNO2680, without repressor (very slow growth).

Experiments designed to demonstrate such a global effect of the SD mutant ribosomes are now in progress. Cells containing either mutant or wild type ribosomes (at 42°) will be incubated with ^{35}S-methionine and the labeled proteins electrophoretically separated on two dimensional O'Farrell gels (19). If the SD mutation does affect binding efficiency and translation, then distinct differences in relative ratios of the proteins in the two gel patterns should be observed. Identical protein patterns should be obtained when cells with the two different plasmids are grown and labeled at 30°.

MUTAGENESIS OF RESIDUE C1402 OF 16S rRNA

There is no direct evidence for base-pairing between rRNA and tRNA. The two RNAs are in close proximity to one another on the ribosome, however, as is evident by the crosslinking of the wobble base of tRNA$_{\text{Val}}$ to residue C1400, near the 3' end of 16S rRNA (10). This region is highly conserved in sequence and is thought to be in a single stranded conformation from studies on kethoxal reactivity (20). Our study of this region of 16S rRNA has focused on a point mutation two bases removed from C1400, at residue C1402. This site is unique in E. coli rRNA in that it is methylated on both the base and the ribose moieties. The function of this residue is unknown but the unusual double methylation near C1400 makes the C1402 mutant an interesting candidate for studying potential tRNA interaction with the ribosome.

The mutation of C1402 → U was created by cytosine deamination with bisulfite in a 121 base single stranded loop-out formed in a heteroduplex of linearized deletion mutant plasmid, pHaeII-82 (lacking bases 1385-1505), and wild type plasmid pKK3535. The two plasmids were linearized by restriction enzyme digestion, mixed, denatured, renatured and treated with bisulfite according to a published procedure (21). Competent BD 817 cells, deficient in the DNA repair enzyme, N uracil glycosylase, were transformed with the multi-copy plasmid and we obtained only a single mutation, C1402 → U, in the region around position C1400.[*]

We were surprised to discover that the C1402 → U mutation was viable. Mutagenesis was performed on plasmid pKK3535, a multicopy plasmid with two active ribosomal promoters. We estimate that approximately 50% of the ribosomes in the cell are synthesized from the plasmid. Thus any mutations that give rise to ribosomes that seriously affect cell growth are selected against. In the case of mutant C1402 the doubling time of transformed cells was 60 minutes, not considerably slower than the 50 minutes doubling time of cells containing wild type plasmid pKK3535.

Mutant plasmid C1402 was cloned into the maxicell strain CSR603 in order to determine the fate of the rRNA transcript. The mutant rRNA was processed normally to mature 16S rRNA and found in 30S subunits capable of

---

[*]R. Zimmermann and co-workers have isolated a series of mutants in this region by bisulfite mutagenesis of a cloned fragment of the rrnB operon in M13 (R. Zimmermann, personal communication).

interacting with 50S subunits, as analyzed by two dimensional gel electrophoresis. However, we have not determined the state of the methylation of the uridine residue at position 1402 in the mutant rRNA.

Until recently our in vivo studies of mutant C1402 have been limited to measurement of growth rate and rRNA processing and assembly. Now it has become possible to assay the function of the mutant ribosomes in vivo. One of us (M.S.) developed an assay for mutant ribosomes which can determine whether certain changes in rRNA can influence translational fidelity. The procedure involves introducing a mutant plasmid into cells with a nonsense mutation in β-galactosidase, and measuring the level of enzyme synthesis in the presence of the mutant ribosomes. Cells carrying an UGA nonsense codon in the lac Z gene and a suppressor tRNA (su^{+9}) (22) were transformed with either pKK3535 (control) or mutant plasmid C1402. The lac Z gene is partially expressed in these cells due to the presence of the suppressor tRNA. In the presence of the mutant plasmid C1402 the level of β-galactosidase is reduced 30 to 70% below that of cells with plasmid pKK3535 or no plasmid. Thus mutant plasmid C1402 caused restriction of reading of a nonsense codon in mRNA by the supressor tRNA. This effect is analogous to that produced by mutations in ribosomal protein S12, which give streptomycin resistance and restrict reading of nonsense codons. While this result does not demonstrate a direct physical interaction between tRNA and rRNA, it does indicate that there is a functional association. More direct evidence must await the development of techniques for the isolation of pure mutant rRNA for reconstitution and study by more specific in vitro functional assays.

SUMMARY

We have presented two examples of rRNA mutants, produced by site directed mutagenesis, which affect the interaction of ribosomes with mRNA and tRNA. These examples demonstrate the potential usefulness of the technique of rRNA mutagenesis to the study of rRNA structure and function. It is now possible to produce mutations at any site in the rRNAs, analyze processing and assembly of mutant rRNA by maxicells, and regulate expression of the mutant rRNA with a temperature sensitive repressor of the pL operon. The functional activity of mutant ribosomes in vivo can be assayed by two dimensional protein gel electrophoresis, assaying of β-galactosidase synthesis from normal mRNA and measuring restriction of translation of β-galactosidase mRNA with a nonsense mutant. Using these procedures we are beginning to gain an understanding about the functional roles of different regions of the rRNA which are involved in interactions with tRNA and mRNA.

REFERENCES

1. P. L. Wollenzien, C. F. Hui, C. Kang, R. F. Murphy, and Cantor, C. R., in: "Molecular Mechanism in Protein Synthesis," E. Bermek, ed., pp. 1-22 (1983).
2. A. Expert-Bezancon, P. L. Wollenzien, J. Mol. Biol. 184:53-66 (1984).
3. R. Brimacombe and W. Stiege, Biochem. J. 229:1-17 (1985).
4. H. Heus, J. vanKimmenade, P. vanKnippenberg, C. Haasnoot, S. deBruin, and C. Hilbers, J. Mol. Biol. 170:939-956 (1983).
5. C. R. Woese, R. Gutell, R. Gupta, and H. F. Noller, Microbiological Reviews 47:621-669 (1983).
6. P. Spitnik-Elson, D. Elson, S. Avita, and R. Abramowitz, Nuc. Acids Res. 13:4719-4738 (1985).
7. R. Gourse, M. Stark, and A. Dahlberg, J. Mol. Biol. 159:397-416 (1982).

8. M. Grunberg-Manago, in: "Ribosomes; Structure, Function and Genetics," G. Chambliss, G. Craven, J. Davies, K. Davis, L. Kahan, and M. Nomura, eds., University Park Press, Baltimore, p. 445-478 (1979).

9. J. Shine and L. Dalgarno, Proc. Natl. Acad. Sci. U.S.A. 71:1342-1346 (1974).

10. J. B. Prince, B. H. Taylor, D. L. Thurlow, J. Ofengand, and R. Zimmermann, Proc. Natl. Acad. Sci. U.S.A. 79:5450-5454 (1982).

11. J. Messing, B. Gronenborn, B. Muller-Hill, P. H. Hofschneider, Proc. Natl. Acad. Sci. U.S.A. 74:3642-3646 (1977).

12. J. Brosius, A. Ullrich, M. A. Raker, A. Gray, T. J. Dull, R. R. Gutell, and H. F. Noller Plasmid 6:112-118 (1981).

13. R. Gourse, Y. Takebe, R. Sharrock, M. Nomura, Proc. Natl. Acad. Sci. U.S.A. 82:1069-1073 (1985).

14. A. Sancar, A. Hack, D. Rupp, J. Bacteriol. 137:692-693 (1979).

15. M. J. R. Stark, R. L. Gourse, and A. E. Dahlberg, J. Mol. Biol. 159:417-439 (1982).

16. H. L. Yang, K. Heller, M. Gellert, G. Zubay, Proc. Natl. Acad. Sci. U.S.A. 76:3304-3308 (1979).

17. B. A. Oostra, A. J. Van Vliet, G. Ab, and M. Gruber, J. Bacteriol. 148:782-787 (1981).

18. A. Dahlberg, Electrophoresis of ribosomes and polysomes, in: "Gel Electrophoresis of Nucleic Acids: A Practical Approach," D. Rickwood and B. D. Hames, eds., IRL Press Ltd., Oxford, pp. 213-225 (1982).

19. P. O'Farrell, J. Biol. Chem. 250:4007-4021 (1975).

20. H. F. Noller, Biochemistry 13:4694-4703 (1974).

21. D. Kalderon, B. A. Oostra, B. K. Ely, and A. E. Smith, Nuc. Acids Res. 10:5161-5171 (1982).

22. L. A. Petrullo, P. J. Gallagher, D. Elseviers, Mol. Gen. Genet. 190:289-294 (1983).

RIBOSOMAL RNA AT THE DECODING SITE OF THE tRNA-RIBOSOME COMPLEX

James Ofengand, Jerzy Ciesiolka, and Kelvin Nurse

Roche Institute of Molecular Biology
Roche Research Center
Nutley, New Jersey 07110

INTRODUCTION

Until recently, the protein components of the ribosome were considered to be the functional entities, the ribosomal RNA (rRNA) being viewed as merely a scaffold upon which the proteins were arranged. This view had its origin in part from the expectation that the proteins, by analogy with soluble enzymes, would play the crucial role in recognition of tRNA as well as in other ribosomal functions. It also was based on the fact that the early affinity labeling experiments identified primarily proteins (Ofengand, 1980). More recently, however, direct evidence for the proximity of rRNA at functional sites on the ribosome has been obtained (reviewed in Ofengand et al., 1984; 1985). The first demonstration that rRNA could be closely associated with a functional site was the crosslinking of 23S RNA by peptidyl transferase center-affinity labels (reviewed in Ofengand, 1980), although the exact site of crosslinking was not identified until very recently (Barta et al., 1984). The second example of rRNA proximity at a functional center, and the first case in which the exact nucleotide was identified, was the discovery of the close contact between C-1400 of *Escherichia coli* 16S RNA and the anticodon of P site bound tRNA at the decoding site (Ofengand et al., 1979; Prince et al., 1982). Subsequently, a nearby residue, C-1409, was shown to be indirectly involved in codon recognition(Li et al., 1982), and the not too distant G-1322 was found to be crosslinkable to S12 (Chiaruttini et al., 1982), a protein in contact with mRNA. The mRNA base-pairing region, A-1531 to A-1542 (Gold et al., 1981), is also not far from C-1400 in the 16S RNA secondary structure. However, G-462 and G-474, which were placed at the decoding site by Wagner et al. (1976) are quite distant. This latter result may be indicative of tertiary folding of rRNA within the subunit.

CROSSLINKING OF THE ANTICODON LOOP OF P SITE BOUND tRNA TO 16S RNA

Structure and Specificity of the Crosslink

The close contact between C-1400 of 16S rRNA and the anticodon loop of tRNA mentioned above was shown by crosslinking P site bound tRNA to 16S rRNA (Schwartz & Ofengand, 1978; Ofengand et al., 1979). As shown in Fig. 1, the crosslink was a cyclobutane dimer (Ofengand & Liou, 1980) between the 5'-anticodon base, cmo^5U-34 in $tRNA_1^{Val}$, the tRNA used for most of these experiments, and C-1400 of 16S RNA (Table 1). This is the

TABLE 1 CROSSLINKING SITES IN SMALL SUBUNIT rRNA

tRNA	tRNA Residue 34	Ribosome	rRNA Crosslink Site	Refs.
E. coli	uridine-5-OCH$_2$COOH	*E. coli*	UACACACC$\downarrow$$_{1400}$G	a,b,c
E. coli	uridine-5-OCH$_2$COOH	Yeast	UACACACC$\downarrow$$_{1626}$G	b,c
E. coli	uridine-5-OCH$_2$COOH	*A. salina*	UACACACC$\downarrow$$_{1644}$G	d
B. subtilis	uridine-5-OCH$_3$	*E. coli*	UACACACC$\downarrow$$_{1400}$G	b,c

Arrow denotes the residue crosslinked to cmo^{5}U-34 of tRNA as
determined by sequence analysis.
(a) Prince et al., 1982;   (b) Ehresmann et al., 1984;
(c) Ehresmann and Ofengand, 1984;   (d) Ciesiolka et al., 1985b.

only case known so far of cyclobutane dimer formation between two distinct
nucleic acid molecules. The requirement for cyclobutane dimer formation
means that cmo^{5}U-34 and C-1400 must either stack or be adjacent so that
the planes of the crosslinking bases are parallel and within ca 4 Å of
each other. Analogous crosslinking has been found in other ribosomes to a
cytidine residue which is the eukaryotic equivalent of C-1400 (Table 1).
In these ribosomes, no other rRNA residue is crosslinked, not even the
5'-adjacent C residue. Also, the U-34 COOH group is not needed for
crosslinking. Crosslinking is codon-dependent (Ofengand & Liou, 1981) and
occurs whether the A site is occupied or empty (Table 2). Crosslinking
also occurs with spinach chloroplast ribosomes (Ofengand et al., 1982) and
Tetrahymena thermophila cytoplasmic ribosomes (unpublished results) but
not with ribosomes from yeast mitochondria, even though P site binding
could be demonstrated (Ofengand et al., 1984). C-1400 lies in the center
of a 17 nucleotide long sequence which has been conserved in all known
small subunit rRNA primary structures (Table 3). This conserved region is
believed to be single-stranded in the ribosome and to lie at the interface
between 50S and 30S subunits (Noller, 1984). No function has been so far
assigned to this region and no ribosomal proteins are known to have any
interaction with it (Brimacombe et al., 1983; Noller, 1984). It seems
highly unlikely, however, that such strong sequence conservation, coupled
with the functional conservation exemplified by the ability of species as
disparate as *E. coli* and *Artemia salina* to crosslink at the same residue,
would have been retained without having some important role in ribosome
structure or function.

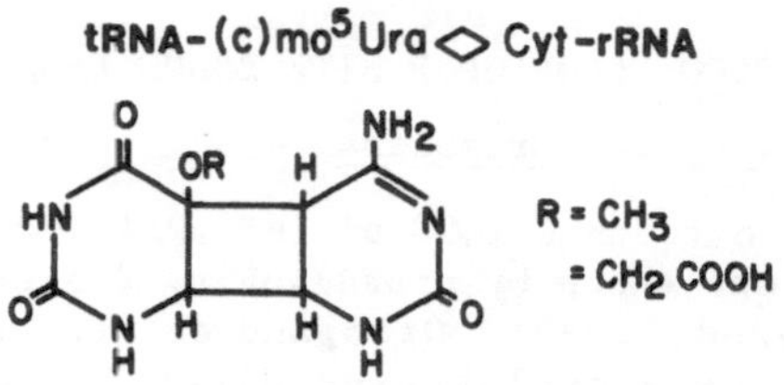

tRNA-(c)mo^{5}Ura $\diamondsuit$ Cyt-rRNA

R = CH$_3$
= CH$_2$COOH

Fig. 1. Structure of the tRNA-16S RNA crosslink. The tRNA base,
cmo^{5}U-34, is shown on the left and the rRNA base, C-1400, is on the right.

TABLE 2   EFFECT OF Phe-tRNA IN THE A SITE ON P SITE CROSSLINKING OF tRNA^{Val}

| Codon | Minus Phe-tRNA | | Plus Phe-tRNA | | |
	tRNA^{Val} bound	% Cross-linking	tRNA^{Val} bound	tRNA^{Phe} bound	% Cross-linking
GU$_2$	0.59	68	0.56	0.02	62
GU$_3$	0.59	66	0.45	0.02	65
GU$_4$	0.57	64	0.34	0.20	59
GU$_5$	0.61	69	0.50	0.48	56
GU$_6$	0.58	69	0.47	0.56	60
GU$_7$	0.55	70	0.44	0.57	60

[^{3}H]tRNA^{Val} was prepared by [^{3}H]ATP exchange with A-76 of the tRNA using tRNA nucleotidyl transferase. Codons were obtained by 3'-polymerization of U onto GpU using polynucleotide phosphoylase, and separation according to size on DEAE-cellulose. P and A site binding and P site crosslinking were performed and assayed essentially as described (Gornicki et al., 1984; 1985). 750 nM [^{3}H]tRNA^{Val}, 150 nM ribosomes, 50 μM codon, 113 nM [^{14}C]Phe-tRNA, and 20 mM Mg^{++} were used. Values for tRNA bound are shown as pmole tRNA per pmole ribosome. Values for tRNA^{Val} crosslinked are given as percent of the tRNA^{Val} bound.

Localization of the decoding site on the ribosomal small subunit

Location of C-1400 on the 30S *E. coli* subunit by immunoelectron microscopy was accomplished by attaching a dinitrophenyl antigen directly to the crosslinking base in the tRNA via a flexible carbon chain leash. Reaction with antibody followed by electron microscopy showed that all the antibodies were bound to the cleft separating the head and neck from the large protrusion (Gornicki et al., 1984). This region must therefore correspond to the decoding site . The location of this site is shown in Fig. 2. It has not yet been possible to obtain information about front-to-back localization of the site. This site is close to the m$_2^6$Am$_2^6$A region and is ca. 50 Å from the 3'-end of rRNA, A-1542. The site for IF3 (involved in mRNA binding to the ribosome) overlaps the C-1400 region (Stöffler & Stöffler-Meilicke, 1984).

An attempt was made to perform a similar localization experiment with *A. salina* ribosomes. However, since the overall crosslinking yield in pmole/pmole ribosomes was less than 1/4th that of *E. coli* ribosomes (Table 4, lines 3 & 4), an attempt was made to purify the crosslinked 40S from uncrosslinked particles by antibody-linked dimer formation via the attached DNP group. The procedure had been used successfully with *E. coli* ribosomes when the DNP group was placed on the amino acid moiety of the tRNA (Keren-Zur et al., 1979). As shown in Table 4, lines 1 and 2, it was also successful with *A. salina*, although the dimer yield was only half that when *E. coli* was used. However, when the DNP group was linked to the crosslinked anticodon base directly, no dimer formation with *A. salina* crosslinked 40S subunits could be detected even though substantial amounts of dimer (15%) were found with *E. coli* 30S particles. This result has so far frustrated our efforts to localize the decoding site on *A. salina* ribosomes by immunoelectron microscopy. It has, however, given some insight into the differences between these two particles. It is evident that the *A. salina* decoding site is in some way more sterically hindered than that of *E. coli*. We cannot yet distinguish between hindrance to dimer formation or to reaction of antibody with the DNP group itself since we have so far only studied the former reaction.

Table 3   CONSERVED SEQUENCE IN rRNA FROM THE SMALL SUBUNIT

	Residue Number[a]			References
	1392	1400	1408	
Prokaryotes				
Escherichia coli	G U A C A C	A C C G C C C	G U C A	Brosius *et al* (1978); Carbon *et al* (1979); Noller (1984)
Proteus vulgaris	. . . . . .	. . . .	. . . .	Carbon *et al* (1981)
Bacillus brevis	. . . . . .	. . . .	. . . .	Kop *et al* (1984)
Mycoplasma capricolum	. . . . . .	. . . .	. . . .	Iwami *et al* (1984)
Mycoplasma (5 other species)	. . . . . .	. . . .	. . . .	Woese *et al* (1980)
Halobacterium volcanii	C	. . . .	. . . .	Gupta *et al* (1983)
Halobacterium morrhua	–	. . . .	. . . .	Leffers & Garrett (1984)
Halobacterium halobium	C	. . . .	. . . .	Magrum *et al* (1978)
Thermoplasma acidophilum	C	. . . .	. . . .	Woese *et al* (1980)
Methanogens (15 species)	C	. . . .	. . . .	Balch *et al* (1979)
Methanococcus vannielii	C	. . . .	. . . .	Jarsch & Böck (1985)
Agrobacterium tumefaciens	. . . . . .	. . . .	. . . .	Yang *et al* (1985)
Pseudomonas testosteroni	. . . . . .	. . . .	. . . .	Yang *et al* (1985)
Anacystis nidulans	. . . . . .	. . . .	. . . .	Tomioka & Sugiura (1983)
Prochloron	. . . . . .	. . . .	. . . .	Seewaldt & Stackebrandt (1982)
Eukaryote cytoplasm				
Saccharomyces cerevisiae	. . . . . .	. . . .	. . . G	Rubstov *et al* (1980); Mankin *et al* (1981); Hogan *et al* (1984)
Tetrahymena thermophila	. . . . . .	. . . .	. . . G	Spangler & Blackburn (1985)
Dictyostelium discoideum	. . . . . .	. . . .	. . . G	McCarroll *et al.* (1983)
Bombyx mori	. . A C . .	. . . .	. . . G	Samols *et al* (1979)
Drosophila melanogaster	. . . . . .	. . . .	. . . G	Jordan *et al* (1980)
Xenopus laevis	. . . . . .	. . . .	. . . G	Salim & Maden (1981); Atmadja *et al* (1984)
Artemia salina	. . . . . .	. . . .	. . . G	Nelles *et al* (1984)
Mouse	. . . . . .	. . . .	. . . G	Raynal *et al* (1984)

276

Organism	Sequence													Reference
Rat	·	·	·	·	·	·	·	·	·	·	·	·	G	Torczynski *et al* (1983); Chan *et al* (1984)
Rabbitt	·	·	·	·	·	·	·	·	·	·	·	·	G	Connaughton *et al* (1984)
Zea mays (corn)	·	·	·	·	·	·	·	·	·	·	·	·	G	Messing *et al* (1984)
Oryza sativa (rice)	·	·	·	·	·	·	·	·	·	·	·	·	G	Takaiwa *et al* (1984)

Chloroplasts

Organism	Sequence													Reference
Zea mays	·	·	·	·	·	·	·	·	·	·	·	·	·	Schwartz & Kössel (1980)
Euglena gracilis	·	·	·	·	·	·	·	·	·	·	·	·	·	Graf *et al* (1982)
Chlamydomonas reinhardii	·	·	·	·	·	·	·	·	·	·	·	·	·	Dron *et al* (1982)
Nicotiana tabacum	·	·	·	·	·	·	·	·	·	·	·	·	·	Tohdoh & Sugiura (1982)

Mitochondria

Organism	Sequence													Reference
Saccharomyces cerevisiae	C	·	·	U A	·	U	·	A	·	U	·	A	· · ·	Sor & Fukuhara (1980); Li *et al* (1982)
Aspergillus nidulans	·	·	·	U A	·	·	A	·	U	·	·	·	A · · ·	Köchel & Küntzel (1981)
Paramecium primaurelia & *tetraurelia*	C	·	·	U	·	·	·	·	U	·	·	A	· · ·	Seilhamer *et al* (1984)
Leishmania tarentolae, *Crithidia fasciculata*, *Trypanosoma brucei*	·	G U A U U G U U	·	·	·	·	·	A C	·	·	·	·	·	de la Cruz *et al* (1985); Sloof *et al.* (1985)
Drosophila yakuba	·	·	·	·	·	·	U	·	·	·	·	·	G · · ·	Clary & Wolstenholme (1985)
Xenopus laevis	·	C	·	·	·	·	·	·	·	·	·	·	·	Roe *et al.* (1985)
Hamster	·	·	·	·	·	·	·	·	·	·	·	·	·	Baer & Dubin (1980)
Aedes albopictus	·	·	·	U · U	·	·	·	·	A	·	·	·	G · · ·	Dubin & HsuChen (1983)
Mouse	·	C	·	·	·	·	·	·	·	·	·	·	·	Bibb *et al* (1981)
Rat	·	C	·	·	·	·	·	·	·	·	·	·	·	Kobayashi *et al* (1981)
Cow	·	C	·	·	·	·	·	·	·	·	·	·	·	Anderson *et al* (1982)
Human	·	·	·	·	·	·	·	·	·	·	·	·	·	Eperon *et al* (1980); Anderson *et al* (1981)
Zea mays	·	·	·	·	·	·	·	·	·	·	·	·	·	Chao *et al* (1984)
wheat	·	·	·	·	·	·	·	·	·	·	·	·	·	Spencer *et al* (1984)

a Numbering system according to *E. coli* (Noller, 1984).
Dots indicate identity with *E. coli*; dashes the absence of a nucleotide.

TABLE 4

RIBOSOME BINDING, CROSSLINKING, AND DIMER FORMATION WITH DNP-MODIFIED tRNA

tRNA	Ribosome	pGUU-dependent tRNA Binding	pGUU-dependent tRNA Crosslinking	% Cross-linking	% Dimers
		pmol/pmol ribosomes			
DNP-Val-tRNA	*E. coli*	0.30	0.30	100	47
	A. salina	0.24	0.15	64	22
AcVal-tRNA$^{EDA-DNP}$	*E. coli*	0.28	0.35	125	15(29)
	A. salina	0.11	0.08	72	<2

Ribosomal P site binding, cross-linking, and subunit separation were performed and assayed as described previously (Gornicki et al., 1984; 1985; Ciesiolka et al., 1985b). 30S-antiDNP-30S complex formation and sucrose gradient centrifugation was modified from Keren-Zur et al. (1979). Incubation with anti-DNP was in 20 mM Hepes, pH 7.5, 100 mM KCl, 2 mM Mg^{++} for 10 min at 37°C followed by 20 min at 0°C. Sucrose gradient separation of 30S dimers from monomers was done in the same buffer, except the value in parentheses which used 20 mM Mes, pH 6.0, 100 mM NH_4Cl, 2 mM Mg^{++} in the sucrose gradient. DNP-Val-tRNA is N-(2,4-dinitrophenyl)valyl-tRNA; AcVal-tRNA$^{EDA-DNP}$ is N-(acetyl)valyl-tRNA modified like NAK in Fig. 4 except that the N_3 is replaced by NO_2 and there are 3, not 5, CH_2 groups between the left-hand NH and CO groups.

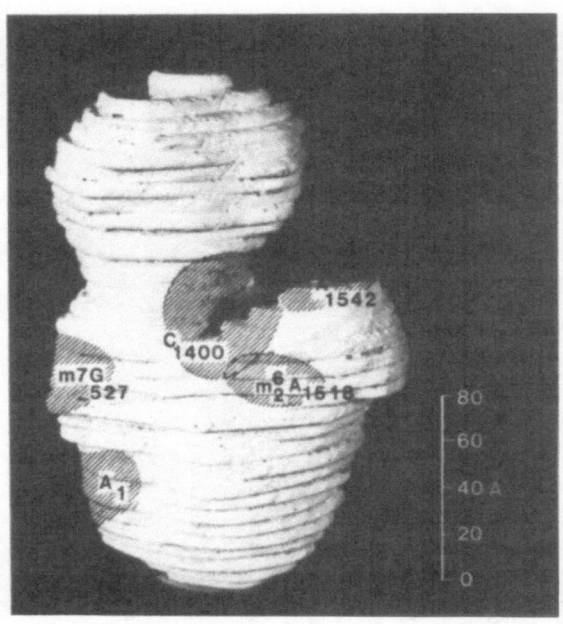

Fig. 2. Model of the *E. coli* subunit with topographically identified rRNA sites. The model is that of Verschoor et al. (1984). C-1400, the ribosomal RNA residue crosslinked to the anticodon of tRNA, was located by immunoelectron microscopy (Gornicki et al., 1984). The locations of the 3'-end, A-1542 (Olson and Glitz, 1979; Shatsky et al., 1979; Lührmann et al., 1981 ; Stöffler-Meilicke et al. 1981), m_2^6A-1517,1518 (Politz and Glitz, 1977; Stöffler and Stöffler-Meilicke, 1981), m^7G-527 (Trempe et al., 1982), and the 5'-end, A-1 (Mochalova et al., 1982) are also shown.

Folding of 16S rRNA

The folding of the 3'-terminal region of 16S RNA shown in Fig. 3, follows from the requirement, as illustrated, that the 5'-end of the Shine-Dalgarno sequence in φX174 H mRNA be within 26 Å of the 3' base of the initiation codon in that mRNA. Since that residue is equivalent in space to the tRNA 5'-anticodon base which in turn is connected to C-1400, ribosomal RNA residues 1403-1530 must fit into a 20 Å linear space. Segments 1409-1491 and 1506-1529 are known to be helical (Noller, 1984). The additional single-stranded residues must also be looped out of the way. As the relative spatial arrangement of the Shine-Dalgarno region and the decoding site of the 30S is not likely to change for different mRNAs, these spatial relationships are probably generally applicable. Other mRNAs appear to span this distance by varying the number of non-base paired nucleotides between their Shine-Dalgarno regions and the initiating AUG and/or by varying the position *along the rRNA* of the 3' base paired residue of the Shine-Dalgarno sequence. Since the distance per residue for a single-stranded RNA is 2.4 times that for one which is base-paired, elimination of base-pairing is a way to add distance without adding nucleotides.

CROSSLINKING OF THE ANTICODON LOOP OF A SITE BOUND tRNA TO 16S rRNA

The folding possibilities shown in Fig. 3 prompted a search for other segments of 16S RNA at the decoding site. In order to do this, non-specific aryl azide photoaffinity probes, linked via alkyl chain leashes of variable length and structure, were attached to the same tRNA

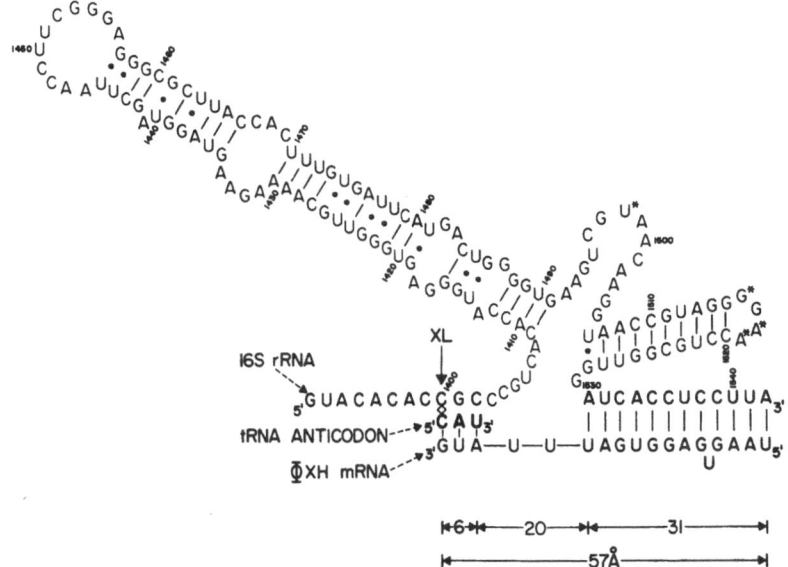

Fig. 3. Folding of the 3'-region of E. coli 16S rRNA postulated on the basis of the location of Shine-Dalgarno and initiation AUG sequences in φX174 H mRNA (Sanger et al., 1977). The tRNA 5'-anticodon base which crosslinks to C-1400 is spatially equivalent to the 3'-G of the initiation codon. The linear spacing of single (Yathindra and Sundaralingam, 1975) and double-stranded (Arnott et al., 1975) residues of the mRNA is to scale, the distance for unpaired residues being 2.4 times that for helical ones. The exact folding and direction of the 16S rRNA segments not directly interacting with tRNA or mRNA (lighter type face) is arbitrary, as is the linear direction of the mRNA. Figure modified from Gornicki et al. (1984).

nucleotide, cmo^{5}U-34, involved in crosslinking to C-1400 (Fig. 4). The protocol for binding and crosslinking is shown in schematic form in Fig. 6 (complex A). Efficient crosslinking (30-50%) at the A site was found when the NAK or SNAP probes were used (Table 5). Crosslinking was probe and

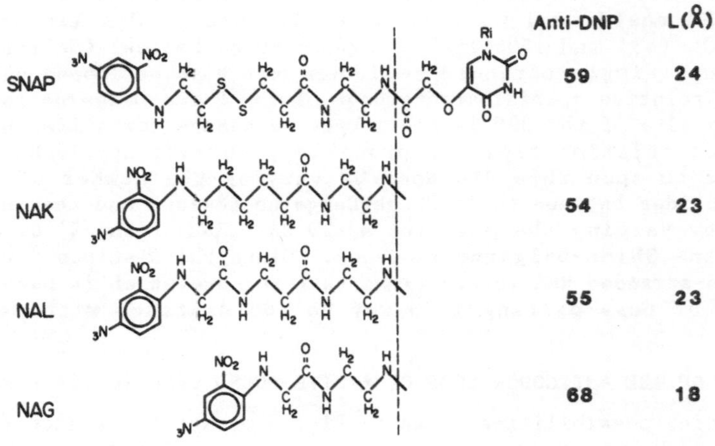

Fig. 4. Structure of the various aryl azide derivatives of tRNA^{Val} used. Attachment was to the COOH group of the cmo^{5}U-34 residue. Reactivity of the modified tRNAs was tested by crosslinking to anti-DNP (Gornicki et al., 1985). The values shown are 100 x (moles crosslinked/-moles bound). L (Å): distance from C$_5$ of cmo^{5}U-34 to N$_3$ in the maximally extended conformation.

TABLE 5 CROSSLINKING OF MODIFIED tRNA^{Val} TO THE RIBOSOMAL A SITE

Modified tRNA	Dependency	Bound	Crosslinked	Percent Crosslinking
		pmoles tRNA/pmole ribosomes		
NAK	Irrad.	0.27	0.13	48
	Codon	0.26	0.10	38
	EFTu	0.27	0.11	41
NAK*	Irrad.	0.20	0.004	2.0
SNAP	Irrad.	0.44	0.13	30
	EFTu	0.29	0.10	35
NAL	Irrad.	0.15 (0.13)	0.01 (0.01)	7 (8)
NAG	Irrad.	0.18	0.021	12
	EFTu	0.18	0.011	6.1
None	Irrad.	0.30	<0.001	<0.2

Binding and crosslinking values in the absence of the stated dependency have been subtracted. Irrad., irradiation; EFTu, elongation factor Tu from *E. coli*. NAK*, Val-tRNA^{NAK} pre-photolyzed under standard conditions in the absence of ribosomes. Values for NAL in parenthesis were obtained by measuring [^{3}H]NAL. Other values were obtained from [^{3}H] or [^{14}C]Val-tRNA. Table modified from Gornicki et al. (1985). NAK, SNAP, NAL, NAG denote the structures shown in Fig. 4.

irradiation dependent in all cases, and where tested was also codon and EFTu-dependent, and blocked by pre-photolysis. Although NAK, SNAP and NAL were almost the same length, the crosslinking yields were different. NAL crosslinking was even less than that of the 5 Å shorter NAG. We attribute the low yields with SNAP and especially with NAL to additional rigidity of the alkyl chain leash due to the presence of the S-S or CONH group. The crosslinking yield with NAK approaches that obtained by reaction with anti-DNP (see Fig. 4). This high yield implies close contact with the ribosomal surface or accessibility to a hydrophobic pocket. The degree of resistance to inhibition of crosslink formation by varying concentrations of mercaptoethanol also led to the conclusion that the A site crosslink was relatively shielded from solvent (Gornicki et al., 1985). Both results are consistent with location of the anticodons of A and P site bound tRNAs in the cleft of the 30S subunit, as directly shown above for the P site tRNA. Crosslinking was to the 30S subunit only and mainly (>70%) to 16S RNA.

The exact site of crosslinking was determined using both the non-cleavable NAK probe and the cleavable SNAP analog (Ciesiolka et al., 1985a). To our surprise, sequence analysis showed that C-1400 was the *only* nucleotide to be crosslinked with both the SNAP and NAK probes. C-1400 was previously shown to be within 4 Å of the cmo^{5}U-34 of P site bound tRNA. In that work, unique crosslinking to C-1400 was rationalized on the basis of the stereochemical constraints of cyclobutane dimer formation. Finding the same specificity for the relatively non-specific probes used here implies that the other nearby nucleotides of the ribosomal RNA are largely inaccessible to such probes at the decoding site. This point is reinforced by the fact that although tRNA^{SNAP} and tRNA^{NAK} differed in crosslinking yield almost 2-fold (Table 5), they both crosslinked to the same nucleotide.

Studies on the chemical reactivity of rRNA nucleotides in the 1392-1408 region do not show unique exposure of C-1400 (Table 6). While there is a change in going from free rRNA to an activated 30S particle, *both* C1399 and C1400 go from a hidden to a partially exposed state. Moreover, G-1401 appears as exposed as C-1400. Why then was only C-1400 crosslinkable in our studies? We speculate that formation of the 70S couple which shields G-1405, or of the 70S-tRNA complex which shields G-1401, may have a similar effect on other nucleotides. The unique reactivity of C-1400 to both aryl azide probes and cyclobutane dimer formation could then be accounted for if there was some special reason for keeping it exposed in the 70S-tRNA complex.

Crosslinking to C-1400 from the A site is stereochemically reasonable if the 3'-stack model of Fuller and Hodgson (1967), which is supported by considerable experimental evidence (Steiner et al., 1984; and references therein), is assumed. This arrangement places the 5'-anticodon bases of the two tRNAs about 22 Å apart. If C-1400 is positioned within 4 Å of the P site 5'-anticodon base (in order to account for cyclobutane dimer formation), then a 23 Å probe at the A site should be sufficient to reach it. In Fig. 5, the N$_3$ of the NAK probe has been placed in the approximate location where C-1400 is to be expected, based on its close contact with the 5'-anticodon base of P site bound tRNA. The latter nucleotide is expected to be adjacent to the 3' base of the A site anticodon by virtue of the 5' to 3' direction of reading mRNA, the antiparallel nature of codon-anticodon base pairing, and stereochemical considerations (Sundaralingam et al., 1975; Spirin & Lim, 1985).

CROSSLINKING OF BOTH P AND A SITE BOUND tRNAs TO THE SAME rRNA OLIGO-
NUCLEOTIDE

Since both A and P site tRNAs were crosslinked to C-1400 when reacted separately, they might also do so when both tRNAs are on the same ribosome. The experimental protocol to test this hypothesis is shown in

TABLE 6 EXPOSURE OF RESIDUES 1392–1408 IN *E. COLI* 16S RNA AND RIBOSOMES

		1392 XL ↓ 1408
"native" RNA (1)	exposed	G_1 U_3 $\underline{A_{1,7}}$ A_1 $\underline{A_{1,7}}$ G_1 $\underline{G_{1,7}}$ A_1
		: : :
	partial	C_3 : C_3 : :
		: : + :
	hidden	A_7 C_3 C_3 G_7 $\overset{\circ}{C}_3$ C_3 C_3 U_3 $\overset{+}{C}_3$ A_7
active 30S at 4°C (1)	exposed	A_1 G_1 A_1
	partial	G_1 U_3 C_3 C_3 C_3 C_3 G_1
	hidden	A_1 A_1 $\overset{\circ}{C}_3$ C_3 C_3 U_3 $\overset{+}{C}_3$
70S vs. 30S	enhanced (2)	G_7
	protected (3)	G_1
70S·P site tRNA vs. 70S (4)	enhanced	G_7
	protected	G_7

Exposure of bases was tested by chemical modification as indicated below.
Subscript numbers indicate atom modified. Underlining or dotted line
connects residues where both N_1 and N_7 were probed. $\overset{\circ}{C}$, m^4Cm; $\overset{+}{C}$, m^5C.
(1) U_3, carbodiimide; C_3 and A_1, dimethylsulfate; G_1, carbodiimide and
   kethoxal (Moazed et al., 1985). A_7, diethylpyrocarbonate (Van Stolk &
   Noller, 1984; Douthwaite et al., 1983). G_7, dimethylsulfate
   (Douthwaite et al., 1983).
(2) G_7, dimethylsulfate (Meier and Wagner, 1985).
(3) G_1, kethoxal (Herr et al., 1979).
(4) G_7, dimethylsulfate (Meier and Wagner, 1984).

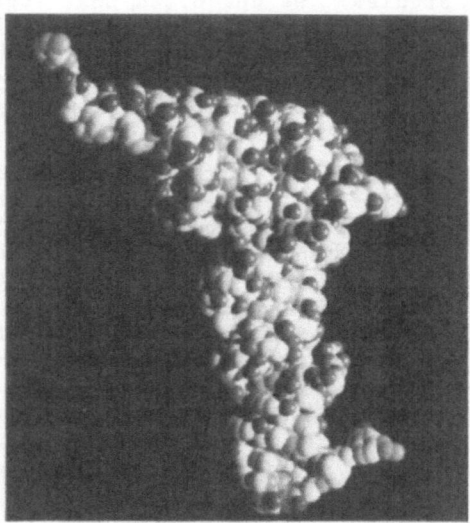

Fig. 5. Orientation of the NAK modification at the anticodon of
tRNA. The three-dimensional structure of yeast tRNA[Phe] was modified by
computer graphics replacement of Gm–34 with NAK-modified cmo[5]U–34. The N_3
was placed in the desired location by rotation about the C–5 oxygen atom.

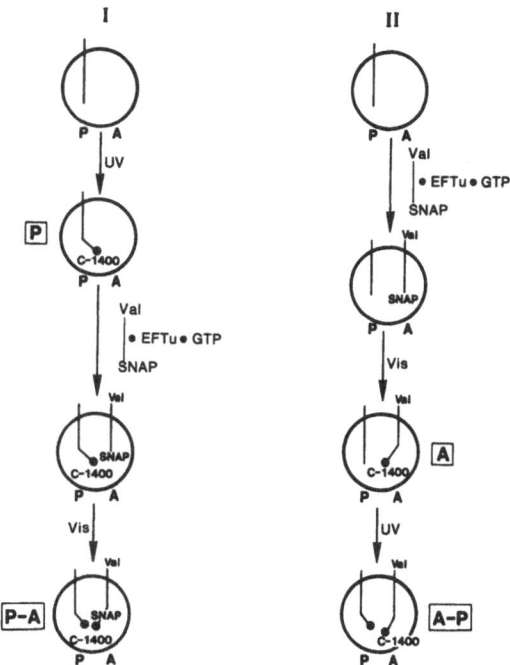

Fig. 6. Procedure for crosslinking both A and P bound tRNAs to the same 16S rRNA molecule. Protocol I yields P and P→A products, while protocol II generates A and A→P crosslinks.

Fig. 6. In scheme I, it was not possible to bind both tRNAs to the ribosome before crosslinking, as done in scheme II, since the near UV light used for P site crosslinking also induced crosslinking by the SNAP probe. Thus, P site crosslinking was done before A site binding. The protocol for complexes P and A is identical to that already used, where it was established that C-1400 was the sole crosslinking site. The purpose of the experiment was, therefore, to determine if a *second* crosslink to the rRNA could form and to which nucleotide. As shown in Fig. 7, a slow-moving oligomer band only appeared when both irradiation steps were included. This slower band was shown to be the expected trimer of two tRNA anticodon oligomers linked to a single rRNA oligomer. This was done by first cleavage of the cyclobutane link with UV light (Ofengand & Liou, 1980), and then mercaptan cleavage of the SNAP-linked oligomer, with electrophoretic isolation of the dimer and monomer fragments at each step. The rRNA fragment must include C-1400, and as it migrated like the authentic rRNA 9mer, UACACACCG, it is unlikely that the 3' oligomer, m^4CmCCG, was attached. The exact nucleotide crosslinked when the trimers are formed is currently under investigation.

The data of Fig. 7 also show that formation of the first crosslink partially inhibits formation of the second, since only 20-30% of the expected amount of trimer was found independent of the order of crosslinking. This result suggests that the first crosslink to C-1400, whether from the A or P site, perturbs the tertiary structure of rRNA in its vicinity. We cannot yet say whether the decrease in the second crosslink is due to a lessened availability of C-1400 or to a larger conformational change such that another nucleotide is being crosslinked. The sequence studies now in progress should answer this question.

0 -

TRIMER -

DIMER -

XC -

Amount of Crosslinked Product

	P	A	P→A	A→P
A site XL	–	7.4	6.6	7.7
Dimer	900	292	1025	1228
Trimer found	–	–	31.6	33.4
% of expected	–	–	21-32	19-29

Fig. 7. Quantitive analysis of the amount of trimer formed. A site XL was determined as pmole [3H]Val-tRNA^{SNAP} crosslinked per pmol 16S rRNA after isolation of the rRNA-tRNA complexes by SDS sucrose gradient centrifugation. RNase T$_1$ digestion to completion was used to generate oligonucleotides which were separated by 7 M urea gel electrophoresis after 5'-[32P]-labeling. Dimer and trimer bands were quantitated by excision from the gel (upper left). Values are given as [32P] cpm x 0.01 and divided by 3 for the trimers and 2 for the dimers. The expected amount was calculated assuming independent crosslinking, and the same fraction (40-60%) of active ribosomes, at each site. The percent of expected varies linearly with the fraction of active ribosomes. The values shown correspond to 40-60% activity.

CONCLUSION

Evidence for the presence of the conserved sequence 1392-1408, and in particular of C-1400, at the decoding site of the ribosome is now quite strong. So far however, intriguing as it may be, this result remains a phenomenological observation. Although speculations on possible functional roles for this close contact have been proposed previously (Ofengand et al., 1979; Ofengand and Liou, 1981; Ofengand et al., 1982) no evidence for any essential function in the decoding process has as yet been obtained. Hopefully, function will follow structure, so that as more is learned about the three-dimensional organization of rRNA around the decoding site, plausible and testable hypotheses about function may emerge.

ACKNOWLEDGEMENTS

We thank Bernard Brooks and Richard Feldmann of the Division of Computer Research and Technology, N.I.H. for the generous gift of their time and skill in the construction of Fig. 5. We thank Christian Oste for performing the experiments summarized in Table 2.

ABBREVIATIONS

cmo^{5}U, 5-(carboxymethoxy)uridine; m$_2^6$A, $\underline{N}^6$-dimethyladenosine; m^{7}G, $\underline{N}^7$-methylguanosine; DNP, 2,4,-dinitrophenyl.

REFERENCES

Anderson, S., Bankier, A.T., Barrell, B.G., deBruijn, M.H.L., Coulson, A.R., Drouin, J., Eperon, I.C., Nierlich, D.P., Roe, B. A., Sanger, F., Schreier, P.M.H., Smith, A.J.H., Staden, R., and Young, I.G., 1981, *Nature* (London) **290**:457-464.

Anderson, S., de Bruijn, M.H.L., Coulson, A.R., Eperon, I.C., Sanger, F., and Young, I.G., 1982, *J. Mol. Biol.* **156**:683-717.

Arnott, S., Chandrasekaran, R., and Selsing, E., 1975, *in: Structure and Conformation of Nucleic Acids and Protein-Nucleic Acid Interactions* (Sundaralingam, M., Rao, S.T., eds.), pp. 577-596, University Park Press, Baltimore, MD.

Atmadja, J., Brimacombe, R., and Maden, B.E.H., 1984, *Nucleic Acids Res.* **12**:2649-2667.

Baer, R., and Dubin, D.T., 1980, *Nucleic Acids Res.* **8**:4927-4941.

Balch, W.E., Fox, G.E., Magrum, L.J., Woese, C.R., and Wolfe, R.S., 1979, *Microbiol. Rev.* **43**:260-296.

Barta, A., Steiner, G., Brosius, J., Noller, H.F., and Kuechler, E., 1984, *Proc. Natl. Acad. Sci. USA* **81**, 3607-3611.

Bibb, M.J., Van Etten, R.A., Wright, C.T., Walberg, M.W., and Clayton, D.A., 1981, *Cell* **26**:167-180.

Brimacombe, R., Maly, P., and Zwieb, C., 1983, *Progr. Nucl. Acid. Res. Mol. Biol.* **28**:1-48.

Brosius, J., Palmer, M.L., Kennedy, J.P., and Noller, H.F., 1978, *Proc. Natl. Acad. Sci. USA* **75**:4801-4805.

Carbon, P., Ehresmann, C., Ehresmann, B., and Ebel, J.-P., 1979, *Eur. J. Biochem.* **100**:399-410.

Carbon, P., Ebel, J.P., and Ehresmann, C., 1981, *Nucleic Acids Res.* **9**:2325-2333.

Chan, Y-L., Gutell, R., Noller, H.F., and Wool, I.G., 1984, *J. Biol. Chem.* **259**:224-230.

Chao, S., Sederoff, R., and Levings III, C.S., 1984, *Nucleic Acids Res.* **12**:6629-6644

Chiaruttini, C., Expert-Bezancon, A., Hayes, D., and Ehresmann, B., 1982, *Nucleic Acids Res.* **10**:7657-7676.

Ciesiolka, J., Gornicki, P., and Ofengand, J., 1985a, *Biochemistry*, (in press).

Ciesiolka, J., Nurse, K., Klein, J., and Ofengand, J., 1985b, *Biochemistry* **24**:3233-3239.

Clary, D.O., and Wolstenholme, D.R., 1985, *Nucleic Acids Res.* **13**:4029-4045.

Connaughton, J.F., Rairkar, A., Lockard, R.E., and Kumar, A., 1984, *Nucleic Acids Res.* **12**:4731-4745.

de la Cruz, V.F., Lake, J.A., Simpson, A.M., and Simpson, L., 1985, *Proc. Natl. Acad. Sci. USA* **82**:1401-1405.

Douthwaite, S., Christensen, A., and Garrett, R.A., 1983, *J. Mol. Biol.* **169**:249-279.

Dron, M., Rahire, M., and Rochaix, J.-D., 1982, *Nucleic Acids Res.* **10**:7609-7620.

Dubin, D.T., and HsuChen, C.C., 1983, *Plasmid* **9**:307-320.

Ehresmann, C., Ehresmann, B., Millon, R., Ebel, J.-P., Nurse, K., and Ofengand, J., 1984, *Biochemistry* **23**:429-437.

Ehresmann, C., and Ofengand, J., 1984, *Biochemistry* **23**:438-445.

Eperon, I.C., Anderson, S., and Nierlich, D.P., 1980, *Nature* **286**:460-467.

Fuller, W., and Hodgson, A., 1967, *Nature* **215**:817-821.

Gold, L., Pribnow, D., Schneider, T., Shinedling, S., Singer, B.S., and Stormo, G., 1981, *Ann. Rev. Microbiol.*, **35**:365-403.

Gornicki, P., Nurse, K., Helmann, W., Boublik, M., and Ofengand, J., 1984, *J. Biol. Chem.* **259**:10493-10498.

Gornicki, P., Ciesiolka, J., and Ofengand, J., 1985, *Biochemistry* (in press)

Graf, L., Roux, E., and Stutz, E., 1982, *Nucleic Acids Res.* **10**:6369-6381.

Gupta, R., Lanter, J.M., and Woese, C.R., 1983, *Science* 221:656-659

Herr, W., Chapman, N.M., and Noller, H.F., 1979, *J. Mol. Biol.* 130:433-449.

Hogan, J.J., Gutell, R.R., and Noller, H.F., 1984, *Biochemistry* 23:3322-3330.

Iwami, M., Muto, A., Yamao, F., and Osawa, S., 1984, *Mol. Gen. Genet.* 196:317-322.

Jarsch, M., and Böck, A., 1985, *Syst. Appl. Microbiol.* (in press).

Jordan, B.R., Latil-Damotte, M., and Jourdan, R., 1980, *FEBS Lett.* 117:227-231.

Keren-Zur, M., Boublik, M., and Ofengand, J., 1979, *Proc. Natl. Acad. Sci. USA* 76:1054-1058.

Kobayashi, M., Seki, T., Yaginuma, K., and Koiko, K., 1981, *Gene* 16:297-307.

Kop, J., Kopylov, A.M., Magrum, L., Siegel, R., Gupta, R., Woese, C.R., and Noller, H.F., 1984, *J. Biol. Chem.* 259:15287-15293.

Köchel, H.G., and Küntzel, H., 1981, *Nucleic Acids Res.* 9:5689-5696.

Leffers, H., and Garrett, R., 1984, *EMBO J.* 3:1613-1619.

Li, M., Tzagoloff, A., Underbrink-Lyon, K., and Martin, N.C., 1982, *J. Biol. Chem.* 257:5921-5928.

Lührmann, R., Stöffler-Meilicke, M., and Stöffler, G., 1981, *Mol. Gen. Genet.* 182:369-376.

Magrum, L.J., Luehrsen, K.R., and Woese, C.R., 1978, *J. Mol. Evol.* 11:1-8.

Mankin, A.S., Kopylov, A.M., and Bogdanov, A.A., 1981, *FEBS Lett.* 134:11-14.

McCarroll, R., Olsen, G.J., Stahl, Y.D., Woese, C.R., and Sogin, M.L., 1983, *Biochemistry* 22:5858-5868.

Meier, N., and Wagner, R., 1984, *Nucleic Acids Res.* 12:1473-1487.

Meier, N., and Wagner, R., 1985, *Eur. J. Biochem.* 146:83-87.

Messing, J., Carlson, J., Hagen, G., Rubenstein, I., and Oleson, A., 1984, *DNA* 3:31-40.

Moazed, D., Stern, S., and Noller, H.F., 1985, (personal communication).

Mochalova, L.V., Shatsky, I.N., Bogdanov, A.A., and Vasiliev, V.D., 1982, *J. Mol. Biol.* 159:637-650.

Nelles, L., Fang, B.-L., Volckaert, G., and Vanden, R., 1984, *Nucleic Acids Res.* 12:8749-8768.

Noller, H.F., 1984, *Ann. Rev. Biochem.* 53:119-162.

Ofengand, J., Liou, R., Kohut, III, J., Schwartz, I., and Zimmermann, A., 1979, *Biochemistry* 18:4322-4332.

Ofengand, J., 1980, in: *Ribosomes: Structure, Function, and Genetics* (Chambliss, G., Craven, G., Davies, J., Davis, K., Kahan, L. and Nomura, M., eds.), pp. 497-529, University Park Press, Baltimore, MD.

Ofengand J., Gornicki, P., Nurse, K., and Boublik, M., 1984, On the structural organization of the tRNA-ribosome complex, in: *The Translational Step and Its Control* (B.F.C. Clark and H.U. Petersen, eds.) pp. 293-315, Munksgaard, Copenhagen.

Ofengand, J., Ciesiolka, J., Gornicki, P., and Nurse, K., 1985, in: Structure, Function and Genetics of Ribosomes (B. Hardesty and G. Kramer, eds.), Springer-Verlag, New York (in press).

Ofengand, J., and Liou, R., 1980, *Biochemistry* 19:4814-4822.

Ofengand, J., and Liou, R., 1981, *Biochemistry* 20:552-559.

Ofengand, J., Gornicki, P., Chakraburtty, K., and Nurse, K., 1982, *Proc. Natl. Acad. Sci. USA* 79:2817-2821.

Olson, H.M., and Glitz, D.G., 1979, *Proc. Natl. Acad. Sci. USA* 76:3769-3773.

Politz, S.M., and Glitz, D.G., 1977, *Proc. Natl. Acad. Sci. USA* 74:1468-1472.

Prince, J.B., Taylor, B.H., Thurlow, D.L., Ofengand, J., and Zimmermann, R.A., 1982, *Proc. Natl. Acad. Sci. USA* 79:5450-5454.

Raynal, F., Michot, B., and Bachellerie, J.-P., 1984, *FEBS Lett.* 167: 263-268.

Roe, B.A., Ma, D-P., Wilson, R.K., and Wong, J.F-H., 1985, *J. Biol. Chem.* **260**:9759-9774.

Rubtsov, P.M., Musakhanov, M.M., Zakharyev, V.M., Krayev, A.S., Skryabin, K.G., and Bayev, A.A., 1980, *Nucleic Acids Research* **8**:5779-5794.

Salim, M.,, and Maden, B.E.H., 1981, *Nature* **291**:205-208.

Samols, D.R., Hagenbuchle, O., and Gage, L.P., 1979, *Nucleic Acids Res.* **7**:1109-1119.

Sanger, F., Air, G.M., Barrell, B.G., Brown, N.L., Coulson, A.R., Fiddes, J.C., Hutchison III, C.A., Slocombe, P.M., and Smith, M., 1977, *Nature* **265**:687-695.

Schwarz, Z., and Kössel, H., 1980, *Nature* **283**:739-742.

Schwartz, I., and Ofengand, J., 1978, *Biochemistry* **17**:2524-2530.

Seewaldt, E., and Stackenbrandt, E., 1982, *Nature* (London) **295**:618-620.

Seilhamer, J.J., Olsen, G.J., and Cummings, D.J., 1984, *J. Biol. Chem.* **259**:5167-5172.

Shatsky, I.N., Mochalova, L.V., Kojouharova, M.S., Bogdanov, A.A., and Vasiliev, V.D., 1979, *J. Mol. Biol.* **133**:501-515.

Sloof, P., Vanden Burg, J., Voogd, A., Benne, R., Agostinelli, M., Borst, P., Gutell, R., and Noller, H. (1985) *Nucleic Acids Res.* **13**:4171-4190.

Sor, F., and Fukuhara, H., 1980, *C.R. Acad. Sci.* Paris, **291**:933-936.

Spangler, E.A., and Blackburn, E.H., 1985, *J. Biol. Chem.* **260**:6334-6340.

Spencer, D.F., Schnare, M.N., and Gray, M.W., 1984, *Proc. Natl. Acad. Sci. USA* **81**:493-497.

Spirin, A.S., and Lim, V.I., 1985, *in*: Structure, Function and Genetics of Ribosomes (B. Hardesty and G. Kramer, eds.) Springer-Verlag, New York (in press).

Steiner, G., Luhrmann, R., and Kuechler, E., 1984, *Nucleic Acids Res.* **12**:8181-8191.

Stöffler-Meilicke, M., Stöffler, G., Odom, O.W., Zinn, A., Kramer, G., and Hardesty, B., 1981, *Proc. Natl. Acad. Sci. USA* **78**:5538-5542.

Stöffler, G., and Stöffler-Meilicke, M., 1981, *International Cell Biology* 1980/81 (H.G. Schweiger, ed.), Springer Verlag, Berlin, Heidelberg, New York, 93-102.

Stöffler, G., and Stöffler-Meilicke, M., 1984, *Ann. Rev. Biophys. Bioeng.* **13**:303-330.

Sundaralingam, M., Brennan, T., Yathindra, N., and Ichikawa, T., 1975, in: *Structure and Conformation of Nucleic Acids and Protein-Nucleic Acid Interactions* (Sundaralingam, M., Rao, S.T., eds.), pp. 101-105, University Park Press, Baltimore, MD.

Takaiwa, F., Oono, K., and Sugiura, M., 1984, *Nucleic Acids Res.* **12**:5441-5448.

Tohdoh, N., and Sugiura, M., 1982, *Gene* **17**:213-218.

Tomioka, N., and Sugiura, M., 1983, *Mol. Gen. Genet.* **191**:46-50.

Torczynski, Bollon, A.P., and Fuke, M., 1983, *Nucleic Acids Res.* **11**:4879-4890.

Trempe, M.R., Ohgi, K., and Glitz, D.G., 1982, *J. Biol. Chem.* **257**:9822-9829.

Van Stolk, B.J., and Noller, H.F., 1984, *J. Mol. Biol.* **180**:151-177.

Verschoor, A., Frank, J., Radermacher, M., Wagenknecht, T., and Boublik, M., 1984, *J. Mol. Biol.* **178**:677-698.

Wagner, R., Gassen, H.G., Ehresmann, C., Stiegler, P., and Ebel, J.-P., 1976, *FEBS Lett.* **67**:312-315.

Woese, C.R., Maniloff, J., and Zablen, L.B., 1980, *Proc. Natl. Acad. Sci. USA* **77**:494-498.

Yang, D., Oyaizu, Y., Oyaizu, H., Olsen, G.J., and Woese, C.R., 1985, *Proc. Natl. Acad. Sci. USA* **82**:4443-4447.

Yathindra, N., and Sundaralingam, M., 1975, in: *Structure and Conformation of Nucleic Acids and Protein-Nucleic Acid Interactions* (Sundaralingam, M., Rao, S.T., eds.), pp. 649-676, University Park Press, Baltimore, MD.

DOUBLE STRANDED RNA IN THE DECODING OF THE mRNA

BY THE BACTERIAL RIBOSOME

V. Eckert, A. Lang, A. Kyriatsoulis and
H. G. Gassen

Institut für Organische Chemie und Biochemie
Technische Hochschule Darmstadt
D-6100 Darmstadt, Petersenstraße 22

SUMMARY

In the process of the decoding of the mRNA by the 70 S
ribosome a hexanucleotide double stranded RNA is formed by the
mRNA and the aminoacyl- and the peptidyl-tRNA. Both, in initiation or termination, however, only either the formylmethionyl-
or the peptidyl-tRNA are present on the ribosome. According to
a theory as outlined by Shine and Dalgarno the 3' end of the
16 S RNA can form a short double strand with the mRNA. In order
to prove these assumptions we prepared oligonucleotides containing the Shine-Dalgarno region and the initiation codon.
These were examined in their capabilities to direct the binding
of fMet-tRNA to the 30 S ribosome. The data indicate that the
double strand formation between 16 S RNA and mRNA acts as a
signal to stimulate the binding of the initiator-tRNA to the
ribosomal subunit.

To elucidate details of the termination reaction the binding constants between the release factors and the codons UAA,
UAG and UGA have been determined. Furthermore, we crosslinked
the codons to the 70 S ribosomal termination complex. The
nucleotides were crosslinked to both the proteins and the 16 S
RNA. The data suggest that the termination codons interact with
the 16 S RNA. A model, which explains the function of the 16 S
RNA in initiation and elongation is proposed.

INTRODUCTION

Ribosome-dependent protein synthesis in bacteria represents
a complex sequence of reactions involving more than 150 compounds. The ribosome functions as an enzyme which catalyzes
peptide bond formation. In contrast to most known enzymes the
ribosome is programmed by the mRNA, the short-lived transcript
of DNA. The active center of the ribosome is modulated with
respect to its substrate specificity by a defined codon, such
that it can differentiate between at least 20 aminoacyl-tRNAs
with high selectivity. If the ribosome is viewed as a programmed
enzyme the following mechanistic problems must be considered

- How is the start signal within the mRNA recognized and the
  nucleotide text framed into coding units;
- How is the mRNA as an information tape shifted by one trinucleotide per peptide bond formed;
- How is the release of the nascent peptide chain from the ribosomal complex triggered by the termination codon?

In the following we point out the dominant role of the
interplay between single-stranded and double-stranded RNA for
the initiation, elongation and termination reaction of ribosome-dependent protein synthesis. (Fig. 1).

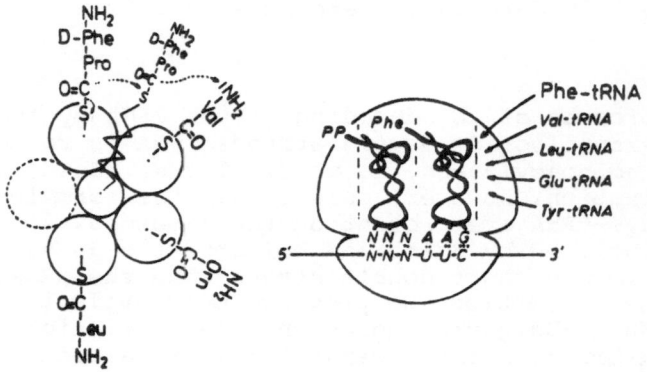

Fig. 1. Comparison of multienzyme complex displaying substrate
specificity and the ribosome as a programmed enzyme. - Left: a
multienzyme complex which catalyzes the synthesis of peptide
antibiotics. In order to explain the role of the pantetheine
carrier the peptide synthetase is drawn to emphasize the similarities with the fatty acid synthetase. Right: the substrate
specificity of the ribosomal aminoacyl-tRNA binding site is
governed by the codon of the mRNA.

SELECTION OF THE START SIGNAL BY A RNA-RNA HYBRID

In order to elucidate how ribosomes are able to distinguish
real initiator triplets from the many internal AUG and GUG
codons in mRNA much attention was given to a comparison of the
base sequences of ribosome binding sites, particularly those
upstream from the AUG. Shine and Dalgarno sequenced the 3' end
of 16 S RNA and found it to be, in part base complementary to
the polypurine tracts which precede mRNA initiator codons.
Therefore they hypothesized that the formation of Watson-Crick
base pairs between the mRNA and the 16 S RNA allows the ribosome to differentiate between initiator and internal AUG or
GUG codons (Shine and Dalgarno, 1974; Steitz and Jakes, 1975).

Now this hypothesis is supported by extensive biochemical evidence and is augmented by an impressive list of approximately 100 mRNA binding sites (Taniguchi and Weissmann, 1978; Backendorf et al., 1980; Steitz, 1980).

We have shown before that the initiator-tRNA interacts with more than tree nucleotides in the decoding process. In previous experiments the pentanucleotide UAUGA was the best affector for the binding of the fMet-tRNA to the 30 S ribosome. Furthermore oligonucleotides such as GAGG which are complementary to the CUCC-region of the 16 S RNA stimulate the binding of the initiator-tRNA to the 30 S ribosomal subunit (Schmitt et al., 1980).

Thus the formation of two short RNA-RNA hybrids may act as positive control signal for the correct framing of the mRNA within the 30 S initiation complex. In order to substantiate this concept we prapared with the aid of RNA ligase a set of oligonucleotides containing the GAGG and the AUG sequences separated by a defined spacer (Fig. 2).

Synthesis of the Pentadecanucleotide

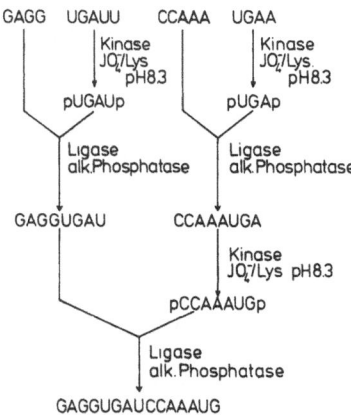

Fig. 2. Synthetic scheme for large scale synthesis of oligonucleotides.

Kinase and ligase reactions were performed in 50 mM Tris-HCl, pH 7,5.10 mM MgCl$_2$, 10 mM mercaptoethanol,10 µg/ml bovine serum albumin, 0,1 mM EDTA at 37° C for analytical scale reactions and at 18 - 20° C overnight for preparative reactions.

Each oligonucleotide was purified by preparative HPLC using a C18-reversed phase column. The compounds were tested in their capacity to stimulate the binding of the fMet-tRNA to the 30 S ribosomal subunit in the presence and absence of the initiation factor 2 (Fig. 3).

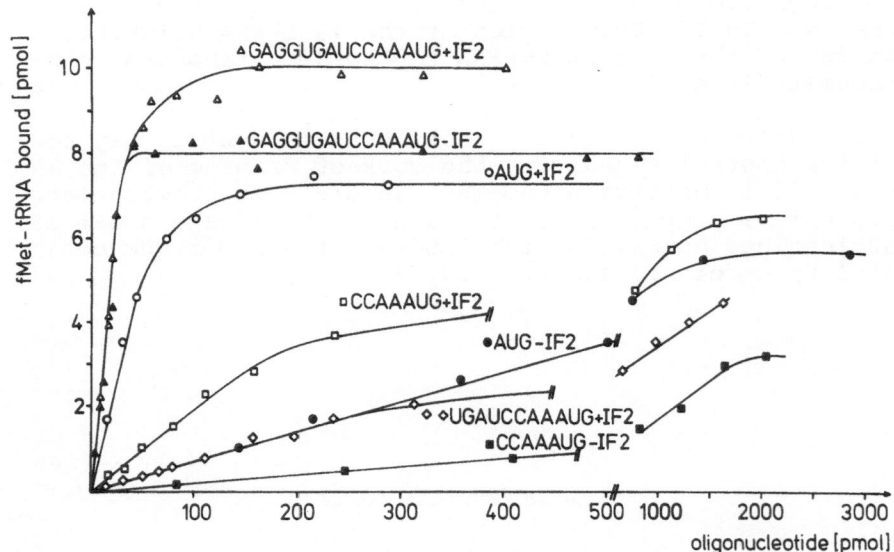

Fig. 3. Stimulation of initiator-tRNA binding to the 30 S ribosome by AUG-containing oligonucleotides.

The oligonucleotide directed binding of [3]H fMet-tRNA to 30 S subunits was measured by adsorption of the complexes to nitrocellulose filters. The reaction mixture contained in a final volume of 100 µl: 50 mM Tris-HCl, pH 7,5, 100 mM NH$_4$Cl, 15 mM Mg(OAc)$_2$, 1 mM GTP, 1 mM DTE, 20 pMol 30 S subunits, 25 pMol [3]H fMet-tRNA saturating amounts of IF-2 (20 µg) and varying amounts of oligonucleotides.

The pentadecanucleotide GAGGUGAUCCAAAUG which is complementary to the 16 S RNA showed in the presence of the initiation factor 2 a titration-type stimulation of the binding of the tRNA to the ribosome. Equilibrium was reached with a two to threefold excess of oligonucleotide over ribosomes. The increased plateau value in the presence of IF-2 points towards a higher amount of active ribosomes. With AUG the initial slope and the plateau value are lower as compared to the pentadecanucleotide. In the absence of the initiation factor a high concentration of AUG is needed to achieve the binding of the fMet-tRNA.

Oligonucleotides lacking the GAGG sequence - i. e. UGAUCCAAAUG and CCAAAUG show a reduced codon efficiency as compared to AUG. This may indicate that there is no base-base interaction between the spacing nucleotides and the 16 S RNA. Possibly the nucleotide chain preceeding the AUG places the initiator codon in a position outside the 30 S decoding site.

With a different set of oligonucleotides we investigated the effect of the length of the spacer between the GAGG and the AUG (Table 1). Uridylates of increasing chain length were placed between the two binding signals. The compounds were prepared as described.

Table 1. Binding of fMet-tRNA to 30-S ribosomes in the presence of initiation factor 2 as stimulated by AUG-containing oligonucleotides

oligonucleotide	amount of tRNA bound [pMol]	saturation %
AUG	5.1	34
$GAGGU_2AUGU_4$	6.0	40
$GAGGU_5AUGU_3$	10.2	68
$GAGGU_7AUGU_4$	13.0	87
$GAGGU_{11}AUGU_3$	8.6	58

In each experiment 15 pmoles of 30-S ribosomal subunits were used.

From the mRNA analogues examined it is evident that a spacer of seven nucleotides shows the best stimulatory activity. Both shorter and longer spacers reduce the effect of the oligo-nucleotides. When the uridylates were replaced by adenylates, the efficiency of the analogues were lower (data not shown). This may indicate that a flexible spacer is preferred over a rigid one.

From the data presented one may conclude the following: The correct association of the mRNA to the 30 S ribosomal sub-unit is stabilized by two RNA-RNA hybrids; 16 S RNA and pre-cistronic sequence and initiation codon of the mRNA and anti-codon of the fMet-tRNA. Critical for an efficient complex formation is the length of the spacer but not its base composition (Tessier et al., 1984).

THE INTERPLAY BETWEEN A TRINUCLEOTIDE AND A HEXANUCLEOTIDE RNA-RNA HYBRID EXPLAINS THE TRANSLOCATION REACTION IN THE ELONGATION CYCLE

The translocation of the mRNA in relation to the ribosome during peptide synthesis represents an example for a mechano-chemical reaction in which the chemical bond energy of GTP is transformed into coordinated motion. The elementary steps of the translocation reaction have been intensively investigated and described in detail (Gassen, 1984).

The translocation can be explained simply by binding equilibria between the tRNA, the mRNA and their binding sites on the ribosome. The influence of the cognate tRNA on the stability of the mRNA ribosome complex was examined using $AUGU_3$ as a model system (Table 2). The respective binding constants have been determined by equilibrium dialysis (Holschuh and Gassen, 1981).

Table 2. Effect of the chain length and of cognate tRNAs on the formation of the 7OS-AUGU$_3$ complex. The association constants obtained are depicted in the elongation cycle in Fig. 4.

Complex	tRNA	K_{Ass} M^{-1}	Number of binding sites
$7OS \cdot AUGU_3$	–	$6.8 \cdot 10^5$	1.3
$7OS \cdot AUGU_6$	–	$8.7 \cdot 10^5$	1.3
$7OS \cdot AUGU_{13}$	–	$9.3 \cdot 10^5$	1.4
$7OS \cdot AUGU_3$	$tRNA_f^{Met}$	$4.6 \cdot 10^7$	1.0
$7OS \cdot AUGU_3$	$tRNA_m^{Met}$	$3.9 \cdot 10^6$	1.0
$7OS \cdot AUGU_3$	$tRNA^{Phe}$	$4.1 \cdot 10^6$	1.2
$7OS \cdot AUGU_3$	$tRNA_f^{Met}$, $tRNA^{Phe}$	$2.2 \cdot 10^8$	1.1
$7OS \cdot AUGU_6$	$tRNA_f^{Met}$, $tRNA^{Phe}$	$2.1 \cdot 10^8$	1.0
$7OS \cdot AUGU_{13}$	$tRNA_f^{Met}$, $tRNA^{Phe}$	$2.2 \cdot 10^8$	1.1
$7OS \cdot AUGU_3$	$fMet-tRNA_f^{Met}$	$1.1 \cdot 10^8$	1.1
$7OS \cdot AUGU_3$	$AcMet-tRNA_m^{Met}$	$5.3 \cdot 10^7$	1.1

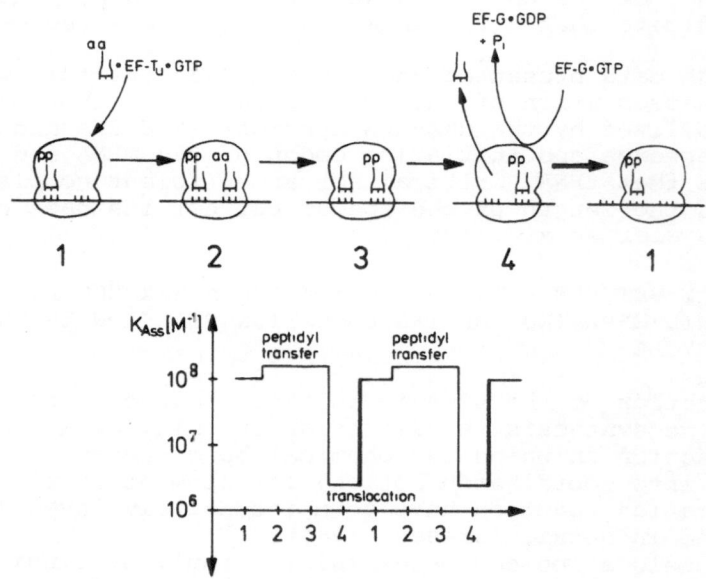

Fig. 4. The mechanism of the translocation reaction. In the upper part of the figure the four steps in translocation are depicted. The drop in the association constants reflect step 4 to 1 namely the translocation of the PP-tRNA mRNA complex versus the ribosome by a codon length.

In the one-tRNA-state a trinucleotide RNA-RNA hybrid is formed between the mRNA and the tRNA. In the two-tRNA-state - hexanucleotide hybrid - there is an interaction between both tRNAs and the ribosome and as a result, the mRNA is tightly bound to the ribosome. In this type of complex the peptidyl moiety is transferred from the P-site located tRNA to the AA-tRNA in the A-site. Mediated by the elongation factor G and GTP the deacylated tRNA dissociates from the ribosome. In the resulting one-tRNA-state the interaction between mRNA and ribosome is loosened and the mRNA-PP-tRNA can be translocated from the aminoacyl site to the peptidyl site. This concept of the translocation reaction is outlined in Figure 4.

IN THE TERMINATION REACTION A COMPLEX IS FORMED BETWEEN THE TERMINATION CODON AND THE 3' END OF THE 16 S RNA

Peptide chain termination is directed by one out of three specific codons, UAA, UAG and UGA and requires soluble protein factors called termination or release factors (Fig. 5).

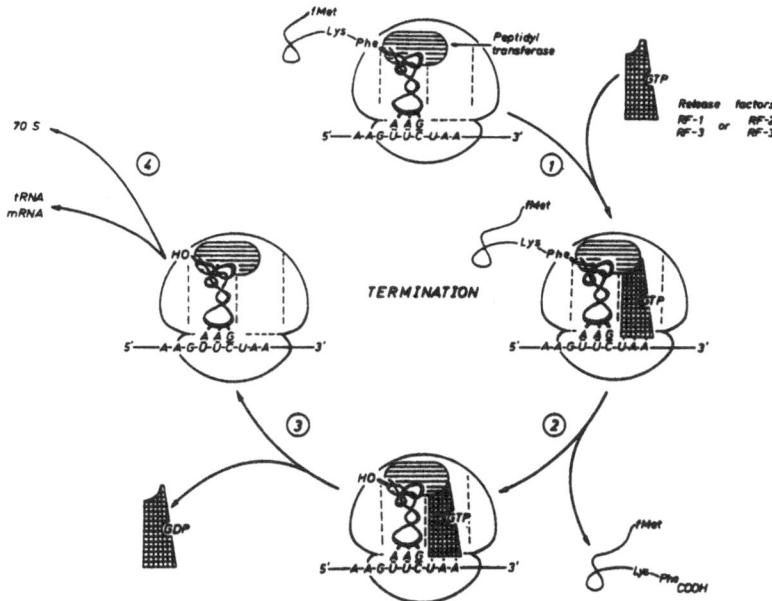

Fig. 5. The mechanism of the termination reaction in bacterial protein biosynthesis.

Very little is known at the present about the molecular mechanism of peptide chain termination and it is even questioned if there are two or three release factors. (Caskey, 1980) An interesting problem to tackle is whether the release factors act as "tRNA-like" protein molecules or whether the recognition of termination codons occurs via a RNA-hybrid between the mRNA and 16 S RNA as postulated by Shine and Dalgarno (Table 3).

To find an answer to these questions we measured the binding of 32P -labelled termination codons to the purified release factors and to factor-ribosome complexes by equilibrium dialysis. The codons show only a low affinity to the protein itself. In the presence of the ribosome, however, the binding constant is increased to 10^5 M^{-1} with specificity between the respective factors and their corresponding codons. In the presence of RF-2 the association constant for the specific codon UGA is raised by one order of magnitude to $2,2 \times 10^5$ M^{-1} as compared to the wrong termination codon UAG. Next we located the binding site of RF-2 on the ribosome (Fig. 6). Binding of the factor was examined with ribosomes which were programmed with $AUGU_7$ and contained fMet-tRNA. The elongation codon U_3 in the A-site lowers the UGA-dependent binding of the release factor. Binding is further reduced if Phe-tRNA occupies the A-site of the ribosome. These data indicate, that the recognition of the release factor by the termination codon takes place at the A-site of the ribosome.

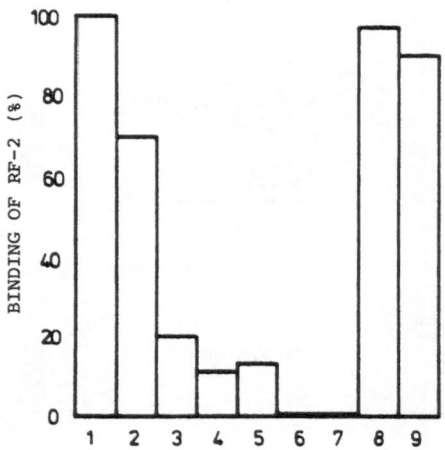

Fig. 6. Evidence for the binding of the termination factor RF-2 to the A-site of the ribosome. The bars represent the amount of RF-2 bound to the 70 S ribosome in %.

1	RF-2 + 70 S + UGAA	2	1 + fMet-tRNA
3	1 + $AUGU_7$ + fMet-tRNA	4	3 + Phe-tRNA
5	4 + EF-Tu · GDPCP	6	4 − RF-2
7	5 − RF-2	8	1 + GTP
9	1 + EF-Tu · GTP		

In order to elucidate, which ribosomal components are involved in the recognition of the termination factor 32P – labelled oligonucleotides were covalently crosslinked to factor-ribosome-complexes by UV-irradiation. The irradiation conditions were selected to cause less than 5 % inactivation of the individual complexes.

Two-dimensional gel electrophoresis was used for the identification of the crosslinked ribosomal proteins. Besides the termination factor the proteins L2, L7, L10, L12 and L20 of the 50 S ribosomal subunit and to a lesser extend S6 and S18 of the small subunit became labelled. This suggests a direct contact between these proteins and the termination codon. In the absence of ribosomes the termination factor could not be crosslinked to the trinucleotide. These findings support the data from the binding experiments, and further demonstrate that the ribosome is required for a factor-oligonucleotide interaction.

Table 3. Association constants for the binding of the termination codons to the RF-2·70 S ribosome complex.

Codon	association constants (M^{-1})		
	K_{CR}	K_{CF-2}	K_{CRF-2}
UGA	$2.4 \cdot 10^4$	10^4	$2.2 \cdot 10^5$
UGAA	$2.7 \cdot 10^4$	10^4	$2.8 \cdot 10^5$
UAA	$1.9 \cdot 10^4$	-	$0.8 \cdot 10^5$
UAG	$1.6 \cdot 10^4$	-	$1.4 \cdot 10^5$

All association constants have been determined from Scatchard plots, using ^{32}P-labelled oligonucleotides. CR = codon·ribose complex, CF-2 = codon·factor complex and CRF-2 = codon·factor ribosome·complex.

Ribosomal RNA was extracted from irradiated particles with phenol/water and the nucleic acids were seperated on a 4 % polyacrylamide gel. The crosslinked RNA was identified by autoradiography. The termination codon UGA became crosslinked only to the 16 S RNA. In the absence of the factor or in the presence of the wrong codon UAG no labelled RNA band could be detected. The crosslinking of 16 S RNA is taken as evidence, that codon and RNA come into direct contact within the termination complex.

Our present data do not allow to determine the part of the 16 S RNA to which the UGA became attached. Caskey et al. have shown that the release factor binding to the ribosome requires an intact 3' terminus of the 16 S RNA. Cleavage of 49 nucleotides off the 3' end prevents the formation of the termination complex. Therefore it is assumed that the UGA binds to the 3' end of the 16 S RNA (Tate et al., 1983, Tate et al., 1983) (compare Fig. 7).

DISCUSSION

As outlined in the introduction, the ribosome is viewed as a mRNA-programmed particle. Since the ribosome itself represents a protein-RNA-complex one is tempted to speculate on the function of the nucleic acids within this particle. At one hand the RNA acts as scaffold for the ribosomal proteins, on the other side parts of the RNA may function in the binding of the various components of the protein synthesizing machinery. The mRNA represents an ideal candidate to study molecular details of RNA-RNA hybrid formation. Structural details of such short RNA-RNA double strands may be derived from the known structure of a codon-anticodon-complex.

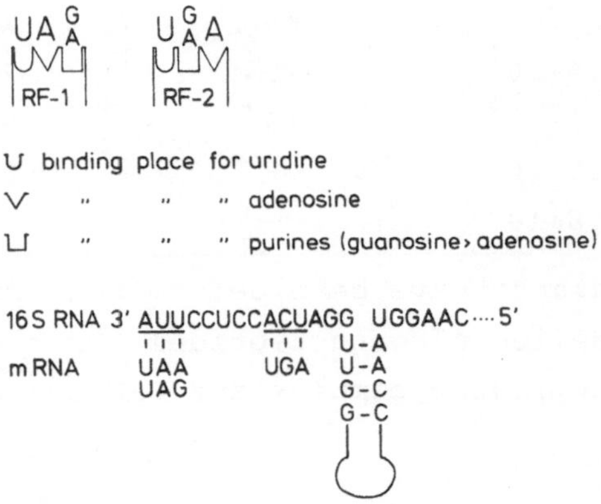

Fig. 7. Interaction between the termination codons and the 16 S RNA in the termination complex.
Both factors interact with two codons, displaying a wobble in either the 3' end (RF-1) or the middle position (RF-2). The termination codons may form a RNA·RNA hybrid with two trinucleotides at the 3' end of the 16 S RNA.

The elongation phase in the protein cycle is characterized by an interplay between the one and the two-tRNA-state. (Fig. 3). During peptide bond formation, two tRNAs form a hexanucleotide double stranded RNA with the mRNA. This arrangement guarantees a stable quarternary complex AA-tRNA·PP-tRNA·mRNA·ribosome. Before translocation can occur the deacylated tRNA has to be removed from the P-site. Next the remaining PP-tRNA·mRNA complex can be moved from A- to P-site.

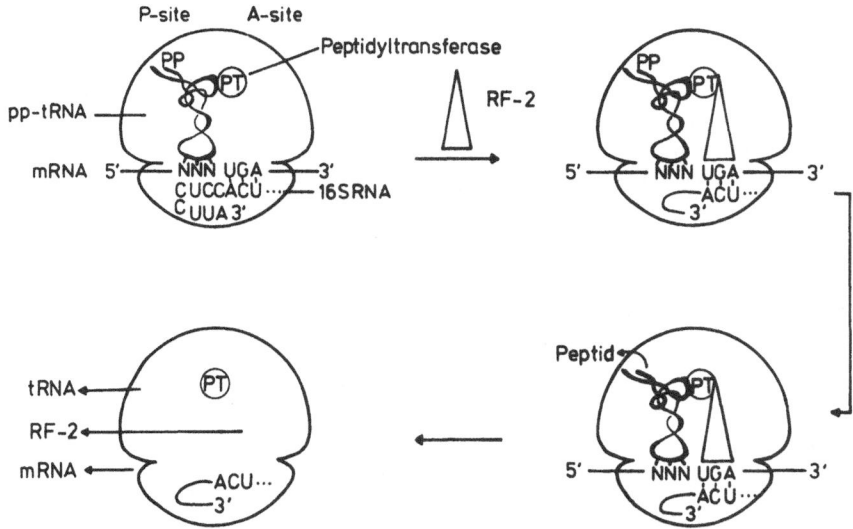

Fig. 8. Model for peptide chain termination displaying a
mRNA-16 S RNA complex. The figure shows a model for the
formation of the termination reaction. Since there is no
terminator-tRNA a RNA·RNA double strand is formed between the
termination codon and the 16 S RNA. This complex is stabilized
by the termination factor. The protein on the other hand modu-
lates the peptidyl transferase center such, that an activated
water molecule hydrolyzes the peptidyl moiety off the tRNA.
The dissociation of the deacylated tRNA destabilizes the 70 S
complex and the mRNA is released.

In either the initiation or the termination reaction only
one tRNA - initiator or peptidyl-tRNA occupies the P-site of
the 70 S ribosome. At this stage the 3' end of the 16 S RNA
should act in a tRNA-like manner. If we conclude from the
codon-anticodon concept, then the mRNA sequence GAGG should
form the codon and the CCUC region of the 16 S RNA the anticodon structure. The same could be true for the termination
reaction, only that different parts of the 3' region of the
16 S RNA have to be used for double strand formation with the
termination codon now located in the P-site (Fig. 8). However,
since in initiation the interaction has to take place upstream
the P-site, yet downstream from the same center for the
termination reaction, a high degree of flexibility would be
required for the terminal 50 nucleotides of the 16-RNA. It
could be one of the functions of either the initiation or the
termination factors to adjust the conformation of the 16 S
RNA such, that it can form the two RNA hybrids in question
(Van Duin et al., 1984).

ACKNOWLEDGEMENT

The work reported was supported by a grant of the Deutsche
Forschungsgemeinschaft.

The help of Mrs. Christine Egner in preparing the manuscript
is appreciated.

REFERENCES

Backendorf, C., Overbeek, G. P., Van Boom, J. H.,
     Van Der Marel, G., Veeneman, G. and Van Duin, J.,
     1980, Role of 16 S RNA in Ribosome Messenger
     Recognition,
     Eur. J. Biochem.,110: 599

Caskey, C. Th., 1980, Peptide Chain Termination,
     Trends Biochem. Sci., 5: 234

Gassen, H. G. 1982, The Bacterial Ribosome: A Programmed
     Enzyme,
     Angew. Chemie, 21: 23

Holschuh, K., Riesner, D. and Gassen, H. G., 1981, Steps
     of mRNA Translocation in Protein Biosynthesis,
     Nature, 293: 675

Holschuh, K. and Gassen, H. G., 1982, Mechanism of Trans-
     location: Binding Equilibria Between the Ribosome,
     mRNA Analoques and Cognate tRNAs,
     J. of Biol. Chem., 257: 1987

Schmitt, M., Manderschied, U., Kyriatsoulis, A.,
     Brinckmann, U. and Gassen, H. G., 1980, Tetra-
     nucleotides as Effectors for the Binding of
     Initiator tRNA to Escherichia coli Ribosomes,
     Eur. J. Biochem., 109; 291

Shine, J. and Dalgarno, L., 1974, The 3' Terminal
     Sequence of Escherichia coli 16 S Ribosomal RNA:
     Complementary to Nonsense Triplets and Ribosome
     Binding Sites,
     Proc. Natl. Acad. Sci. USA, 71: 1342

Steitz, J. A. and Jakes, K., 1975 How Ribosomes Select
     Initiator Regions in mRNA: Base Pair Formation
     Between the 3' Terminus of 16 S rRNA and the
     mRNA During Initiation of Protein Syntheses in
     Escherichia coli,
     Proc. Natl. Acad. Sci. USA, 72: 4734

Steitz, J. A., 1980, RNA-RNA Interaction During Polypeptide
     Chain Initiation, in: "Ribosomes, Structure, Function
     and Genetics",
     Craven, G. R., Davies, J., Davies, K., Kahan, L.
     and Nomura, M., ed., University Park Press, Baltimore

Taniguchi, T. and Weissmann, C., 1978, Inhibition of
     Qß RNA 70 S Ribosome Initiation Complex Formation
     by an Oligonucleotide Complementary to the 3'
     Terminal Region of Escherichia coli 16 S Ribosomal
     RNA,
     Nature, 275: 770

Tate, W. P., Ward, Ch. D., Trotman, N. A. C., Lührmann,R.
and Stöffler, G. 1983, The Shine and Dalgarno
Hypothesis for Termination: The 3´Terminus of the
16S rRNA of the Escherichia coli Ribosome can be
Modified or Base-Paired with a Complementary Oligonucleotide without Affecting Termination in Vitro,
Biochem. Int., 7: 529

Tate, W. P., Hornig, H. and Lührmann, R., 1983, Recogni-
tion of Termination Codon by Release Factor in the
Presence of a tRNA Occupied A-site,
J. Biol. Chem., 258: 10360

Tessier, L.-H., Sondermeyer, P., Faure, T., Dreyer, D.,
Benavente, A., Villeval, D., Courtney, M. and
Lecoy, J.-P., 1984, The Influence of mRNA
Primary and Secondary Structure on Human IFN-
γ-Gene Expression in Escherichia coli,
Nucleid Acids Research, 12: 7663

Van Duin, J., Ravensbergen, C. J. C. and Doornbos, J., 1984)
Basepairing of Oligonucleotides to the 3´end of 16S
Ribosomal RNA is not Stabilized by Ribosomal Proteins,
Nucl. Acids Res., 12: 5079

CONFORMATIONAL DYNAMICS INVOLVED IN RNA SELF-SPLICING

Thomas R. Cech, Francis X. Sullivan, Tan Inoue,* John M. Burke,** Michael D. Been, N. Kyle Tanner and Arthur J. Zaug

Department of Chemistry and Biochemistry, University of Colorado, Boulder, CO 80309

INTRODUCTION

In some species of the ciliated protozoan, *Tetrahymena*, the genes for the large rRNA (ribosomal RNA) are interrupted by an IVS (intervening sequence or intron) approximately 400 base pairs in size (Wild and Gall, 1979). The genes are transcribed by RNA polymerase I to give a pre-rRNA that contains the IVS (Din et al., 1979; Cech and Rio, 1979). The pre-rRNA is then subject to an RNA processing reaction called RNA splicing, by which the IVS is excised and the exons or flanking sequences are ligated. RNA splicing is required for the expression of most mRNA genes and some rRNA and tRNA genes in the nuclei of eukaryotes, and is also involved in the expression of some mitochondrial and chloroplast genes.

Unlike nuclear pre-mRNA and pre-tRNA splicing, splicing of the *Tetrahymena thermophila* pre-rRNA takes place in vitro in the absence of protein (Cech et al., 1981; Kruger et al., 1982). The folded RNA molecule provides a binding site for a free guanosine nucleotide which cleaves the pre-rRNA at its 5' splice site and becomes covalently attached to the 5' end of the IVS (Bass and Cech, 1984). The 3' hydroxyl group at the end of the 5' exon then attacks the 3' splice site, releasing the IVS and ligating the exons. Both steps in splicing occur by transesterification, an exchange of phosphate esters that requires no external energy such as is provided by ATP or GTP hydrolysis in other systems (Cech, 1983). Following its excision from the pre-rRNA, the IVS can undergo self-catalyzed conversion to a circular form with release of the first 15 nucleotides of the IVS. Thus, cyclization is a cleavage-ligation reaction which, like RNA splicing, proceeds through a transesterification mechanism (Zaug et al., 1983).

Self-splicing has been reported for several other IVSs. The best-studied examples are the first IVS of the mitochondrial cytochrome b pre-mRNA in *Neurospora crassa* and the IVS of the mitochondrial pre-rRNA in yeast (Garriga and Lambowitz, 1984; Tabak et al., 1984; Van der Horst and Tabak, 1985). In both cases, the mechanism of self-splicing is analogous to that of the *Tetrahymena* pre-rRNA, including the use of guanosine in the first transesterification reaction.

---

*Present address: Salk Institute, La Jolla, CA 92037, USA
**Permanent address: Department of Chemistry, Williams College, Williamstown, MA 01267, USA

The three self-splicing IVSs described above belong to the larger class of Group I introns, originally categorized by their conserved sequence elements and core secondary structure (Michel et al., 1982, 1983; Davies et al., 1982; Waring et al., 1983). Biochemical analysis of the structure of the *Tetrahymena* rRNA IVS provided additional evidence for the Group I intron secondary structure (Cech et al., 1983). While some of the structural features of the IVS may be static, we will discuss evidence that two regions of the molecule undergo structural transitions essential for the self-catalyzed reactions.

SELECTION OF DIFFERENT REACTION SITES

The first example of a conformational switch involves sequences at the reaction sites near the 5' end of the IVS. The 5' splice site is preceded by an oligopyrimidine sequence, CUCUCU. The major and minor cyclization sites are preceded by UUU and CCU, respectively (Zaug et al., 1983, 1984). In altered versions of the IVS that have insertions or deletions at the major cyclization site, cyclization usually proceeds to a new site; in most cases, the new site is preceded by a tripyrimidine sequence (Been and Cech, 1985).

Evidence that a single active site within the IVS binds the oligo-pyrimidine sequences that precede all of these sites is as follows:

(1) Both the 5' splice site and the various cyclization sites are subject to attack by guanosine; the 5' splice site is normally attacked by free guanosine, and the cyclization sites by the guanosine residue at the 3' end of the IVS RNA. However, in a linear IVS RNA molecule with a blocked 3' end, the cyclization site is subject to attack by free guanosine (Zaug and Cech, 1985). In a "two-thirds molecule" that contains the IVS still joined to the 5' exon, both the 5' splice site and the major cyclization site are susceptible to attack by the 3'-terminal guanosine residue of the IVS (Inoue et al., 1985a). These results support the hypothesis that both splicing and cyclization take place in the same active site.

(2) Oligonucleotides can function as 5' exons, cleaving the pre-rRNA at the 3' splice site and becoming covalently attached to the 3' exon (Inoue et al., 1985b). Oligonucleotides can also attack the cyclization junction of the circular IVS RNA, forming a linear IVS RNA with the oligo-nucleotide covalently attached to its 5' end (Sullivan and Cech, 1985). In both the intermolecular exon ligation reaction and the reverse cyclization reaction, CpU is the most active of the 16 dinucleotides, while UpCpU is the most active of the trinucleotides tested. These results could be explained by two sites with very similar sequence preferences, but are most simply explained by a single oligopyrimidine binding site.

If a single site within the IVS binds the oligopyrimidine sequences for both splicing and cyclization, then structural refolding is required to bring the different target sequences into the active site. More specifically, we envision that the CUCUCU preceding the 5' splice site binds to the active site. The nearby guanosine-binding site (G-site) aligns the 3'-hydroxyl group of guanosine in the correct orientation to attack the phosphorus atom at the 5' splice site, thereby helping to catalyze trans-esterification. After splicing, the exons separate from the IVS and the CUCUCU sequence no longer occupies the active site. A local conformation change brings $U_{13}U_{14}U_{15}$ (preceding the major cyclization site) into the active site. This rearrangement is thermodynamically driven; it is simply the most stable pairing interaction that can be made with the available sequences. The guanosine residue at the 3' end of the IVS now has a free

3' hydroxyl group; it can bind to the G-site and attack the phosphorus atom following $U_{13}U_{14}U_{15}$, resulting in cyclization.

A critical test of the "one active site" model has not been made. Such a test might involve altering the sequence of the oligopyrimidine binding site and showing that the alteration affects the choice of both the 5' splice site and the cyclization site in the same way. In the meantime, it is interesting to speculate where within the IVS the binding site might be located. In the pre-rRNA structural models of Waring et al. (1983) and Michel and Dujon (1983), the UCU preceding the 5' splice site is bound to $G_{22}G_{23}A_{24}$ within the internal guide sequence (Davies et al., 1982). A binding site sequence of GGA is in perfect agreement with the data of Sullivan and Cech (1985). Thus it is plausible that a portion of the internal guide sequence, previously postulated to be involved in the selection of the 5' splice site (Davies et al., 1982), is also involved in the selection of the various cyclization sites.

CATALYSIS OF TRANSESTERIFICATION

The second proposed example of structural dynamics involves two sequence elements, **9L** and **2**, which are conserved among all Group I introns (Burke and RajBhandary, 1982; Michel et al., 1982; called **R** and **S** by Davies et al., 1982). Elements **9L** and **2** are homologous to the left half of *box*9 and to *box*2, respectively, in intron 4 of the yeast mitochondrial cytochrome b gene, sequences originally defined as sites of *cis*-dominant splicing-defective mutations (De La Salle et al., 1982; Anziano et al., 1982). Five bases of element **9L** are postulated to pair with five bases of element **2** (Michel et al., 1982; Davies et al., 1982) as shown in Figure 1 to form part of the core structure of a Group I intron.

We have investigated the effect of mutations in these bases on the self-splicing activity of the *Tetrahymena* rRNA IVS (Burke et al., 1985). Mutation of two bases of element **9L** (Fig. 1) eliminates splicing activity under normal splicing conditions (10 mM $MgCl_2$) at both 30° and 42°C. When this mutation is combined with a compensatory two-base change in element **2** (the **9L/2** double mutant in Fig. 1), activity is restored at 42°C. Unlike the wild-type RNA, the RNA from the **9L/2** double mutant has little splicing activity at 30°C. The higher temperature requirement for splicing may be related to the more stable base-pairing produced by the compensatory base changes. Thus, an attractive interpretation of the data is that splicing requires that pairing occur, but that it not be too strong. That is, the pairing may be involved in a required conformational switch. RNA structure analysis supports the conclusion that the **9L·2** pairing is a weak or unusual interaction in the wild-type IVS RNA (Inoue and Cech, 1985; Tanner and Cech, 1985a).

An analysis of the activity of the **9L/2** double mutant RNA indicates that these sequence elements are involved in catalyzing the various transesterification reactions rather than in the choice of reaction sites.

	Wild Type	9L Mutant	9L/2 Double Mutant
9L	5'GACUA3'	5'GAGUC3'	5'GAGUC3'
	· · · · ·	· · ·	· · · · ·
2	3'CUGAU5'	3'CUGAU5'	3'CUCAG5'

Fig. 1. Wild type **9L·2** pairing and variants produced by in vitro mutagenesis of the *Tetrahymena* rRNA IVS (Burke et al., 1985).

Although the **9L/2** double mutant pre-rRNA splices at 42°C, the rates of both the first step in splicing (attack of guanosine at the 5' splice site) and the second step in splicing (exon ligation) are approximately an order of magnitude slower than for the wild-type RNA (Burke et al., 1985). Furthermore, the excised IVS RNA is defective in cyclization and in site-specific hydrolysis, a reaction which indicates activation of the phosphate at the 3' splice site (Zaug et al., 1984; Inoue et al., 1985a). Thus, changing the sequence of two base pairs in a presumed helical region of the IVS has a wide-ranging effect on the ability of the IVS to catalyze transesterification, but it does not affect the specificity of the reaction. At present there is no firm basis for postulating how a conformational switch involving the weak pairing of elements **9L** and **2** might contribute to catalysis.

Under certain solution conditions, such as high $MgCl_2$ concentration or in the absence of monovalent cation, only a portion of the population of linear IVS RNA molecules undergo cyclization at the normal rate ($t_{1/2} \approx$ 1 min); the remaining portion cyclizes at a much slower rate (Zaug et al, 1985; Tanner and Cech, 1985b). When the same sample of RNA is incubated under other conditions, essentially all the molecules cyclize rapidly with a single order rate constant. One interpretation is that a portion of the molecules can be trapped in a nonproductive conformation, and that conformational isomerization is then rate-limiting for cyclization. Consistent with this view, solution conditions that promote cyclization of the whole population are those that would be expected to "loosen up" the molecule and facilitate conformational changes, i.e., lowering the Mg^{2+}/Na^{2+} ratio, raising the temperature, or partially denaturing the molecules with formamide or by increasing the pH to pH 9. In the future it should be possible to determine whether there are two conformations of the IVS RNA as implicated by the kinetic experiments and, if so, whether they are related to the conformations proposed to be controlled by the **9L·2** pairing.

ACKNOWLEDGEMENTS

This work was supported by grants from the National Institutes of Health (GM28039) and the American Cancer Society (NP-374) to T.R.C. and from the Research Corporation (C-1805), the National Science Foundation (DMB-8502691) and the Williams College Faculty Research Fund to J.M.B.

REFERENCES

Anziano, P. Q., Hanson, D. K., Mahler, H. R., and Perlman, P. S., 1982, Functional domains in introns: Trans-acting and cis-acting regions of intron 4 of the *cob* gene, *Cell*, 30:925.

Bass, B. L., and Cech, T. R., 1984, Specific interaction between the self-splicing RNA of *Tetrahymena* and its guanosine substrate: Implications for biological catalysis by RNA, *Nature*, 308:820.

Been, M. D., and Cech, T. R., 1985, Autocyclization of the *Tetrahymena* ribosomal RNA intervening sequence: Sequence requirements for splice site choice, manuscript submitted.

Burke, J., and RajBhandary, U. L., 1982, Intron within the large rRNA gene of *N. crassa* mitochondria: A long open reading frame and a consensus sequence possibly important in splicing, *Cell*, 31:509.

Burke, J. M., Kaneko, K. J., Irvine, K. D., Oettgen, A. B., Zaug, A. J., and Cech, T. R., 1985, Role of conserved sequence elements **9L** and **2** in self-catalyzed splicing of the *Tetrahymena* ribosomal RNA intervening sequence, manuscript submitted.

Cech, T. R., 1983, RNA splicing: Three themes with variations, *Cell*, 34:713.

Cech, T. R., and Rio, D. C., 1979, Localization of transcribed regions on the extrachromosomal ribosomal RNA genes of *Tetrahymena thermophila* by R-loop mapping, *Proc. Natl. Acad. Sci. USA*, 76:5051.

Cech, T. R., Tanner, N. K., Tinoco, I., Jr., Weir, B. R., Zuker, M., and Perlman, P. S., 1983, Secondary structure of the *Tetrahymena* ribosomal RNA intervening sequence: Structural homology with fungal mitochondrial intervening sequences, *Proc. Natl. Acad. Sci. USA*, 80:3903.

Cech, T. R., Zaug, A. J., and Grabowski, P. J., 1981, In vitro splicing of the ribosomal RNA precursor of *Tetrahymena*: Involvement of a guanosine nucleotide in the excision of the intervening sequence, *Cell*, 27:487.

Davies, R. W., Waring, R. B., Ray, J. A., Brown, T. A., and Scazzocchio, C., 1982, Making ends meet: A model for RNA splicing in fungal mitochondria, *Nature*, 300:719.

De La Salle, H., Jacq, C., and Slonimski, P. P., 1982, Critical sequences within mitochondrial introns: Pleiotropic mRNA maturase and cis-dominant signals of the *box* intron controlling reductase and oxidase, *Cell*, 28:721.

Din, N., Engberg, J., Kaffenberger, W., and Eckert, W., 1979, The intervening sequence in the 26 S rRNA coding region of *T. thermophila* is transcribed within the largest stable precursor for rRNA, *Cell*, 18:525.

Garriga, G., and Lambowitz, A. M., 1984, RNA splicing in *Neurospora* mitochondria: Self-splicing of a mitochondrial intron in vitro, *Cell*, 39:631.

Inoue, T., and Cech, T. R., 1985, Secondary structure of the circular form of the *Tetrahymena* ribosomal RNA intervening sequence: A new technique for RNA secondary structure analysis using chemical probes and reverse transcriptase, *Proc. Natl. Acad. Sci. USA*, 82:648.

Inoue, T., Sullivan, F. X., and Cech, T. R., 1985a, New reactions of the ribosomal RNA precursor of *Tetrahymena* and the mechanism of self-splicing, manuscript submitted.

Inoue, T., Sullivan, F. X., and Cech, T. R., 1985b, Intermolecular exon ligation of the rRNA precursor of *Tetrahymena*: Oligonucleotides can function as 5' exons, manuscript submitted.

Kruger, K., Grabowski, P. J., Zaug, A. J., Sands, J., Gottschling, D. E., and Cech, T. R., 1982, Self-splicing RNA: Autoexcision and autocyclization of the ribosomal RNA intervening sequence of *Tetrahymena*, *Cell*, 31:147.

Michel, F., and Dujon, B., 1983, Conservation of RNA secondary structures in two intron families including mitochondrial-, chloroplast-, and nuclear-encoded members, *EMBO J.*, 2:33.

Michel, F., Jacquier, A., and Dujon, B., 1982, Comparison of fungal mitochondrial introns reveals extensive homologies in RNA secondary structure, *Biochimie*, 64:867.

Sullivan, F. X., and Cech, T. R., 1985, Reversibility of Cyclization of the *Tetrahymena* rRNA intervening sequence: Implication for the mechanism of splice-site choice, *Cell*, in press.

Tabak, H. F., Van der Horst, G., Osinga, K. A., and Arnberg, A. C., 1984, Splicing of large ribosomal precursor RNA and processing of intron RNA in yeast mitochondria, *Cell*, 39:623.

Tanner, N. K., and Cech, T. R., 1985a, A Self-catalyzed cyclization of the intervening sequence RNA of *Tetrahymena*: Inhibition by intercalating dyes, *Nucleic Acids Res.*, in press.

Tanner, N. K., and Cech, T. R., 1985b, Self-catalyzed cyclization of the intervening sequence RNA of *Tetrahymena*: Inhibition of intercalating dyes, *Nucleic Acids Res.*, in press.

Van der Horst, G., and Tabak, H. F., 1985, Self-splicing of yeast mitochondrial precursor RNA, *Cell*, 40:759.

Waring, R. B., Scazzocchio, C., Brown, T. A., and Davies, R. W., 1983, Close relationship between certain nuclear and mitochondrial introns, *J. Mol. Biol.*, 167:595.

Wild, M. A., and Gall, J. G., 1979, An intervening sequence in the gene coding for 25 S ribosomal RNA of *Tetrahymena pigmentosa*, *Cell*, 16:565.

Zaug, A. J., and Cech, T. R., 1985, Oligomerization of intervening sequence RNA molecules in the absence of proteins, *Science*, in press.

Zaug, A. J., Grabowski, P. J., and Cech, T. R., 1983, Autocatalytic cyclization of an excised intervening sequence is a cleavage-ligation reaction, *Nature*, 301:578.

Zaug, A. J., Kent, J. R., and Cech, T. R., 1984, A labile phosphodiester bond at the ligation junction in a circular intervening sequence, *Science*, 224:574.

Zaug, A. J., Kent, J. R., and Cech, T. R., 1985, Reactions of the intervening sequence of the *Tetrahymena* ribosomal ribonucleic acid precursor: pH dependence of cyclization and site-specific hydrolysis, *Biochemistry*, in press.

RNA CATALYZED LARIAT FORMATION FROM YEAST MITOCHONDRIAL PRE-RIBOSOMAL RNA

Henk F. Tabak, Annika C. Arnberg* and Gerda van der Horst

Section for Molecular Biology, Laboratory of Biochemistry
University of Amsterdam, Kruislaan 318, 1098 SM Amsterdam
*Laboratory of Biochemistry, Rijks Universiteit, Nijenborgh
16, 9747 AG Groningen, The Netherlands

INTRODUCTION

Mitochondria of <u>Saccharomyces cerevisiae</u> contain a circular DNA of around 80 kb. It comprises a limited set of genes which code for 2 ribosomal RNAs (rRNA), approximately 25 tRNAs, an RNA involved in tRNA processing, 6 proteins which are part of respiratory chain complexes, a protein (var1) associated with the mitochondrial ribosome and proteins (maturases), which are involved in RNA splicing (for review see Dujon, 1981). In the strain <u>S.cerevisiae</u> KL14-4A three of these genes are interrupted by intervening sequences (introns): the genes coding for large rRNA, apocytochrome <u>b</u> and subunit I of cytochrome <u>c</u> oxidase (see Fig.1). Removal of introns from precursor RNAs requires the presence of a RNA splicing machinery. The mitochondrial splicing process differs in a number of aspects from splicing of precursor RNAs synthesized by RNA polymerase II:

i) mitochondrial introns possess conserved nucleotide blocks, which by inducing specific folding of intron RNA are thought to facilitate the splicing process (Michel and Dujon, 1983; Waring and Davies, 1984). Mutations in such sequence elements have a cis-acting phenotype (Lamouroux et al. 1980). In contrast nuclear introns in higher eukaryotes do not contain sequence elements of this type that contribute to RNA splicing (Wieringa et al. 1984).

ii) some mitochondrial introns contain long open reading frames in phase with the upstream exon. Translation of unspliced RNA by mitochondrial ribosomes leads to synthesis of hybrid proteins (maturases), which are involved in and contribute to intron removal (Lazowska et al. 1980). This is supported by the observation that nucleotide alterations leading to a deficient reading frame can be complemented in trans by the wildtype allele (Lamouroux et al. 1980).

A simplified version of the mitochondrial splicing process for reading frame containing introns is given in Fig.2. Precursor RNA is translated by mitochondrial ribosomes leading to a protein, coded by the upstream exon and the intronic reading frame. This maturase contributes to splicing and destroys in this way its own mRNA (autoregulation). The second wave of mitochondrial ribosomes translates the exon encoded protein from the mature mRNA. In some cases proteins that affect mitochondrial splicing are encoded in the nuclear genome and are imported into the organelle after synthesis by celsap ribosomes.

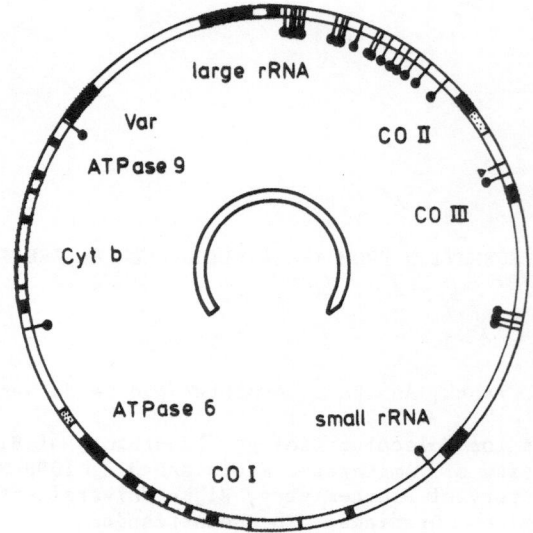

Fig.1. Gene organization of yeast mtDNA.
The map shown is that of <u>S.cerevisiae</u> strain KL14-4A. Shaded regions
represent genes or split genes, while stippled blocks outside genes
indicate long reading frames with unidentified function. Small
filled circles represent tRNA genes. The arrow inside shows the
direction of transcription.

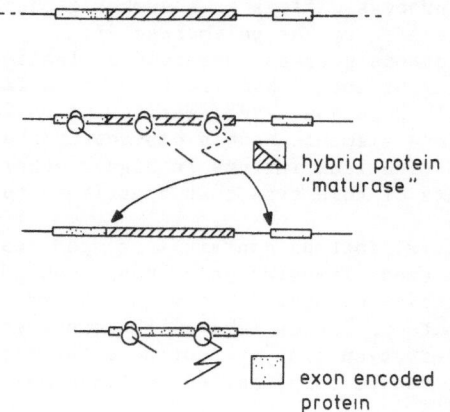

Fig.2. Model for yeast mitochondrial splicing of precursor RNAs
containing introns with coding potential.
Stippled blocks represent exons. The region between these blocks
represents the intron. The hatched intron part represents the
reading frame coding for the maturase. The model is based on
Lazowska et al. 1980.

They can act on a specific mitochondrial intron and may act similarly to the mitochondrial intron encoded proteins (Dieckman et al. 1982; Pillar et al. 1983). Thus, mitochondrial splicing may be guided by nuclear or mitochondrial DNA encoded "maturases". These concepts have been developed largely by genetic studies on splicing deficient mutants. Recently this picture turned out to be more complicated by the finding that some mitochondrial introns can undergo self-splicing in vitro.

SPLICING OF THE PRECURSOR TO LARGE RIBOSOMAL RNA

The gene coding for large rRNA is interrupted by a single intron of 1143 bp (Bos et al. 1978; Dujon, 1980), which contains a reading frame coding for a protein that is not involved in RNA processing however (Jacquier and Dujon, 1985; Macreadie et al. 1985). Studies on transcripts from the rRNA gene by Northern blot analysis revealed the presence of RNA species that exclusively hybridized with an intron specific DNA probe and corresponded in length with that of the intron (Bos et al. 1980). These RNAs were purified by extraction from agarose gels and further investigated by primer extension with reverse transcriptase and electron microscopy. They appeared to consist of an excised linear intron RNA carrying a guanosine residue at its 5'-end which is not encoded in mtDNA and a circular form of the intron. From the RNA sequence in the region of circle closure it followed that three 5'-intron encoded nucleotides were absent (Tabak et al. 1984). The properties of these intermediates are typically those of the splicing pathway of Tetrahymena nuclear pre-ribosomal RNA. This precursor has been shown to be spliced in an RNA catalyzed reaction in which the folding of the intron plays a crucial role (reviewed by Cech, 1983). The same is true for the mitochondrial rRNA precursor. Part of the intron containing gene was cloned behind the SP6 promoter in the SP65 vector DNA. This allows in vitro synthesis of an artificial precursor RNA by SP6 RNA polymerase from this recombinant plasmid, which is completely free of yeast mitochondrial protein contamination. This substitute for natural pre-ribosomal RNA shows self-splicing in a simple mixture containing salts and a guanosine nucleotide leading to intron excision, intron circularization and exon ligation (Van der Horst and Tabak, 1985). The splicing pathway is illustrated in schematic form in Fig.3A. Apart from the self-splicing products expected on the basis of the extensive analysis of splicing of the Tetrahymena pre-ribosomal RNA, we have also observed formation of a number of other RNA products.

At high GTP concentration nucleophilic attack can also open the 3'-splice junction resulting in covalent addition of the guanosine to the 5'-end of the downstream exon in a process which is comparable to opening of the 5'-splice-junction (Fig.3B). Transcription on the SP promoter starts with GTP. This 5'-terminal G residue in the precursor RNA can be exchanged with free GTP present in the self-splicing mixture in an RNA catalyzed reaction (Fig.3C). Finally, electron microscopic analysis of self-splicing RNA products, revealed the presence of branched circles: lariats (Fig.3D). An example of such an electron micrograph is shown in Fig.4. Apart from the circularized introns described before also lariats are present. We are currently investigating their structure but have already found that they do not consist of the excised intron only. We suspect that the reactions summarized in Fig.3B,C and D do not fit into the splicing pathway of pre-ribosomal RNA in mitochondria but we believe that these partial reactions can offer important clues for further investigation into the mechanism of RNA catalyzed reactions.

Such side reactions may be prevented in vivo by the presence of proteins involved in precursor RNA processing. We have observed that two in vitro self-splicing introns from the multiply split gene coding for subunit I of cytochrome c oxidase are not noticeably spliced in mitochondria in the

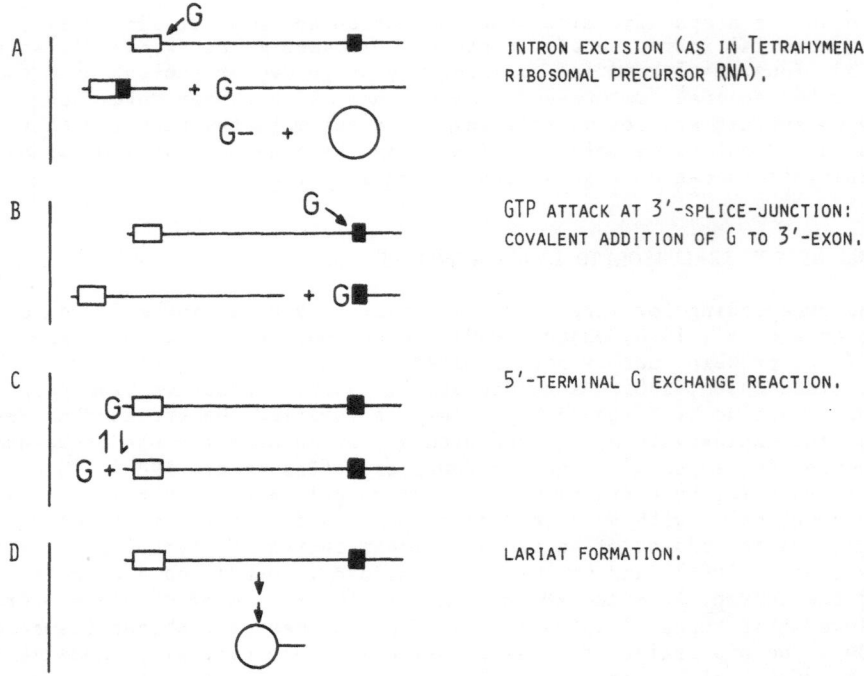

A   INTRON EXCISION (AS IN TETRAHYMENA RIBOSOMAL PRECURSOR RNA).

B   GTP ATTACK AT 3'-SPLICE-JUNCTION: COVALENT ADDITION OF G TO 3'-EXON.

C   5'-TERMINAL G EXCHANGE REACTION.

D   LARIAT FORMATION.

Fig.3. RNA catalyzed reactions of the yeast mitochondrial precursor to large rRNA.

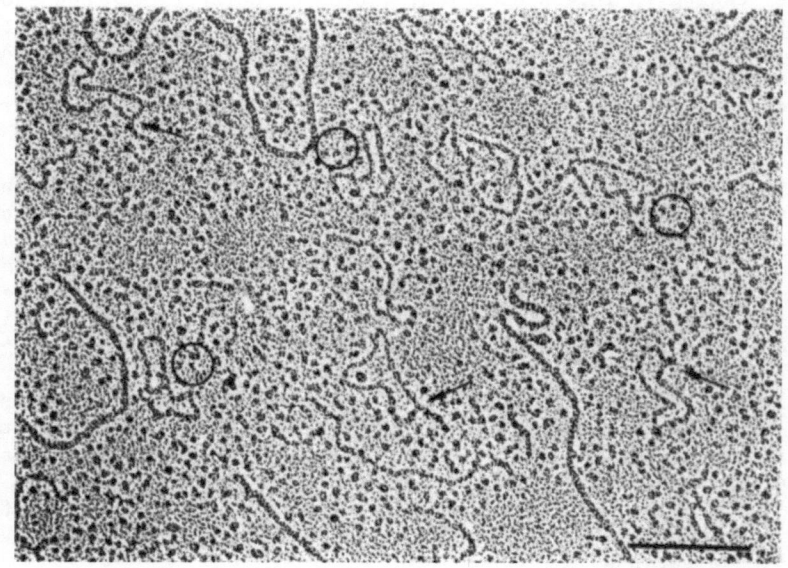

Fig.4. Electron micrograph of self-splicing RNA products.
Linear, circular and lariat structures can be observed in the same field. Lariats are indicated by arrows, circles by circular symbols. The black bar represents 0.2 μm.

absence of mitochondrial protein synthesis (Van der Horst and Tabak, 1985; Tabak et al. 1985). The situation may be compared with the behaviour of RNase P the enzyme involved in precursor tRNA maturation in bacteria. It consists of an RNA and protein moiety. At high Mg^{2+} concentration the RNA component alone is capable of pre-tRNA cleavage while under conditions that are more or less comparable to the in vivo situation the presence of the protein component is also required (Guerrier-Takada et al. 1983).

IMPLICATIONS OF LARIAT FORMATION IN AN RNA CATALYZED REACTION

Lariats were first identified as intermediates in splicing of pre-mRNAs transcribed by RNA polymerase II from nuclear split genes (Ruskin et al. 1984; Padgett et al. 1984; Domdey et al. 1984). The development of cell-free extracts has made it possible to further characterize the nuclear splicing mechanism, which shows enormous complexity and requires the assembly of a spliceosome particle involving a number of proteins and RNA components. Since internal intron sequences, as mentioned before, hardly contribute to the splicing process, it further confirmed the notion that nuclear splicing is a protein-catalyzed reaction. The fact that lariats are products of a protein dependent splicing machinery as well as of self-splicing, challenges this concept and it is quite possible that RNA components like U1 and U2 small nuclear RNAs may play a more active role than anticipated thusfar. One could envisage the possibility that U1 and U2 RNAs by base-pairing to 5'- and 3'-splice- junctions respectively (Mount and Steitz, 1984) may contribute to RNA splicing as RNA catalysts for instance by straining the phosphodiester bonds present in the junctions between exon and intron to prepare them for hydrolysis.

Thusfar, the various splicing themes (Cech, 1983) displayed by pre-tRNA splicing, pre-mRNA splicing and self-splicing of Tetrahymena nuclear and fungal mitochondrial precursor RNAs did not seem to show any resemblance. The finding of a typical RNA product, the lariat, common to two splicing systems raises the question whether the various splicing systems now in existence could have arisen from a common ancestor. It is tempting to postulate that self-splicing RNAs were the first to appear on the evolutionary scene. In one branch of evolution specific RNA segments may have separated from the intron to which they originally belonged to begin an individual way of life in what finally evolved into a spliceosome. After centralizing the splicing machinery in such a particle, introns were permitted to lose the RNA sequence elements that in the past took care of their own excision. Small nuclear RNAs like U1 and U2 may be descendants of such key elements originally present in nuclear self-splicing introns and may be comparable to intron RNA sequences which still exist in self-splicing introns and function as internal splice-guides to bring splice-junctions in close spatial proximity (Bos et al. 1980; Davies et al. 1982). In self-splicing precursor RNAs, the other branch of evolution, the original situation remained fixed and key RNA sequences essential for RNA catalysis are still present in the intron RNA itself.

REFERENCES

Bos, J.L., Heyting, C., Borst, P., Arnberg, A.C. and Van Bruggen, E.F.J., 1978, An insert in the single gene for the large ribosomal RNA in yeast mitochondrial DNA, Nature, 275:336-338.
Bos, J.L., Osinga, K.A., Van der Horst, G., Hecht, N.B., Tabak, H.F., Van Ommen, G.J.B. and Borst, P., 1980, Splice point sequence and transcripts of the intervening sequence in the mitochondrial 21S ribosomal RNA gene of yeast, Cell, 20:207-214.
Cech, T., 1983, RNA splicing: three themes with variations, Cell, 34:713--716.

Domdey, H., Apostel, B., Lin, R.J., Newman, A., Brody, E. and Abelson, J., 1984, Lariat structures are in vivo intermediates in yeast pre-mRNA splicing, Cell 39:611-621.

Dujon, B., 1980, Sequence of the intron and flanking exons of the mitochondrial 21S rRNA gene of yeast strains having different alleles at the omega and rib-1 loci, Cell 20:185-197.

Dujon, B., 1981, Mitochondrial genetics and function, in: Life cycle and Inheritance, J.H. Strathern, E.W. Jones and J.R. Broach, eds., Cold Spring Harbor Laboratory, Cold Spring Harbor, N.Y., 505-635.

Guerrier-Takada, C., Gardiner, K., Marsh, T., Pace, N. and Altman, S., 1983, Cell, 35:849-857.

Jacquier, A. and Dujon, B., 1985, An intron-encoded protein is active in a gene conversion process that spreads an intron into a mitochondrial gene, Cell, 41:383-394.

Krämer, A., Keller, W., Appel, B. and Lührmann, R., 1984, The 5' terminus of the RNA moiety of U1 small nuclear ribonucleoprotein particles is required for the splicing of messenger RNA precursors, Cell, 38:299-307.

Lamouroux, A., Pajot, P., Kochko, A., Halbreich, A. and Slonimski, P.P., 1980, Cytochrome b messenger RNA maturase encoded in an intron regulates the expression of the split gene: II. Trans- and cis-acting mechanisms of mRNA splicing, in: The Organization and Expression of the Mitochondrion Genome, C. Saccone and A. Kroon, eds., North-Holland, New York, 152-156.

Lazowska, J., Jacq, C. and Slonimski, P.P., 1980, Sequence of introns and flanking exons of wildtype and box3 mutants of cytochrome b reveals and interlaced splicing protein coded by an intron, Cell, 22:333-348.

Macready, I.G., Scott, R.M., Zinn, A.R. and Butow, R.A., 1985, Transposition of an intron in yeast mitochondria requires a protein encoded by that intron, Cell, 41:395-402.

Michel, F. and Dujon, B., 1983, Conservation of RNA secondary structure in two intron families including mitochrondrial-, chloroplast- and nuclear encoded members, EMBO J.,2:33-38.

Mount, S.M., Pettersson, I., Hinterberger, M., Karmas, A. and Steitz, J.A., 1983, The U1 small nuclear RNA-protein complex selectively binds a 5' splice-site in vitro, Cell, 33:509-518.

Mount, S.M. and Steitz, J.A., 1984, RNA splicing and the involvement of small ribonucleoproteins, Mol. Cell. Biol., 3:249-297.

Padgett, R.A., Konarska, M.M., Grabowski, P.J., Hardy, S.F. and Sharp, P.A., 1984, Lariat RNAs as intermediates and products in the splicing of messenger RNA precursors, Science, 225:898-903.

Pillar, T., Lang, B.F., Steinberger, I., Vogt, B. and Kaudewitz, F., 1983, Expression of the "split gene" cob in yeast mtDNA: Nuclear mutations specifically block the excision of different introns from its primary transcript, J. Biol. Chem., 258:7954-7959.

Ruskin, B., Krainer, A.R., Maniatis, T. and Green, M.R., 1984, Excision of an intact intron as a novel lariat structure during pre-mRNA splicing in vitro, Cell, 38:317-331.

Tabak, H.F., Van der Horst, G., Osinga, K.A. and Arnberg, A.C., 1984, Splicing of large ribosomal precursor RNA and processing of intron RNA in yeast mitochondria, Cell, 39:623-629.

Van der Horst, G. and Tabak, H.F., 1985, Self-splicing of yeast mitochondrial ribosomal and messenger RNA precursors, Cell, 40:759-766.

Waring, R.B. and Davies, R.W., 1984, Assessment of a model for intron RNA secondary structure relevant to RNA self-splicing: a review, Gene, 28:277-291.

Wieringa, B., Hofer, E. and Weissmann, Ch., 1984, A minimal intron length but no specific internal sequence is required for splicing the large rabbit β-globin intron, Cell, 37:915-925.

VIROIDS: STRUCTURE FORMATION AND FUNCTION

Gerhard Steger, Volker Rosenbaum, and Detlev Riesner

Institut für Physikalische Biologie
Universität Düsseldorf
Düsseldorf, FRG

INTRODUCTION

Among the nucleic acids, the single-stranded RNA molecules have the highest potential to form a large variety of structures and to undergo quite different structural transitions. The determination of their structure and structural transitions is a prerequisite for understanding their function in replication, transcription, translation, and regulation. In many cases the problem of RNA structure is particularly difficult for in addition to the secondary structure formed by Watson-Crick base pairs and possibly wobble base pairs, a complicated tertiary structure exists which cannot be described by a few prototypes of interactions.

During the last years an attractive problem of RNA structure and function came up with the work on viroids. Viroids are an independent class of plant pathogens which are distinguished from viruses by the absence of a protein coat and by their small size of 200 to 400 nucleotides (cf. reviews: [1-5]). Whereas viroids do replicate autonomously in the host cell, virusoids - in size and structure related to viroids - are genomic RNAs of a plant virus. Both, virusoids and viroids, are single-stranded, circular RNA molecules. The structural and dynamic properties of viroids and virusoids have been studied by thermodynamic, kinetic and hydrodynamic methods. From the experimental data and from model building after the first sequence determination it was concluded that the secondary structure of viroids consisted of an unbranched series of short double helices and small internal loops. There is no indication for an additional tertiary structure. In the mean time, both experimental data and sequences are known for several viroid species as potato spindle tuber (PSTV), citrus exocortis (CEV), chrysanthemum stunt (CSV), coconut cadang-cadang (CCCV), avocado sunblotch (ASBV), and for the virusoids from Solanum nodiflorum mottle virus (SNMV) and velvet tobacco mottle virus (VTMoV). We show in this article, how the properties of the structure and the structural transitions of viroids were derived from their nucleotide sequences by applying free energy minimization calculations[14]. The pre-positions for a theoretical treatment of viroids are better than for any other RNA molecule: Only Watson-Crick and wobble basepairing, as mentioned above, have to be assumed without an uncalculable tertiary structure[6-10], and no odd nucleotides, for which the helix parameters would not be known, have been found. We will calculate the curves of the conformational transitions and compare them with the experiments.

As a conclusion of biological relevance it will be discussed if viroids follow an unifying concept in their structure and dynamics clearly different from other RNA molecules, and whether such a concept follows from theory and experiment. Furthermore, it will be asked whether the functional difference between viroids and virusoids correlates with major differences in their structural and dynamic features. From the comparison of viroids with virusoids and computer-generated RNA molecules the unique features of viroids will become more evident.

Although the mechanism of viroid structure formation is quite different from that known from other RNA species, it has been found in all viroids studied. From this point of view one has to expect a direct functional relevance of the structural and dynamic features. Therefore, we will discuss some experimental results of viroid replication[11], pathogenesis[11], and origin[12] with regard to the conservative features of viroids.

METHODS

A theoretical analysis has to derive viroid properties from a thermodynamical set of parameters for helix growth and helix nucleation of RNA and from the nucleotide sequence of a particular viroid. The native structure of a viroid is obtained from searching for the structure of lowest energy. Intermediate structures during the process of structure formation or denaturation will be found by the same procedure if the temperature dependence of the structural parameters is incorporated.

Elementary Thermodynamic Parameters

Several sets of parameters for helix growth and helix nucleation are available from the literature. These had been obtained from studies on oligonucleotides and polynucleotides. The results of the calculations presented below do depend to a high degree upon the source of the elementary parameters. "Small changes in free energy assignments for unpaired bases do not effect predicted secondary structures in ssRNA"[13] is only true for "native" secondary structures, but small changes may have large effects on consecutive melting of helices or even on melting of a whole secondary structure. For a detailed discussion of the parameters used throughout our calculations refer to the literature[14].

The equilibrium constant s for the helix growth is assigned to the formation of a base pair adjacent to an already existing base pair.

$$- RT \ln s = \Delta G_{Bp} = \Delta H_{Bp} - T \cdot \Delta S_{Bp}$$

with ΔG, ΔH, ΔS denoting the changes in Gibbs free energy, enthalpy and entropy. These values are available for all combinations of neighbors of A:U, G:C, and wobble G:U base pairs. All values refer to 1 M NaCl, absence of $MgCl_2$, neutral pH. The results may be extrapolated to other solvent conditions. It should be mentioned that there are values which are particularly suitable to calculate stabilities[14,15], i.e T_m-values, and values which should be used if particularily emphasis is put on ΔH-values[14,16], for example to calculate the width of a transition.

The first base pair of a new helix may form four different types of loops: hairpin loops, internal loops, bulge loops, and bifurcations. The values of ΔG (at 25°C) and ΔS for these loops are given in the literature[14,16-19].

Calculation of thermal denaturation curves

Structure and structural transitions are calculated along the same line. The free energy for the formation of the double helical stretch m with k base pairs is:

$$\Delta G_m = \sum_{i=2}^{k} \Delta H_i - T \cdot \left(\sum_{i=2}^{k} \Delta S_i - \Delta S_1 \right)$$

with ΔH_i and ΔS_i, respectively, for the ith base pair and the loop entropy ΔS_1 due to the formation of the first base pair. ΔG of a total secondary structure sums up all helices m, the entropy of the remaining loops ΔS_r, and if the primary sequence is circular the entropy of the open circle without any base pairs, ΔS_0:

$$\Delta G_{\text{secondary structure}} = \sum_{m=1}^{\text{number of helices}} \Delta G_m + T \cdot (\Delta S_r - \Delta S_0)$$

The hypochromicity of a structural transition is the sum of the relative hypochromicities[20,21] of all base pairs involved.

1) "Cooperative Helix" Approximation

The secondary structure models under consideration, particularly the possibilities for rearrangement, were taken from the base pairing matrix according to Tinoco et al.[22]. As outlined earlier[23], it may be approximated that short double helical stretches as occuring in viroid are either formed completely or are dissociated totally. The most probable order of melting of the individual helices is then obtained by calculating their T_m-values and taking into account for a particular helix the conformational state of the neighboring regions. As dissociation of a helix results in an increase of the loop size, the neighboring helices are often destabilized. Cooperative T_m-values for neighboring helices are obtained in many cases. In summary, if the stability of a helix is estimated, possibilities for helix opening, dissociation of neighboring helices, and for rearrangement have to be checked.

2) Nussinov-Algorithm

At present the most accurate method available for calculating structural transitions is based on an exact algorithm for finding the secondary structure of lowest free energy. The algorithm was originally developed by Nussinov and Jacobson[24,25], a well applicable program was reported by Zuker and Stiegler[26], and the particular requirements for the application to the viroid structure, i.e. allowing for circular strands and calculation of bifurcations, were introduced by Steger et al.[14]. With this method, secondary structures of lowest free energy were determined at varying temperatures. The hypochromicity was calculated as the difference in absorption between the structures at different temperatures. As in the other model, only the structure of lowest free energy was considered at every temperature and not the total partition function which would include also the less probable structures.

STRUCTURE FORMATION

Experimental Background

1) Optical Equilibrium Melting Curves

Melting curves are typical for viroid or virusoid RNA species. In Fig. 1 the thermal denaturation curve of PSTV is depicted. It was composed from equilibrium- and kinetic measurements and exhibits clearly a very sharp main transition and two broader transitions at higher temperature[6,23,27]. Very similar results were obtained with CEV and CSV[27]; the thermodynamic parameters are summarized in Table I. With increasing ionic strength the width of the main transition increases only by some tenth of a degree.

In low ionic strength the thermal denaturation of cadang-cadang-RNA appears similar to that of PSTV. There is one transition of cadang-cadang-RNA at temperatures higher than the main transition which appears similar to the high temperature transition of PSTV. This transition at high temperatures could be resolved not only by evaluating relaxation amplitudes as for PSTV but already from the equilibrium measurements. CCCV RNA1 small and large only show minor differences whereas the main transition in ccRNA2 is remarkably broader. In contrast to PSTV, CEV, and CSV, however the main transition broadens remarkably with increasing ionic strength and splits in at least three transitions at 1 M Na+-ions[28].

The thermal denaturation of virusoids is clearly different from that of viroids. The transition occurs at about 10°C lower temperatures and is 2 or 3 times broader. A second transition at higher temperatures is not detectable[28]. The virusoids from VTMoV and from SNMV behave very similar.

In accordance with the thermodynamic properties the results from hydrodynamic studies show that PSTV, CEV, CSV and the four RNAs from cadang-cadang form homologous structures with an unique persistence length of about 300 Å whereas both virusoids are either more globular or more flexible than the viroids[10].

2) Kinetics

In kinetic experiments, a denaturation process is characterized not only by its overall hypochromicity but also by its complete, exponential time course after a stepwise temperature increase or decrease (cf. slow[29] and fast[30] temperature jump technique). If more than one process contribute, more than one exponential with their corresponding amplitudes are measurable. Slow and fast temperature jump techniques were optimized for the requirements of the study on viroids, i.e. high temperatures and exceedingly little RNA material[27,28,31,32].

The main transition (around 50°C) proceeds in the time range of sec to min. The kinetics confirmed the high co-operativity observed in the equi-

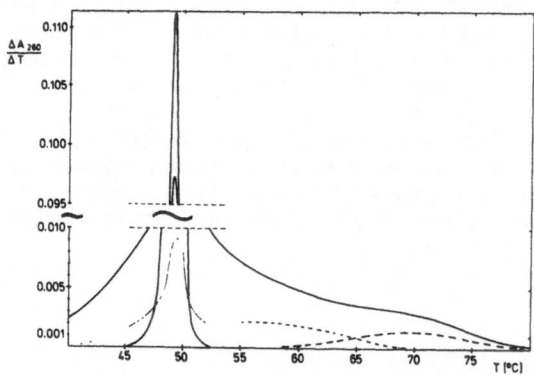

Fig. 1 Composite differentiated denaturation curve of PSTV.
Buffer conditions were 8 mM Na-cacodylate, 0.1 mM EDTA, pH 6.8. The dashed and dotted lines refer to transitions which occur in the time range of ms and were studied with the temperature jump technique. Numbers I, II, and III refer to the denaturation of the stable hairpins which are formed during the main transition.

librium melting curve. 80-90% of the change in base-pairing was found in a single exponential, the remaining contribution in the msec range. Although at temperatures above the main transition the amplitudes decrease drastically (cf. Fig. 1), in PSTV two additional transitions could be followed quantitatively by fast kinetic methods. Similarly, two transitions in addition to the main transition were found in CEV and CSV[27] and only one transition in CCCV RNA1[28]. From ΔH-values the number of base pairs dissociating in the corresponding transition was derived. The G:C content of the dissociating base pairs was estimated from the wavelength dependence of the hypochromicity. As a characteristic result from kinetic experiments, the size of the loop which has to be closed for helix formation, may be estimated. In summary, following results were obtained: T_m, G:C content, number of base pairs, size of the loop. It will be shown below that the transitions could be attributed to well defined hairpin structures.

3) Gel Electrophoresis

The gel electrophoretic properties of viroids have been studied for analytical as well as for preparative purposes[8,33,34]. When viroids undergo the conformational transitions discussed above, they change drastically their electrophoretic mobility. It is caused by the denatured structure of the open circle which migrates much slower than any other RNA of similar molecular weight[35]. Consequently, the nicked form of a denatured viroid runs faster than the circular form. The transition may be made visible in a gel if an urea or a temperature gradient is formed perpendicular to the electrophoresis direction. In Figure 2 a gel electrophoresis of circular and nicked viroids in a temperature gradient is shown. The extraordinary difference in the electrophoretic mobility of native and denatured viroids is evident. Furthermore, at least two transitions may be detected in the transition curve of circular viroids.

4) Influence of Ionic Strength and Other Solvent Conditions

Quite generally, helix-coil transitions of nucleic acids are shifted to higher temperature by increasing ionic strength, and to lower temperature by addition of several organic solvents. These effects have been elucidated with several nucleic acids and would not have to be repeated for viroids if the quantitative data would not be specific for viroids. They are of enormous practical importance for other experiments as gel electrophoresis, hybridization, purification procedures etc.

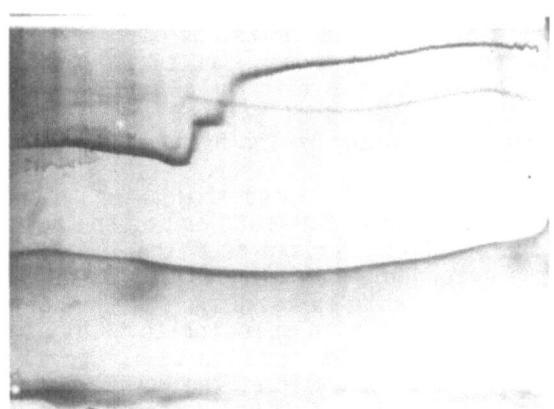

Fig. 2 Denaturation curve of PSTV in 5% polyacrylamid gel-electrophoresis Buffer conditions were 8.9 mM Tris-Borat, pH 8.3. The temperature gradient is linear from 35 to 60°C. The faster migrating band at high temperature represents denatured nicked molecules.

The dependence of the T_m-value of the main transition upon the logarithm of the ionic strength[23] is nearly linear between 0.005 M and 1 M NaCl with a slope of 13.2°C/log c_{Na}^+. In 1 M NaCl the T_m-value is of particular interest, because the theoretical thermodynamic treatment refers to this high ionic strength. The co-operativity decreases only slightly with higher ionic strength. An exception in this respect is CCCV1 which is very co-operative at low ionic strength but exhibits three well resolved transitions in 1 M NaCl[28]. The ionic strength dependence of 13.2°C/log c_{Na}^+ is very similar to the ionic strength dependence of homogeneously double-stranded, polymeric RNA of the same G:C content[36], whereas isolated hairpins of a few base pairs have a lower dependence[21]. Thus, the interruption of a homogeneous double-strand by internal loops as it is the case in viroids decreases neither the ionic strength dependence nor the co-operativity of the helix-coil transition. The conformational transitions of viroids are very sensitive to Mg^{2+}-ions. As little as 10^{-4} M Mg^{2+}-ions bind to viroids and raise the T_m-value by about 30°C.

The dependence of the T_m-values of the transitions of viroids upon the concentration of urea is linear between 0 M and 8 M urea[37]. One should mention that the linear dependence of T_m (c_{urea}) as measured with viroids is more the exception than the rule among nucleic acids.

Secondary Structures

In order to compare the structures of different viroids, they have been characterized by some parameters which may be used as quantitative measures for structure and which are listed in Table I.

1) Viroids

Although the sequence homologies between PSTV, CEV, and CSV are about 60%, it is evident that the structural principle of an unbranched series of short double helices and small internal loops is true for all three viroids (cf. Fig. 3). From Table I it can be seen that also the quantitative values of the structural parameters are similar. The degree of base pairing is near 70%. The G and C are equally distributed between helical and non helical regions. The thermodynamic parameter $\Delta G/N$ is characteristic for an average stability in the molecule. The real thermodynamic parameter, of course, is ΔG, and $\Delta G/N$ is listed only to compensate for different numbers of nucleotides N.

Cadang-cadang RNA consists of four different RNA species which can be derived from the smallest species by partial or total gene duplication. The sequence homology with PSTV, CEV, or CSV is only 15% and is restricted to a well-defined region in the secondary structure (boxed in Fig. 3). The principle of the secondary structure is the same as described above for PSTV, CEV, and CSV. A lower degree of base pairing of only 62% may be compensated by a higher G:C content in the base paired regions.

In contrast to all viroids mentioned above, the secondary structure of ASBV includes a bifurcation at the left end of the molecule (cf. Fig. 3). When the structure was first derived from the Tinoco-plot in analogy to other viroids, the extended structure was presented[38]. The branched form is, however, more favorable by 15 kJ/mol. Consequently, one would expect that this form is present in solution to 99.8% and the extended form to 0.2% only. It is evident that as soon as bifurcations have to be taken into account, only the application of the Nussinov-algorithm resulted in the most favorable structure. In other respects, ASBV is also clearly different from the other viroids. The G+C content in the sequence and the content of G:C base pairs is much lower and results in a lower stability parameter $\Delta G/N$.

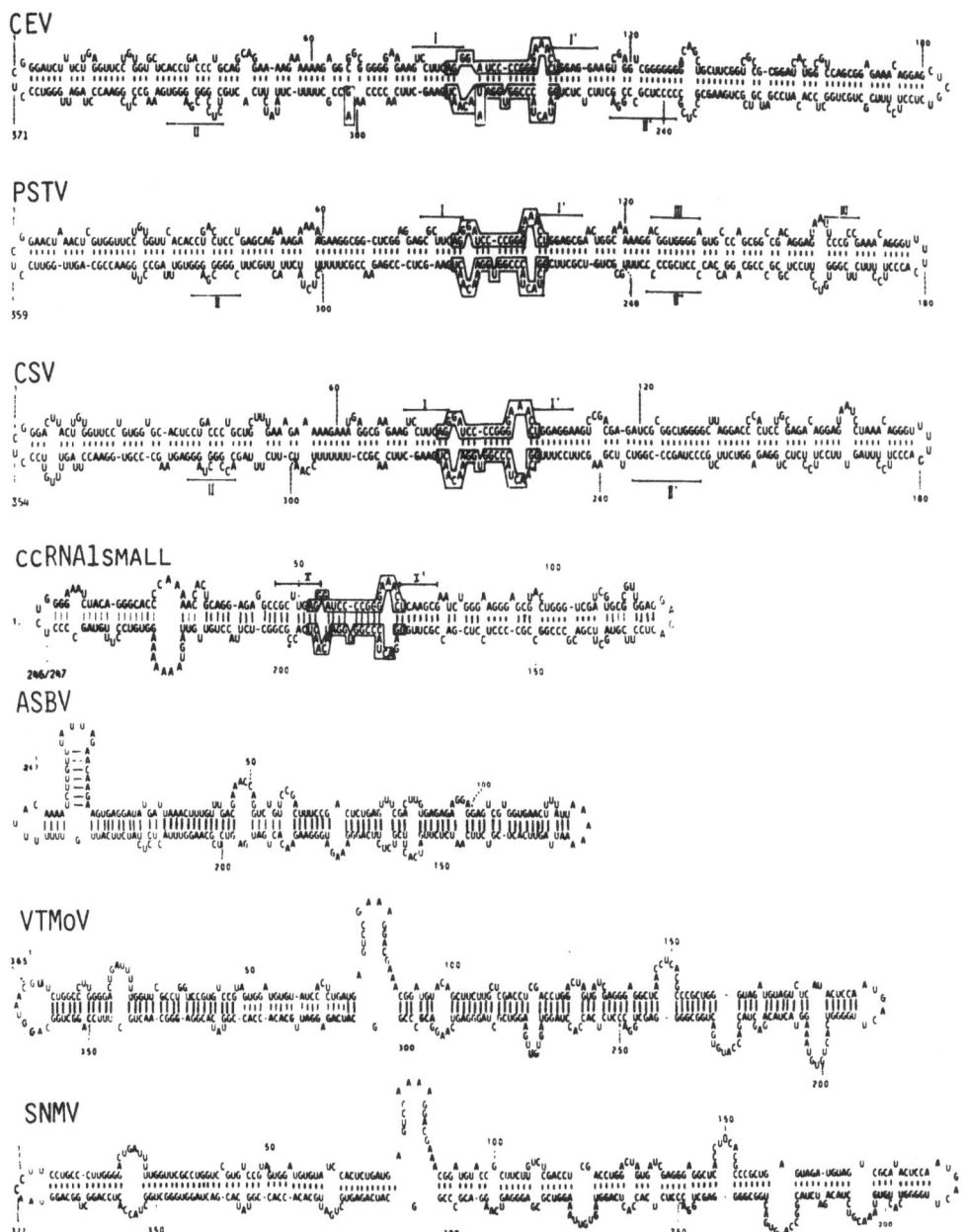

Fig. 3 Sequences and secondary structures of the viroids PSTV, CEV, CSV, CCCV, and ASBV, and the virusoids VTMoV and SNMV.
The sequences in the boxed regions are identical in all viroids, except ASBV. CCCV is a mixture of RNA1 and its exact duplicate RNA2. RNA1 as well as RNA2 were isolated as a "small" form and a "large" form in which the segments 100-123 and 124-147 are duplicated.
The hairpins which are formed during the main transition are depicted with roman numbers.

Table I Structure and Structural Transitions of Viroids, Virusoids, and Random Sequences.

The data on the secondary structures (degree of base pairing, ratio of G:C base pairs relative to G:C, A:U, and G:U base pairs, and number of bifurcations in the native structure) are from calculations with the model of "co-operative helices" and from the Zuker-program. The parameter listed under "thermodynamics" are derived from experiments except ΔG/N (difference in free energy per nucleotide during formation of the structure in the ground state from the completely coiled circle) and except the results on random sequences. For random, i.e. shuffled, sequences the number of nucle-otides, N, and the G:C-content, f_{G+C}, are taken from PSTV. The values for random sequences are average values of five different sequences. T_m- and $\Delta T_{1/2}$-values refer to a ionic strength of 0.011 M Na$^+$, pH 6.8.

RNA species	Sequence		Secondary structure				Thermodynamics				
	N	f_{G+C}	BP/N	helix-length	$\frac{G:C}{G:C+A:U+G:U}$	bi-furcations	ΔG kJ/mol	$\Delta G/N$ kJ/mol	T_m °C	$\Delta T_{1/2}$ °C	mechanism of denaturation
PSTV	359	0.58	0.690	5.0	0.57	0	600.7	1.67	49.5	0.9	formation of
CEV	371	0.60	0.684	4.9	0.58	0	601.0	1.62	50.5	1.0	stable hair-
CSV	354	0.53	0.700	5.2	0.52	0	569.5	1.61	48.5	1.1	pins during
ccRNA1l	302	0.61	0.622	4.5	0.67	0	462.9	1.53	48.1	1.2	co-operative
ccRNA1s	247	0.60	0.616	4.5	0.68	0	377.2	1.53	48.8	1.4	main transition
ASBV	247	0.38	0.664	5.1	0.34	1	278.3	1.13	37.5	1.5	?
SNMV2	377	0.56	0.658	5.4	0.57	1	539.2	1.43	38.2	2.8	–
VTMoV2	365	0.56	0.642	4.9	0.56	1 or 2	482.8	1.32	38.1	2.0	–
random sequence	359	0.58	0.59 ±0.01	4.9 ±1.3	0.57	3±1	457 ±22	1.27 ±0.06	36	5	–

2) Virusoids

Similarly to ASBV, virusoids contain one or two bifurcations in the secondary structure. The two structures of VTMoV having either one bifurcation or two are different in ΔG only by 0.1 kJ/mol; this means that both structures are nearly equally probable. For VTMoV the branched structure is better by -130 kJ/mol and for SNMV by -90 kJ/mol than the unbranched structures published earlier[39]. Even by comparing the quantitative parameters concerning the native structures of viroids and virusoids, mainly the values of $\Delta G/N$ of 1.43 or 1.33 kJ/mol, are significantly lower for virusoids than for viroids with the exception of ASBV.

3) Random sequences

The significance of the parameters listed in table I may be seen best when they are compared with those of circular single-stranded RNA molecules with the same length and G:C-content as a typical viroid but with a random sequence, generated by a computer. These are shuffled sequences in the terminology of Fitch[40]. In this case, N and f_{G+C} were taken from PSTV. For five random sequences their optimized secondary structures were obtained by the Zuker-program. They all contained bifurcations. The structural and thermodynamic parameters given in Table I are average values obtained from the calculation on the five structures mentioned above. The average degree of base pairing with 59% is lower than those of viroids and virusoids. The difference in $\Delta G/N$ with 1.27 ± 0.1 seems more significant. This value also overlaps with the corresponding value of VTMoV and is even higher than that of ASBV.

Summarizing the values of the degree of base pairing, G:C-content of base-paired regions, number of bifurcations and $\Delta G/N$, virusoids are more similar to random sequences than to viroids. Among the viroids ASBV is a clear exception, primarily in the very low G+C content which leads to the low stability $\Delta G/N$.

STRUCTURAL TRANSITIONS

The complete mechanism of conformational transitions of viroids as elucidated by experimental and theoretical methods is described in the following paragraph. Despite the fact that the different theoretical models assume different degrees of co-operativity of the helices, their results are nearly identical and in good agreement with experiments (for example cf. Fig. 5). First, the results for PSTV, CEV, and CSV are described because these species show sequence homology as high as 60% and very similar experimental features. CCCV RNA1 will be considered later.

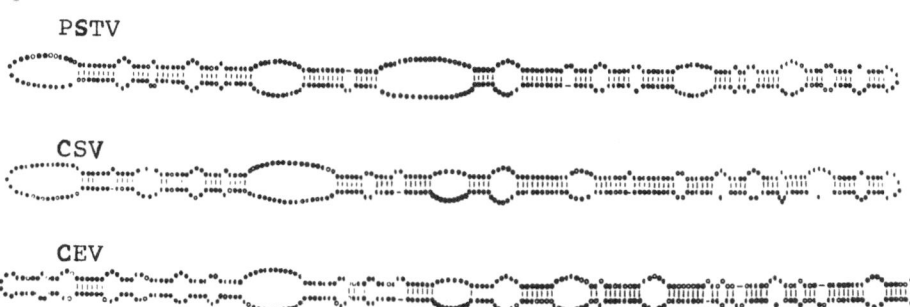

Fig. 4. Secondary structures of PSTV, CEV, and CSV calculated with the "co-operative helix" model for a temperature few degrees below their main transition (about 70°C). Nucleotides which are conservative in all three viroids are shown as points, others as circles.

Fig. 5 Denaturation curves of PSTV

1) measured in 1 M NaCl, 1 mM Na-cacodylate, 0.1 mM EDTA, pH 6.8.

2) calculated with the model of "co-operative helices"

3) calculated with the Zuker-Nussinov-algorithm

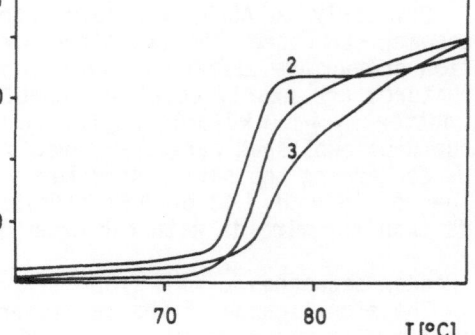

1) Premelting regions

If the stabilities of all helices are calculated at increasing temperatures, two characteristic regions in the viroid molecules show lowest stability. These premelting regions have been compared in Fig. 4 for CEV, CSV, and PSTV. For an easier comparison, conservative and variable nucleotides are marked by different symbols. Thus, it is evident that the premelting regions are in conserved parts of the molecule. The left premelting region contains the oligo-purine sequence which was suspected to act as a functional signal in replication[33], however, without experimental proof so far. The second premelting region is directly adjacent to one of the most stable helices located in the middle of the molecule. This borderline in the thermal stability is conserved in all three viroids. It divides the molecules in a less stable left half and a more stable right half. It will play a crucial role in the co-operative mechanism of denaturation. The right half of the three viroids significantly shows less sequence homology. Despite the sequence variation, however, the secondary structure of this part of the molecules is uniformly more stable than that of the left half.

2) Main transition

At the temperature of the main transition, the linear native secondary structure of the viroids PSTV, CEV, and CSV becomes unstable against a branched structure of two or three very stable hairpins which are not part of the native structure. This denaturation/rearrangement is a result of a thermodynamic coupling of denaturation of the linear structure and formation of the hairpins. The high activation energy for the process, specially the disruption of the basepairs of the center part of the native structure, may be diminished by a sliding mechanism or strand migration.

In case of PSTV, two or three hairpins may be formed in the main transition (cf. Fig. 6): There is no clear theoretical or experimental energy difference between both possibilities. With CEV and CSV, there are formed two hairpins. One of the hairpins - or in case of PSTV possibly two hairpins - melt at temperatures a few degrees above the main transition, the most stable hairpin melts at about 15 to 20°C higher.

The calculated temperatures for the main transition at 1 M NaCl are within the range of error of the experimental T_m-values. The experimental G:C content of the main transition, determined from hypochromicities at 260 and 280 nm, is about 5% higher than the calculated one. This could be a result of the assumption of equal hypochromicity of A:U and G:U base pairs or the disregard of hypochromicity contributions of small loops and length dependence of hypochromicity.

3) Transition of cadang-cadang RNA

The main transition of CCCV RNA1 large in 10 mM NaCl resembles the transition of the other viroids. From experimental equilibrium and kinetic measurements it is visible that the transition of CCCV RNA is also a co-operative rearrangement of the linear native structure to a single hairpin, which is nearly identical to the first hairpin of PSTV, CSV, and CEV. The denaturation of this hairpin is resolvable even with equilibrium measurements at about 10°C above the main transition.

The experimental enthalpy of the main transition demands a co-operativity near 100%, but any theoretical calculation results in transitions with only minor co-operativity. Investigations of denaturation of CCCV RNA at higher ionic strength support these calculations. At 100 mM NaCl the main transition splits into two transitions, and at 1 M NaCl there are at least three transitions with very low hypochromicity.

4) Transition of ASBV
The transition of ASBV in 10 mM NaCl is about 10°C lower than the main transition of all other mentioned viroids. This T_m-value is nearly identical to that of virusoids. The half width of the transition is in the range of cadang-cadang RNA. In contrast to virusoids, the low T_m-value and the high half width are due to the low G:C-content of the secondary structure and the small size of ASBV, but the transition itself is highly co-operative. From the equilibrium melting curves it might be concluded that there is a second transition with low hypochromicity a few degrees above the main transition. Kinetic experiments were not made. Theoretical calculations with the model of cooperative helices are complicated because there are a lot of possible hairpins which are only a little more stable than the native structure but no part of it. The conservative hairpin which is present in all other mentioned viroids is not present in ASBV. Therefore, in case of ASBV, the existence of the typical stable hairpin of viroids is not proven. ASBV may belong to another class of viroids which have evolved separately from the other viroids[38].

5) Random sequences
The sharp main transition of viroid molecules is a result of the denaturation/rearrangement of their native secondary structure with formation of the stable hairpins at the denaturation temperature. These hairpins - up to three in case of PSTV - are up to 11 base pairs long with a G:C content up to 90%. The possibility to form such hairpins even in a

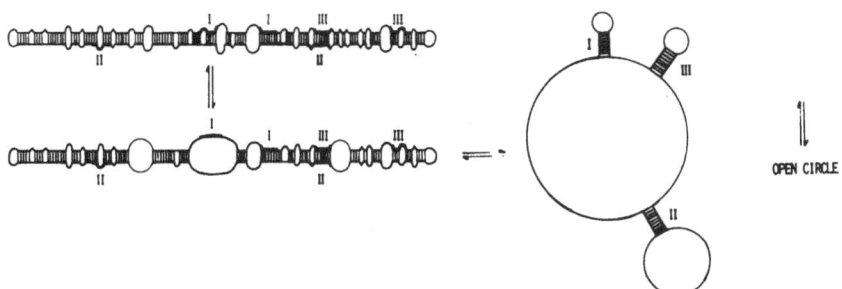

Fig. 6   Mechanism of denaturation and renaturation of PSTV
Schematic structures are represented at temperatures of ca. 25°, 70°, 75°, and above 95°C. The formation of two or three stable hairpins (I, II, III) is possible with PSTV during the main transition. The depicted hairpins are formed by base pairing from following regions: I) 79-87/110-102, II) 227-236/328-319, and III) 127-135/168-160. Analogous hairpins to hairpin I and II of PSTV are found in CEV and CSV. Hairpin I is found in CCCV only.

shuffled sequence is extremely low for statistical reasons. Therefore, the random sequences denature stepwise in temperature intervals of at least 5°C and not in a viroid-like manner.

6) Ionic Strength Dependence of Denaturation

In thermodynamic respects the description of the denaturation is excellent, as long as only transitions in high ionic strengths are treated. Because of the lack of ionic strength dependent parameters, all theoretical models are not able to produce significant results for ionic strengths other than 1 M. An introduction of ionic strength dependence after calculation of the T_m-values is only correct if all used parameters, i.e. elements of the secondary structure, have the same dependence. The stability of small hairpins depends less upon the ionic strength as compaired with internal helices; therefore the stabilities of different parts of the viroid molecule may shift against each other with varying ionic strength resulting in lower co-operativity[14]. Such effects of short hairpins with large loops are of greater influence in case of shorter molecules like CCCV RNA: A loss of only one helix in the denaturation process leads to drastically reduced co-operativity.

FUNCTION OF STRUCTURE FORMATION

Viroids are unique in Structure and Dynamics

The characteristics of viroid denaturation are high co-operativity as in double-stranded RNA and low stability as in other single stranded RNAs. The main transition of viroids is not a mere dissociation of base pairs, but involves the concerted opening of all base pairs and formation of stable hairpins. The similarity of all viroids in respect to their dynamic features is not a consequence of high sequence homology, but only very specific nucleotide sequences may guarantee the dynamic features of viroids. The possibility of forming viroid-analogous hairpins in statistical sequences with length and G:C-content of viroids is extremly small because of length, stability, and special arrangement of the viroid hairpins. Cooperative denaturation is also not possible for random sequences. Their secondary structure is not linear but contains bifurcations. Their calculated denaturation behavior is similar to that of virusoids.

Viroid structure may be described as an optimal compromise between stability and flexibility. On one side, the native structure is nearly intact up to the main transition; on the other side, the opening of the structure of viroids is much easier than for example in a completely double stranded RNA. The fact, that viroid denaturation starts at so called "premelting regions", and that stable hairpins are formed, may be correlated with detailed biological features.

Premelting and Virulence

Different strains of PSTV show from mild to lethal symptoms on tomato plants[41]. The predominant sequence differences of these strains are in the "premelting region" from nucleotides 42-60 and 300-319. Decreasing stability, i.e. easier opening of this region correlates with increasing virulence[41,42]. The authors suggested from these results, that this region interacts with host factors, and that this interaction leads to the expression of symptoms. In CEV, the sequence variations are located predominantly in two regions, and one region overlaps with the premelting region[43]. In contrast to PSTV, the stability of the premelting region was found higher in sequence variants with severe symptoms than in those with mild symptoms.

Function of Stable Hairpins

The formation of stable hairpins together with their location in the highly conserved region gave raise to several hypotheses on their functional relevance. Because of a striking sequence homology between the lower part of the conserved region (254-277 in PSTV) and the 5'end of eukaryotic snRNA U1 viroids had been discussed to interfere with the normal splicing process[33,44]. The interaction of the viroid instead of the snRNA U1 with the splicing site of the unprocessed mRNA would be facilitated if the opposite part in the viroid structure would switch over into the stable hairpin leaving the lower part with less internal base-pairing[45]. There is, however, no experimental evidence for these models. The switch from the native to the hairpin containing structure was discussed quite generally as switch between different functions[46]. For example, the native structure could resemble protein binding sites, possibly for replication, and the hairpin could be generated transiently during replication and serve as a favorable structure for cutting oligomeric intermediates to monomeric species.

There are experimental indications for the involvement of the stable hairpin region in viroid maturation[47]. Plasmids carrying two units of viroids tandemly are highly infectious, whereas with plasmids carrying only the monomeric viroid cDNA no infectivity was obtained. The authors argued that after the viroid precursor-RNA was transcribed from the plasmid DNA, the region of the stable hairpin is relevant to cut the viroid-precursor in that region to the exact monomeric length, probably by cellular endonucleases. These results are in accordance with the finding, that M13-cloned single-stranded PSTV DNA is infectious if an eleven nucleotide long sequence from the conserved region is present twice[48].

CONCLUDING REMARKS

In summary, we were able to show that the viroids studied form a class of RNA molecules with very homogeneous features concerning structure and structural transitions, in that sense comparable with tRNA. Viroids may clearly be differentiated from all other RNAs, especially virusoids, which closer resemble the features of random RNAs. Other progress in viroid structure research has to come from the in vivo structure, i.e. details of the complexes of viroids and viroid intermediates in their subcellular location, the nucleus. Because of the problem of isolation of pure in vivo complexes in amounts for thermodynamic studies, experimental methods for examination of crude extracts have to be used to get thermodynamic parameters for the interaction of viroids with other components. Possible methods can be the above mentioned gel electrophoresis with temperature gradients and hydrodynamic methods with fluorescence detection[11].

REFERENCE

1. T.O. Diener, Viroids: Structure and function, Science, 205:859 (1979).
2. H.J. Gross, and D. Riesner, Viroids: a class of subviral pathogens, Angew. Chem. Int. Ed. 19:231 (1980).
3. H.L. Sänger, Biology, structure, functions, and possible origin of viroids, in: "Encyclopaedia of Plant Physiology, New Series", vol 14B, B. Parthier and D. Boulter, ed., Springer, Berlin, Heidelberg, New York (1981).

4. D. Riesner, M. Colpan, T.C. Goodman, L. Nagel, J. Schumacher, G. Steger, and H. Hofmann, Dynamics and interaction of viroids, J. Biomolec. Struct. Dynamics, 1:669 (1983).
5. D. Riesner and H.J. Gross, Viroids, Ann. Rev. Biochem., 54:531 (1985).
6. D. Riesner, K. Henco, U. Rokohl, G. Klotz, A.K. Kleinschmidt, H. Domdey, P. Jank, H.J. Gross, and H.L. Sänger, Structure and structure formation of viroids, J. Mol. Biol., 133:85 (1978).
7. J.M. Sogo, T. Koller, and T.O. Diener, Potato spindle tuber viroid. X. Visualization and size determination by electron microscopy, Virology, 55:70 (1973).
8. H.L. Sänger, G. Klotz, D. Riesner, H.J. Gross, and A.K. Kleinschmidt, Viroids are single-stranded covalently closed circular RNA molecules existing as highly base-paired rod-like structures, Proc. Nat. Acad. Sci. USA, 73:3852 (1976).
9. U. Wild, K. Ramm, H.L. Sänger, and D. Riesner, Loops in viroids, Eur. J. Biochem., 103:227 (1980).
10. D. Riesner, J.M. Kaper, and J.W. Randles, Stiffness of viroids and viroid-like RNA in solution, Nucl. Acids Res., 10:5587 (1982).
11. T.C. Goodman, L. Nagel, W. Rappold, G. Klotz, and D. Riesner, Viroid replication: Equilibrium association constants and comparative activity measurements for the viroid-polymerase interaction, Nucl. Acids Res., 12:6231 (1984).
12. H.J. Gross, H.L. Sänger, J. Schumacher, G. Steger, and D. Riesner, Viroids: Their structure and possible origin, in: "Endocytobiology", vol. II, H.E.A. Schenk, W. Schwemmler, ed., de Gruyter, Berlin.
13. R. Nussinov and I. Tinoco, Small changes in free energy assignments for unpaired bases do not affect predicted secondary structures in single stranded RNA, Nucl. Acids Res., 10:341 (1982).
14. G. Steger, H. Hofmann, J. Förtsch, H.J. Gross, J.W. Randles, H.L. Sänger, and D. Riesner, Conformational transitions in viroids and virusoids: Comparison of results from energy minimization algorithm and from experimental data, J. Biomolec. Struct. Dynamics, 2:543 (1984).
15. D. Pörschke, O. C. Uhlenbeck, and F. H. Martin, Thermodynamics and kinetics of the helix-coil transition of oligomers containing GC base pairs, Biopolymers, 12:1313 (1973).
16. J. Gralla and D.M. Crothers, Free energy of imperfect nucleic acid helices, J. Mol Biol., 73:497 (1973).
17. J. Gralla and D.M. Crothers, Free energy of imperfect nucleic acid helices. III. Small internal loops resulting from mismatches, J. Mol. Biol., 78:301 (1973).
18. T.R. Fink and H. Krakauer, The enthalpy of the 'bulge' defect of imperfect nucleic acid helices, Biopolymers, 14:433 (1975).
19. T.R. Fink and D.M. Crothers, Free energy of imperfect nucleic acid helices. I. The bulge effect, J. Mol. Biol. 66:1 (1972).
20. J.R. Fresco, L.C. Klotz, and E.G. Richards, New spectroscopic approach to the determination of helical secondary structure in ribonucleic acids, Cold Spring Harb. Symp. Quant. Biol., 28:83 (1963).
21. S.M. Coutts, Thermodynamics and kinetics of GC base pairing in the isolated extra arm of serine-specific tRNA from yeast, Biochem. Biophys. Acta, 232:94 (1971).
22. I. Tinoco, O.C. Uhlenbeck, and M.D. Levine, Estimation of secondary structure in ribonucleic acids, Nature, 230:362 (1971).
23. J. Langowski, K. Henco, D. Riesner, and H. L. Sänger, Common structural features of different viroids: Serial arrangement of double helical sections and internal loops, Nucl. Acids Res., 5:1589 (1978).
24. R. Nussinov, G. Pieczenik, J.R. Griggs, and D.J. Kleitman, Algorithms for loop matchings, SIAM J. Appl. Math., 35:68 (1978).
25. R. Nussinov and A. Jacobson, Fast algorithm for predicting the secondary structure of single-stranded RNA, Proc. Natl. Acad. Sci. USA, 77:6309 (1980).

26. M. Zuker and P. Stiegler, Optimal computer folding of large RNA sequences using thermodynamics and auxiliary information, Nucl. Acids Res., 9:133 (1981).
27. K. Henco, H.L. Sänger, D. Riesner, Fine structure melting of viroids as studied by kinetic methods, Nucl. Acids Res., 6:3041 (1979).
28. J.W. Randles, G. Steger, and D. Riesner, Structural transitions in viroid-like RNAs associated with cadang-cadang disease, velvet tobacco mottle virus, and Solanum nodiflorum mottle virus, Nucl. Acids Res., 10:5569 (1982).
29. F.M. Pohl, Einfache Temperatursprung-Methode im Sekunden- bis Stundenbereich und die reversible Denaturierung von Chymotrypsin, Eur. J. Biochem., 4:373 (1968).
30. M. Eigen and L. DeMaeyer, Theoretical basis of relaxation spectroscopy, in "Techniques of Organic Chemistry", Vol. VIII/2, A. Weissberger, ed., John Wiley, New York (1963).
31. K. Henco, G. Steger, and D. Riesner, Melting curves on less than 1µg of nucleic acid, Anal. Biochem., 101:255 (1980).
32. D. Riesner, M. Colpan, and J.W. Randles, A microcell for the temperature-jump technique, Anal. Biochem., 121:186 (1982).
33. H.J. Gross, G. Krupp, H. Domdey, M. Raba, P. Jank, Ch. Lossow, H. Alberty, K. Ramm, and H.L. Sänger, Nucleotide sequence and secondary structure of citrus exocortis and chrysanthemum stunt viroid, Eur. J. Biochem., 121:249 (1982).
34. P. Palukaitis and R.H. Symons, Purification and characterization of the circular form of CSV, J. Gen. Virol., 46:477 (1980).
35. M. Colpan, J. Schumacher, W. Brüggemann, H.L. Sänger, and D. Riesner, Large-scale purification of viroid RNA using Cs_2SO_4 gradient centrifugation and HPLC, Anal. Biochem., 131:257 (1983).
36. G. Steger, H. Müller, and D. Riesner, Helix-coil transition in double-stranded viral RNA: fine resolution melting and ionic strength dependence, Biochem. Biophys. Acta, 606:274 (1980).
37. K. Henco, D. Riesner, and H.L. Sänger, Conformation of viroids, Nucl. Acids Res., 4:177 (1977).
38. R.H. Symons, ASBV - primary sequence and proposed secondary structure, Nucl. Acids Res., 9:6527 (1981).
39. J. Haseloff and R.H. Symons, Comparative sequence and structure of viroid-like RNAs of two plant viruses, Nucl. Acids Res. 10:3681 (1982).
40. W.M. Fitch, Random sequences, J. Mol. Biol. 163:171 (1983).
41. M. Schnölzer, B. Haas, K. Ramm, H. Hofmann, and H.L. Sänger, Correlation between structure and pathogenicity of PSTV, subm.
42. H.L. Sänger, Minimal infectious agents: the viroids, in: "The Microbe, Part I, Viruses", B.W.J. Mahy and J.R. Pathison, ed., Cambridge University Press (1984).
43. J. Visvader and R.H. Symons, Eleven new sequence variants of CEV and the correlation of sequence with pathogenicity, Nucl. Acids Res. 13:2907 (1985).
44. T.O. Diener, Are viroids escaped introns?, Proc. Natl. Acad. Sci. USA, 78:5014 (1981).
45. D. Riesner, G. Steger, J. Schumacher, H.J. Gross, J.W. Randles, and H.L. Sänger, Structure and Function of viroids, Biophys. Struct. Mech., 9:145 (1983).
46. P. Keese and R.H. Symons, Domains in viroids: evidence of inter-molecular rearrangements and their contribution to viroid evolution, Proc. Natl. Acad. Sci. USA, 78:5014 (1981).
47. T. Meshi, M. Ishikawa, Y. Watanabe, J. Yamaya, Y. Okada, T. Sano, and E. Shikata, The sequences necessary for the infectivity of hop stunt viroid cDNA clones, Mol. Gen. Genet., 200:199 (1985).
48. M. Tabler and H.L. Sänger, Cloned single- and double-stranded DNA copies of PSTV RNA and co-inoculated subgenomic DNA fragments are infectious, EMBO J., 3:3055 (1984).

AUTHOR INDEX